# DINNER WITH
# KING TUT

## ALSO BY SAM KEAN

*The Disappearing Spoon*

*The Violinist's Thumb*

*The Tale of the Dueling Neurosurgeons*

*Caesar's Last Breath*

*The Bastard Brigade*

*The Icepick Surgeon*

# DINNER WITH KING TUT

### How Rogue Archaeologists Are Re-creating the Sights, Sounds, Smells, and Tastes of Lost Civilizations

## SAM KEAN

Little, Brown and Company
New York Boston London

Copyright © 2025 by Sam Kean

Hachette Book Group supports the right to free expression and the value of copyright. The purpose of copyright is to encourage writers and artists to produce the creative works that enrich our culture.

The scanning, uploading, and distribution of this book without permission is a theft of the author's intellectual property. If you would like permission to use material from the book (other than for review purposes), please contact permissions@hbgusa.com. Thank you for your support of the author's rights.

Little, Brown and Company
Hachette Book Group
1290 Avenue of the Americas, New York, NY 10104
littlebrown.com

First Edition: July 2025

Little, Brown and Company is a division of Hachette Book Group, Inc. The Little, Brown name and logo are trademarks of Hachette Book Group, Inc.

The publisher is not responsible for websites (or their content) that are not owned by the publisher.

The Hachette Speakers Bureau provides a wide range of authors for speaking events. To find out more, go to hachettespeakersbureau.com or email hachettespeakers@hbgusa.com.

Little, Brown and Company books may be purchased in bulk for business, educational, or promotional use. For information, please contact your local bookseller or the Hachette Book Group Special Markets Department at special.markets@hbgusa.com.

ISBN 9780316496551
Library of Congress Control Number: 2025934396

Printing 1, 2025

LSC-C

Printed in the United States of America

# Contents

*Introduction*   *3*

1: Africa—75,000 Years Ago   9

2: South America—7500 BC   43

3: Turkey—6500s BC   81

4: Egypt—2000s BC   117

5: Polynesia—1000s BC   165

6: Rome—AD 100s   201

7: California—AD 500s   243

8: Viking Europe—AD 900s   277

9: Northern Alaska—AD 1000s   315

10: China—AD 1200s   347

11: Mexico—AD 1500s   381

*Conclusion*   *419*

*A Thank-You and a Bonus*   *421*

*Acknowledgments*   *423*

*Works Cited*   *425*

*Index*   *437*

The archaeologist is digging up not things, but people.
—Michael Balter, *The Goddess and the Bull*

# DINNER WITH
# KING TUT

# INTRODUCTION

~~~

I've always had a gripe with archaeology. On the one hand, it seems like the most thrilling field in science. It illuminates the deep history of humankind in a way no other discipline can—unveiling who we human beings are, where we came from, and what unites us today with every culture in our past. It reveals no less than our fundamental selves.

On the other hand, whenever I actually visited an archaeological dig, I felt my shoulders sag in disappointment. The work looked so... dull. No matter where I was in the world, it was just scores of sunburned men and women sprawled in the dirt, dusting off broken pot shards with toothbrushes. And not for a few hours, but day after day, year after year. It was such a letdown, the most godawful tedium I could imagine. In no other field is there such a disconnect between the stirring conclusions reached and the sheer monotony of the daily grind.

Imagine my delight, then, when I first discovered experimental archaeology.

As the name implies, experimental archaeology puts ideas about the past to the test, either in the lab or out in nature. Many practitioners even subject *themselves* to experimentation, safely or not, to replicate different aspects of our ancestors' lives—their food, clothing, shelter, body art, and more. Put another way, instead of just digging things up and passively theorizing about them, experimental archaeologists *do* things— actively re-create the past. They brew Viking beer. They make mummies. They drive chariots, play Aztec ballgames, revive ancient yeast and bake the tangy sourdough that King Tut ate. They build rickety ships and plunge out onto the open sea with all the verve of Indiana Jones, to trace the epic journeys of our ancestors. Indeed, some proponents call the field not experimental archaeology but *experiential* archaeology or even *living* archaeology. It doesn't re-create the past as much as resurrect it.

I especially appreciate how sensory-rich the field is compared to traditional archaeology. Excavations and artifacts allow us to sketch a decent picture of what the past *looked* like, but by and large, conventional dirt archaeology neglects the other senses. Not experimental archaeology. You can smell the crab-like odor of a deer hide as you tan it, or taste the salty pinch of fermented Roman fish sauce. You can hear the boom of medieval cannons, or feel the bone-wearying fatigue of spending several hours grinding your own grain into flour — as well as the compensatory satisfaction of tearing into a fresh loaf of bread afterward, knowing it tastes twice as good because you earned your appetite.

Another exciting aspect of the field is how few rules there are; insight can spring from anywhere. Some adherents are hardcore lab geeks, very data- and tech-driven. Others are traditional archaeologists who sensed something missing in their previous work and longed to connect with lost societies on a more intimate level. Some are grouchy, live-off-the-land survivalists with tales of fending off grizzlies and tanning moose hides. (I've now met multiple *Naked and Afraid* contestants.) Some, frankly, are screwball enthusiasts who nevertheless got me thinking big thoughts about the human condition, and kept me thinking for days. Not everyone you'll meet in these pages calls themselves an experimental archaeologist, or even an archaeologist, period. But they all share a commitment to the messy fun of re-creating the lives of our ancestors.

Their work fascinated me so much that I decided to embrace the spirit of the field and get active myself. There's no better way to learn than by doing, and thankfully, these men and women were more than happy to let me tag along on their adventures and guide me on experiments of my own, however ill-advised. Under their tutelage, I've made a DIY mummy, fired a giant catapult, even tattooed someone. I've eaten guinea pig, llama, caterpillar, whale, and walrus. I've been spattered with urine, blood, blubber oil, and worse, in countries all over the globe. I loved every second of it.

## Introduction

To be sure, I didn't emerge from this book as any sort of survival expert. I failed far more than I succeeded, even at simple tasks. (There's nothing more humbling than realizing that, if I were teleported to pretty much any era in the past, I would have starved to death in about half an hour.) But in each case, the process proved more revealing than the results. I still can't hurl a spear straight or make a decent loaf of acorn bread—and God help me if I ever have to perform another trepanation, a form of primitive neurosurgery (really) that's been saving lives for 10,000 years. But I learned a vast amount about ancient life anyway. It was like running my fingers over a flat surface and suddenly feeling a third dimension ripple up. This work made the people of ancient Rome and Africa and Polynesia and Egypt come alive for me in ways they never had before, and keeps them alive for me now.

And yet, despite its delights, experimental archaeology remains controversial within the larger field. In some cases, this is simply healthy skepticism. Can we really re-create, say, ancient Egyptian bread or million-year-old tools? As you'll see within these pages, the answer, increasingly, is yes. And even when we can't re-create things with 100-percent, atom-by-atom fidelity (an impossible standard anyway), the tacit knowledge we gain fills important gaps in our understanding of the past, knowledge that's impossible to glean from examining artifacts alone. However laudable our brains, we *Homo sapiens* are builders and makers deep down. To truly grasp something, you have to use your hands.

Still, much of the disdain for experimental archaeology stems from something deeper than mere skepticism. Some academics despise the field as a rogue upstart, and dismiss even well-designed experiments as "theater." One foe insists that the only reason people go into experimental archaeology is "the satisfaction of character deficiencies." Perhaps. But many indigenous groups have found experimental archaeology invaluable for reconnecting with cultural traditions, traditions that have nearly been wiped out over the past few centuries. It seems callous to dismiss that as "character deficiencies." Moreover, my sympathies will always lie with those who subject their ideas to testing and bring some practical knowledge to bear on their research. As one archaeologist told me, "Anyone who's doing work on [excavating] foundations should learn how to lay stone; anyone who's doing [research on] beermaking should be brewing it." It says something curious about the state of archaeology that these are considered radical notions.

A word on the book's structure. Unlike my previous books, *Dinner with King Tut* combines fiction and nonfiction. The nonfiction consists of both archaeological research and first-person reported accounts from my trips to visit scientists, historians, practitioners of lost arts, and plain old eccentrics around the globe. In parallel with those accounts, each chapter also re-creates a day in the life of a specific man or woman in a specific time and place, starting in sub-Saharan Africa 75,000 years ago, and moving forward toward the present. Along the way, we'll track vicious game with South American hunters, help build (and rob) pyramids in Egypt, endure bitter cold in remote Alaskan villages, and navigate between impossibly far-flung Polynesian islands, among other adventures. But even these fictional sections are built on fact. That is, while the characters in each chapter are invented, they are representative of their time and place, and the details of their lives—what they ate, where they slept, what they smelled and heard and saw—are based on real archaeological research. In short, everything that happens to them could and did happen to people in ancient times, and in a modest way, these stories aim to do some experimental archaeology of their own in re-creating the life experiences of ancient people.

In combination, the novelistic and nonfictional elements allow you to not only experience what ancient people did, but get inside their minds and grasp their worldviews—what made them laugh and cry, what terrified and humiliated them, their entire gestalt.

Beyond relating some incredible tales, I hope this book can illuminate what human beings have lost in modern times, and encourage us to preserve traditions that face extinction or eradication across the world. There's a real cost to being cut off from our heritage. People once smelled hay and dung and blood and cooking smoke constantly. They once had to build or make most everything around them, and they couldn't shy

## Introduction

away from the ugly aspects of life—death very much included. Today, we live in a far more sanitized world, and have little tangible knowledge about where our food and other goods come from. We're narrower as people, and live in a way that would feel alien to virtually every ancestor we have. That seems unhealthy, even destabilizing. Human beings are social creatures, with social needs and instincts, and we're drifting further and further from the types of societies that once nurtured and fulfilled us.

But experimental archaeology—living archaeology—can reconnect us. When you make your own clothes and build your own hearth, butcher your own food and navigate your own path through the wilderness using nothing but your wits, well, there's something deeply human about that. The stories here are meant to be instructive and fun, but they're also a form of time travel—passing traditions down, connecting one generation to the next. Indeed, as close to time travel as we humans can ever hope to get.

# AFRICA — 75,000 YEARS AGO

~~~

Our species, *Homo sapiens,* arose roughly 300,000 years ago in Africa, and by 75,000 years ago, a full suite of modern behavior was on display, including tool use, trade, artistic and symbolic behavior, and more. These ancestors of ours — every bit as intelligent as us today, and every bit as prone to folly — lived in small bands as nomadic hunter-gatherers, roaming from place to place as the seasons passed to collect and hunt food.

No modern human group lives exactly like people did way back when, obviously, but some live much closer to the old hunter-gatherer ways than others, including the famous San people of southern Africa (formerly known as the Bushmen). Today the San are largely confined to the Kalahari, a harsh desert, but historically they ranged much more widely throughout southern Africa, including into relatively lush areas. This chapter combines aspects of the age-old San lifestyle with archaeological evidence of how people in those lusher areas lived.

Contrary to popular belief, the hunting-gathering life was not a constant struggle to survive. Much of the time, in fact, life was quite pleasant — rich and fulfilling, with people having as much or even more leisure time than we do today. Still, when things went bad for people back then, they went very bad indeed. As the saying goes, it doesn't matter how well you eat for eleven months of the year if you don't eat at all during the twelfth…

Kayate always sleeps lightly at night, waking with every hyena howl or *pop* of wood in the fire. Tonight, though, it's the absence of noise that awakens him. He hasn't heard the baby cry for hours.

He rises onto an elbow, his bed of dead sedge grass crinkling beneath him, and scours the darkness inside the cave. He can just make out the firelit forms of his older sister Namkabe, age seventeen, and her husband Xate. On his sister's bare breast lies an unnamed girl, six weeks old. She looks dead. Kayate holds his breath to still his vision, an old hunter's trick, and tries to perceive the rising and falling of her papery chest. He watches her hands, too. Her fingers are splayed open, and he wills them to clench shut, like an eagle's talons. According to his clan's lore, babies sometimes dream of being birds of prey.

At last, the girl's tiny arm jerks. She's alive. Kayate exhales and lies back down.

But he cannot relax. A new day will be dawning soon, and if he and his brother-in-law cannot find food today, the girl will starve—and the rest of them might, too. It's October, the dry season, and the landscape has been scorched by the severest drought in his clan's memory. Or what remains of his clan; most people have either collapsed dead or split off with their own families, to limit the number of mouths to feed. Even desperation foods, like the bitter *tsama* watermelons, have withered. There's another whole month to endure before the rainy season starts, before any hope of relief. Kayate's family's last meal was a scavenged antelope skin; they'd burned off the hair and gnawed the leathery hide.

At least they have shelter this week. The cave is ten paces wide and twenty paces deep, burrowed into the side of a cliff. If he stretches his arms when standing, Kayate can just about reach the ceiling seven feet above. It's got a hardpacked dirt floor and uneven stone walls pockmarked with tiny holes.

Still, it's hardly cozy, and even if he could relax tonight, sleep seems unlikely. Normally inside a cave, he'd lay down some bedding of ash, fresh grass, and leaves to keep bugs away, but he and his family arrived here only yesterday, and with so little ash built up, he can feel the ticks wriggling up his legs. The drought also withered the grass and green leaves whose scent drives mosquitos off. Every few seconds, he has to slap his leg or arm.

Between the bites and his gnawing worries, he decides to rise. He should stir the fire anyway; it's getting low.

Unfortunately, in his foggy state, he forgets about the ostrich eggs at his bedside. It's normally his sister's job to fetch water, which they store in ostrich-egg canteens; they're the size of melons and hold six cups of water each. To make them, Namkabe drills an inch-wide hole in each shell with a chert blade, empties the yolks and whites for a meal, and deodorizes the interior with fragrant herbs. A grass tuft acts as a stopper. The shells even sweat like humans do, keeping the water inside cool.

For all their wonderful properties, however, the canteens are still eggs, still fragile. Kayate wanted to spare his sister the burden of fetching water this morning, so he'd concealed the eggs next to his bed last night—a fact he's since forgotten, until he takes a step and hears a sickening crunch.

The precious water soaks into the mud floor. Kayate hears his brother-in-law sit up. Xate doesn't bother keeping his voice down, despite his sleeping wife and daughter.

—What did you break?

Kayate whispers that nothing's wrong, but Xate jumps up and stalks over. Upon seeing the broken shells, he starts picking up shards and pelting Kayate with them. They sting where they strike.

—You stupid shit! If it weren't for me, you'd starve.

Kayate bristles at this. Xate is a fierce hunter, no question. But he's been scarfing down over half the food he brings in, despite the fact that his wife is struggling to nurse. It's a flagrant violation of the clan's moral code to distribute all game equally—to share and share alike, no matter what. In normal times such selfishness could mean banishment; in a standard drought, execution. But in this extreme drought, with his tribe disintegrated, there's little Kayate can do. Xate stands several inches taller, and he's far more muscular. Not for the first time, Kayate wishes his parents had married Namkabe to anyone else in the world.

Xate chugs the contents of the other canteen, then wanders off to piss. As he does so, Namkabe slips over with her daughter. She rubs Kayate's shoulder with her free hand, assuring him that Xate doesn't mean what he says. Xate yells his answer from the corner.

—I do, too!

Kayate notices the little girl hanging limp in his sister's arms and feels a pinch of worry again; they've got to find food today. So however distasteful the idea, he asks Xate if he can help knap tools for today's hunt. They need to replace some broken spearpoints, as well as make hide-scrapers and hand-knives for cutting flesh. And they need to do so soon; knapping can be noisy, and they've got to finish before the antelope begin stirring at dawn.

Knapping is normally relaxing work—sitting around a fire, honing tools and chatting. Not today. While they chink and shape the rocks with their hammerstones, Namkabe sits in the corner, futilely trying to nurse. Xate blames her for her lack of milk, which leaves Kayate fuming. Xate then turns his abuse on Kayate himself. Being a decade older, Xate can knap rocks quickly and precisely, and every time he hears Kayate make a poor strike, he growls at him to quit wasting rock.

In fact, Xate is so absorbed in criticizing Kayate that he neglects to pay attention to his own work. Just before dawn, Kayate hears a noise like animal hide being sliced open. Then a gasp. A spray of red fluid lashes his feet. He looks up to see a quarter of Xate's pinky dangling by the skin at an impossible angle.

Xate freezes, but the rest of the cave dissolves into chaos. Namkabe rushes forward, baby in arm. Kayate snatches a piece of hide to press against the wound, then eases Xate onto a grass bed. He thinks of washing the wound with water, then grimaces to remember the smashed canteen.

The hide soon soaks through with blood. Xate recovers from his shock quickly—Kayate can tell because he hisses at him to stop wasting good hide. Still, it's clear that Xate cannot hunt today; his hand will be useless. The knowledge sits on Kayate like a boulder. The entire hunt—and perhaps his family's survival—now depends on him.

The first beds in human history date back 200,000 years, to a cliffside cave on the border between South Africa and Eswatini. There, Lyn Wadley, who practices traditional and experimental archaeology at the University of Witwatersrand, has excavated several carefully arranged layers of ash and plant matter that served as ancient mattresses. The ash layer probably arose when the occupants "cleaned" the cave by burning the refuse inside (a treatment that several places I lived during college would have benefited from). Atop the ash lay bundles of grass. Later, by 75,000 years ago, cave-dwellers were mixing in layers of sedge, a grass-like plant collected from riverbanks, as well as a "top sheet" of broad-leaves from the aromatic Cape quince tree.

The discovery of this bed fell under the purview of conventional archaeology—digging stuff out of the dirt. But Wadley wasn't content to stop there. She has an imagination, and grew curious about what the beds felt like. Were they soft? Itchy? Warm? So she re-created them and found two volunteers to sleep overnight on them in the cave. By all accounts the beds were comfortable, although neither fellow got much sleep. For one thing, the cave floor was so sloped that they kept rolling off. The lack of blankets and pillows didn't help—nor did sleeping in the wild, with unknown critters creeping about in the dark.

Still, they provided padding for weary bones, and the ash and plant layers repelled insects, a major nuisance for prehistoric people. The Cape quince (a.k.a., the wild apricot) gave the bedding a fresh, fruity odor that kept away mosquitos. The ash, meanwhile, helped fight ticks, as Wadley showed with another experiment: She lives in the country, so one day, she marched out to some tall grass and collected a handful of brown ticks. Then she gathered ash from her fireplace, molded it into a donut ring in a bucket, and dumped the ticks in the center. Over the next day, several ticks tried to escape their prison by burrowing through. Most failed: ash is soft and provides little traction, making it difficult to wriggle through. It also clogged their breathing pores, effectively suffocating them. A few hardy ticks did manage to escape, but their mouths were so choked with ash that they wouldn't have been able to bite anyone.

Sadly, Wadley's experiments never made it into the scientific literature. When she submitted a paper about the discovery of the world's oldest bed, the editors loved all the traditional archaeology—the digging-through-dirt stuff. But they refused to include her experiments on the re-created bedding or the ash ring, even as informal observations. This sort of snobbery would have offended the likes of Charles Darwin, who ran many such casual experiments in his backyard, experiments that often sparked future research.

This is the burden that experimental archaeologists face. The field can provide real, rich insights into the lives of long-lost people—making us vividly aware of how Kayate and his kinsmen might have struggled to sleep. But some scientists don't want to hear about it.

Organic materials like beds rarely survive the destructive weathering of nature. By necessity, then, most traditional archaeological research focuses on Kayate's main chore that morning, knapping stone tools. Indeed, stone tools provide the greatest insight we have into the lives and minds of ancient people. Our ancestors have been crafting them for three million years—ten times longer than *Homo sapiens* has existed—and they make up the vast majority of surviving artifacts from prehistory. To learn how ancient people made such tools, I visited master knapper Metin Eren at Kent State University in Ohio.

Eren is a burly, bearded fellow with often unruly hair. On my second day at his lab, upon seeing his hair lying flat, I compliment him on what I assume is a fresh haircut. After a baffled look, he smiles and admits, "No, I just combed it." If I had to describe Eren in one word, it would be *gung-ho:* he's up for pretty much any experiment at any time. (Just wait until we get to the poop-knife.)

Eren first realized how effective stone tools were as an undergraduate at Harvard, when he and some friends ordered beef for dinner one night and found themselves staring down at plates of gristle instead, lovingly overcooked into rubber. Blunt cafeteria knives couldn't hack through the stuff, but Eren happened to have some flint blades on hand that he'd knapped that day in an archaeology course. He fished them out, and *voilà*—they sliced right through. Eren later became a master knapper in graduate school, spending eight to nine hours every day breaking rocks, going through whole truckloads. These days, he says, "I can make anything from the past three million years"—any stone tool, from any place on Earth.*

My tutorial takes place in Eren's lab at Kent State, a delightful mess of spears, antlers, arrowheads, animal skulls, and rocks, rocks, rocks—several oil drums' worth, plus hundreds of more pounds scattered on the floor. From the pile, Eren hands me some chert. My piece has a chalky outer surface (the cortex) and a shiny black interior. We sit on plastic office chairs to start. (Eren plops right down on top of several apple-sized stones, but either doesn't notice or doesn't mind.) Then we drape a bundle of leather pads over our thighs, to protect our legs from cuts and bruises. My bundle also includes the chopped-off leg of an old pair of Eren's corduroys, for extra padding. "Nothing goes to waste here," he says.

My first lesson involves disabusing me of what Eren calls the *2001 fallacy*, after the monkeys banging rocks at the beginning of *2001: A Space Odyssey*. In my head, I figured we'd do something similar: bash stones together more or less at random and just hope for the best. Sometimes a tool would emerge, sometimes gravel. But no. As strange as it sounds, breaking rocks takes finesse.

To knap a stone tool, you hold one rock in your hand (the hammerstone) and strike another rock sitting in your lap (the future tool). The hammerstone should be a hard, dense mineral and can range in size from one to ten pounds, depending on the task. It should also be quite smooth; a

---

* Given his skills, people often say to Eren, "Dude, if the apocalypse comes, you'll be ready." Eren just shakes his head. "If the apocalypse comes," he says, "I'm not gonna sit around making stone tools. I'm breaking into my neighbor's house and stealing all their metal knives."

## Africa — 75,000 Years Ago

river cobble is perfect. Master knappers get pretty attached to their favorite hammerstones—Eren has shipped his to jobs all over the world, despite the cost—and they're always scouting for more. Eren once got detained by the police in Dallas for backing his truck up to a grocery store's landscaping plot and raiding some fine-looking hammerstones there.

For the actual tool, knappers prefer glassy, silicon-based rocks like flint, chert, or obsidian, which all fracture cleanly when hit. (You can also knap glass bottles. Eren knows people who knap porcelain toilets.) There's no surefire way to tell if a rock will produce a worthwhile tool until you start knapping, but good rock rings solidly when struck. If the rock produces a dull thud, you can toss it: it's infested with flaws that will lead to unpredictable breaks.

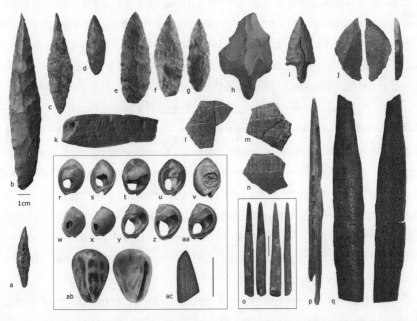

Human beings and their ancestors have been shaping stones into tools for millions of years. Here's a collection of blades and scrapers (a–j), ochre (k, ac), engraved ostrich eggs (l–n), bone tools (o–q), and shell beads (r–ab) from Africa. (Credit: Eleanor Scerri.)

Despite the name, you don't swing the hammerstone—you're not pounding nails into a barn. Rather, its weight alone inflicts the damage: You hold it a foot or so above the rock you're knapping, then simply let it drop, with your fingers guiding it down until it strikes. Moreover, you don't aim the blow at the middle of the rock. Instead, the hammerstone just catches an edge. I find one aspect of working with them especially counterintuitive: that the piece you knock off the target rock—usually a thin flake—peels off the *bottom*, and lands in your lap. In other words, knapping happens largely upside-down. You can't actually see the surface that produces the flake you want, forcing you to rely as much on touch as sight.

To demonstrate the precision involved, Eren takes a hunk of chert from an oil drum and inks a black dot onto it with a Sharpie. He then flips the rock 90 degrees to expose another face, and uses the marker to draw a dotted curve a few inches long. He claims that when he strikes the dot with his hammerstone, a flake of that exact shape will fall off. I know he's a master, but I'm dubious he can be so exact. Silly me. He settles the rock in his lap and gently drops the hammerstone. *Chonk*. The flake he holds up has the rough shape of a quahog clam, with a smooth bulb that flattens out to a sharp edge. And it's the exact size he predicted; he practically bisected the dotted line.

After Eren's demonstration, it's my turn. He inks a dot onto my rock while I adjust my safety glasses and try to remember everything. My first few, tentative blows achieve nothing, which embarrasses me. Who can't break rocks? I feel like the *Space Odyssey* chimps before the monolith. Finally, I gird my loins and land a proper blow—*crack*. Unfortunately, instead of liberating a graceful clam shell, I've shattered the rock in half. Eren says, "Well, okay." Taking the two halves, he finds the Sharpie dot he drew, then points out the scratch mark where I landed the blow. They're an inch apart.

For the next few minutes, Eren continues grimacing as I veer between prissy, too-soft strikes and he-man blows that sting my hands. None produce usable tools. I'm quickly learning to appreciate the skill of Stone Age knappers, not to mention their sophisticated knowledge of the 3D geometry involved.

Finally, on my dozenth blow, it all comes together. Just by the sound—*chonk*—I can tell it's a beaut: it felt pure and smooth in a way the other strikes hadn't. Madeleine-like, the blow also evokes a childhood memory.

## Africa — 75,000 Years Ago

I played baseball as a kid and rarely hit for power; I was far more likely to dribble a groundout to second. But one time, in a cosmic fluke, I connected perfectly with a pitch. The ball absolutely *popped* off my bat, soaring down the left-field line and bouncing over the fence for a double. It was stunning, effortless power. What I remember most, though, is that I barely felt anything touch the bat, as if I'd swatted a Ping-Pong ball. This rock-strike feels like that. The *chonk* rings out, and the chert absorbs every ounce of force from the hammerstone with no rebound or sting. I lift the chert to find a slender shard in my lap—a shoehorn with a razor edge.

Eren leaps off his chair. "Oh, yeah!" He holds the shoehorn up, running his finger along the rim. He pronounces it a perfect hide-scraper. "Welcome to the Upper Paleolithic! You just joined modern *Homo sapiens*."

After this, Eren teaches me some subtler aspects of knapping, including thinning rocks into blades using billets of moose antler, which he just has lying around. (Antlers are softer than rock, and therefore crack flint differently.) He also introduces me to perhaps the most coveted mineral in the ancient world, obsidian, a black volcanic glass that fractures into thin flakes and forms extremely sharp edges with minimal effort. Even a schmuck like me can produce an obsidian blade so fine it's translucent—I can see the ghosts of my fingers through it when I hold it up to the sun. To test its cutting prowess, Eren hands me a thick swatch of leather, which I initially, comically, try to tear with my hands. No chance. But obsidian makes easy work of it, dissecting it in one smooth, silken slash. It feels more like tearing newspaper than cutting hide. If knapped properly, obsidian blades are far sharper than even surgical scalpels, and a few modern surgeons have performed operations with them on overly enthusiastic experimental archaeologists. For my part, I find obsidian discomfiting to use; it's almost too good at cutting—seductively, dangerously sharp.*

---

* Why operate with steel blades, then, if they're duller than obsidian? Steel edges are straighter—which doesn't matter much if you're butchering meat but is vital for surgery. Steel also doesn't chip, and you can churn out steel blades in factories, whereas making obsidian still requires master knappers. Perhaps most importantly, steel keeps its edge longer, and you can sharpen steel blades without losing much metal.

After an afternoon of knapping, my pants are covered in rock dust, and waste chips are piling up at my feet. I'll admit that, despite Eren's guidance, I mostly mishit. But every time I strike true, I am ridiculously, childishly proud of the hide-scraper or little chopper that results. There's a famous saying that any sufficiently advanced technology is indistinguishable from magic. In this case, the opposite is true: this is Stone Age technology, but no less magical. Looking over my growing pile of choppers and scrapers, I can't deny my glee. I started with a pile of useless rock and, using just my hands, created *tools*.

Of course, the very thing that makes stone tools useful—sharp points, glistening edges—also makes them dangerous, both to handle and produce. This becomes clear in the office outside Eren's lab. On a whiteboard, he's drawn a clumsy blob-man in green marker, with two Xs for eyes. But my amusement fades when I notice lines and numbers radiating away from different body parts, and I realize what I'm looking at. Eren teaches stone-knapping to students, and the numbers on the blob-man document the lab's history of injuries—689 in five years, almost 140 per annum. Most are minor, sliced fingers and such, but other injuries are both alarming and mysterious. Two clumsy souls managed to injure their own "ass." Two others sliced or crushed their testicles. Someone managed to hurt their lungs. (By inhaling a chip?) There's even a case of acid reflux—which, after reading the butcher's bill here, I'm starting to understand.*

Even master knappers aren't immune. Eren once sliced his pinky so deep that he could see bone. Indeed, he's thought long and hard about

---

Honing obsidian requires flaking off bits of rock, and it reduces to a stub quickly. Nevertheless, obsidian is still sharper overall, which is pretty darn cool—that a material our ancestors used a million years ago can still outperform modern technology in some ways.

* After a knapping session a few years ago, an ugly growth erupted beneath Eren's eye—possibly, he feared, from a tiny flake lodged there. Still, he found the prospect intriguing. He's a fan of a macabre subset of scientific papers—the "bizarre medical injury" genre—and he dreamed of landing a byline in the *New England Journal of Medicine*. Alas, a biopsy of the growth found no trace of flint inside.

the dangers of knapping, and especially what those dangers reveal about the past. "There must've been some grisly injuries in the Paleolithic," he says, compounded by a high risk of infection. Yet people made billions of stone tools over millions of years. It's a testament to how valuable such tools were, that people kept at it despite the risk—when a clan's entire fate could turn on one badly timed blow...

Kayate lies at the entrance of the cave, fingering his spear and watching the sun peek over the distant hills. A fly buzzes around him in the dawn light, gaining strength as the day warms, until it blunders into a spider web and gets hopelessly tangled. Kayate watches it struggle for a moment, then returns his gaze to the landscape below.

The cave sits atop a steep slope the height of ten men, and getting up and down it can be exhausting. But this position has one distinct advantage. It's a perfect blind, situated directly over a mineral lick at the base of the cliff. Gradually, animals begin arriving in twos and threes for their weekly salt fix. Warthogs. Shaggy wildebeests. Zebra. Kayate smiles to see baboons with their quick hands. Most importantly, their staple game arrives, African antelope. Graceful impala, tiny quivering duikers, majestic mohawked kudu. All are variations on a theme—tawny fur with black and white tufts, plus sharp horns. One or two kills here over the next week, and Kayate's family has a good shot of surviving to the rainy season.

Kayate watches the antelope's ears for signs of fright. Drooping ears mean a relaxed animal; raised ears indicate possible flight. He notices two male kudu with giant spiraling horns eyeing each other territorially, circling and snorting. The drought has raised tension among the animals, too. Kayate flinches when they lower their heads and smash their horns together. The clack echoes like a thunderclap, and the ears of every animal spring up. Luckily, the kudu play nice after this, and the animals go back to nibbling salt.

Suddenly, Xate appears at Kayate's side, startling him. Injured or not, he's a stealthy hunter. He points his blood-caked hand at one of the males and whispers.

—Go for the kudu. That one.

Normally they'd go after juveniles or the elderly—easy pickings. Going after a beast in its prime is risky. But Kayate agrees with his brother-in-law. The elderly antelope have already died off this season, and the juveniles look as scrawny as his sister's child. Only the two males have bulges of muscle and fat, and their vantage point now gives him a good chance at a kill. Kayate rises to a crouch. The sure-kill zone on the torso of a kudu is just a foot in diameter, but if he can pierce its heart, they won't even need to run the animal down to finish it off.

After a deep breath, Kayate springs up and lets fly.

It's not a bad toss; it flies relatively true. But with the skirmish earlier, the kudu is on edge, and it flinches at Kayate's leap. The spear strikes it obliquely and lodges beneath its skin for a second, dangling like a porcupine quill. But once the kudu darts off, the spear clatters to the red dirt. All the other animals scatter, too.

Merely injuring an animal is not a disaster; it's normal, in fact. But Kayate groans. The kudu now requires tracking, which won't be easy in his weakened state. He just hopes the wound is deep. With luck, the stone spearpoint broke off inside.

But today is not a lucky day. Kayate scrambles down the slope, his bare toes gripping the crumbling red rock. Despite his injury, Xate insists on following, muttering that *he* would have killed the kudu outright. Down below, Kayate finds the spearpoint intact on the wooden shaft, which means a shallow wound. Xate clicks his tongue. At least there's a trail of blood to follow.

Kayate scrambles back up the cliff to assemble his hunting kit. He snags a pouch made of hide and tosses this morning's scrapers and handknives inside, to skin and butcher the kudu. He wraps some grass rope around his waist as well; a kudu is too heavy to drag back whole, and rather than leave the precious meat for scavengers, he'll string it up in a tree.

He turns to leave, only to see his sister standing in front of him. She's holding a turtle carapace with a paste of red powder and turtle fat inside.

—You need ochre, she says. The sun will be too strong.

—I don't have time.

But when he tries to push past her, she blocks him. It's futile to protest. Kayate sighs and raises his arms to let her rub the paste on.

Over the millennia, his people have used ochre for an astonishing number of purposes. The reddish-brown mineral makes glue stronger, and helps tan animal skins to prevent decay. It makes a fine pigment for rock art and face paint, and has astringent and antiseptic properties to treat wounds. It also acts, as Namkabe knows, as a mineral sunblock. More poetically, its rich red color evokes blood—a propitious sign for a hunt, and a solemn reminder that taking an animal's life is a grave and sacred thing.

Kayate feels her small hands patting his back. Why on earth did Xate, the better hunter, get injured and not him?

She seems to sense his dour mood.

—You will do well today. You have the most important thing a hunter needs.

He asks what she means.

—Do you remember when we were children, at the camp near the eagle nests? The adults offered a prize for the boy who could capture the most mice. Everyone else ran around like fools, trying to snatch them with their hands or spear them with twigs. Then I saw you, standing as still as a tree. What was it you told me you were thinking?

Kayate frowns. Why is she asking him this?

—I simply thought, if I were a mouse, what would make me come out of my nice, warm hole?

—So what did you do?

—I filled their holes with sand so they couldn't breathe, then scooped them up when they burrowed out.

—And you caught a dozen that way! Far more than anyone else. Just by putting yourself in the animal's mind. That's why you'll do well.

Kayate shakes his head. Those were mice. This is serious hunting.

When she finishes his face, Kayate grabs his toolkit and hurries down the cliffside. He finds Xate there examining the injured kudu's hoofprints, which run parallel to the trail of blood. Kudu have cloven hoofs, a pair of three-inch-long teardrops that taper to points. Unfortunately, there's nothing distinct about this kudu's prints—no chipped hoof or other defect to set them apart. Xate remarks that even he would have trouble tracking it. Kayate ignores the barb, grabs his spear, and begins jogging north alongside the prints, glad to leave his brother-in-law behind.

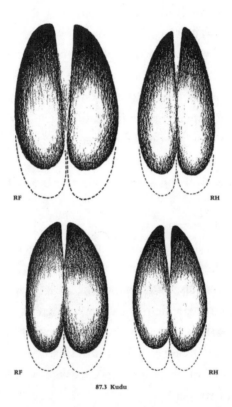

Kudu prints. By studying such tracks, hunter-gatherers could determine a creature's age and whether it was suffering from any injuries. (Copyright: Louis Liebenberg.)

The landscape ahead is a grass savanna, dotted with flat-topped acacia trees and dead bushes breaking through the red earth like skeletal hands. It's normally a lush green, but the drought has withered the grass and turned it yellow. After a half mile, Kayate turns and looks back, fixing the cave's location in his mind. To do so, he uses a huge rock formation behind it that looks like an anteater. His people fix their positions like this constantly on hunts; when your life depends on ranging far and wide, then returning to the exact same spot before dark, you need a firm grasp on your surroundings. As he fixes his position, he sees Namkabe waving from the cave entrance, limp child in arm. Kayate raises his arm in return, then jogs off alone.

There are a thousand distractions as he lopes along. Giraffes and elephants grazing in the distance, ostriches streaking by, the squeak-whistle of the red bishop bird, six-foot-tall termite chimneys skittering with activity. But for the most part, Kayate keeps his eyes low, scouring the ground for spoor, the physical signs of the kudu he's tracking.

In mud or wet sand, tracking is trivial: you follow the footprints. On a savanna, the signs are subtler — trampled tufts of grass, broken twigs, smears of saliva on leaves or foliage. Moreover, there might be a quarter mile between successive signs, making it easy to lose the trail. The key to a successful hunt, then, is not merely following an animal's movements but *anticipating* them. Kayate has to parse each clue and put himself in the animal's mind to guess where it's going.

Today, during one tricky stretch, Kayate passes a pile of leaves, and something catches his eye. Intense sunshine can bake leaves brown on top in hours, even as the underside remains green. Here, though, he sees a few leaves lying green-side up — meaning they were recently disturbed. But was it the wind, his kudu, another animal? Just beyond the pile he notices some high grass angling to the left. One patch hasn't sprung back up; it's been trampled recently. Kayate jogs over — and notices a drop of red on the blades. He dabs the blood with his finger; it's still tacky. His heart begins to race. The kudu must have turned here — perhaps heading for a small watering hole that way. Kayate twists his head, fixes his position relative to home again, and jogs off in the new direction.

At the watering hole, he briefly considers drinking before pushing the idea out of his mind. His people believe — deeply, unshakably — that hunters cannot take sustenance or drink water mid-hunt. During a chase, they say, humans and animals are cosmically, sympathetically linked: if the hunter eats and drinks, that gives the beast strength, too. Conversely, the hunter's suffering weakens the animal in lockstep. Drinking water is therefore out of the question. He needs to focus on the pursuit.

He soon finds kudu tracks beneath a nearby tree—too many tracks, in fact; other antelope have been resting here as well. Even worse, there's no blood. He fears the kudu's wound is scabbing over.

The only thing he can do is methodically eliminate the wrong tracks. On inspection, half are actually impala tracks, which lack the symmetry of kudu prints. Among the kudu, one set of prints reveals a chipped hoof, another an obvious growth. With a few prints, he can see that the dirt wall forming the side has started to cave in, meaning the track is old. Whittling things down from there takes more work, but one by one, he eliminates every track he can.

Unfortunately, two fresh, viable sets remain, and they veer in different directions. But he realizes that one set is on the east side of the tree, directly in the hot sun; he doubts the kudu would have rested there. And when he drops to his knees to examine these hoofprints more closely, he sees what looks like a tiny black seed inside one. It's not pressed down into the earth, so it must have been deposited after the print was made. Kayate picks it up, breaks it with his fingernail, and sniffs it. It's just as he suspected—a mouse turd. He's overjoyed. Mice only emerge at night here, which means these prints must have been made yesterday. He can safely ignore them.

He pops back up to his feet and scans the horizon. There's an archipelago of trees stretching off in the direction of the western prints—trees a wounded kudu would use to rest and cool itself. Everything fits. Thrilled to be inside the animal's mind, Kayate turns, fixes his position again, and takes off.

The sun is now cooking the earth, but with each mile, Kayate grows more exhilarated. He's already fantasizing about the first morsels of food he'll eat when he runs the kudu down—the silky liver, the tangy glands in the throat. Before long, the heat and the fantasy push him into a delirious fugue.

Even when he begins faltering, it doesn't bother him initially. He's already put himself inside the kudu's mind, and in his delirium, he now feels their bodies fusing, too. Somehow, his stumbling merges with the kudu's stumbling, until he can feel the trembling of its legs inside his own. He's all but *become* the kudu. He starts to laugh.

Over the next hour, however, reality intrudes. Every rivulet of sweat—on his arms, legs, chest, face—wicks off more ochre, until just bare skin remains. As the sun begins broiling him, his breath grows labored, his spittle goes gummy, and he can barely peel his tongue off the roof of his mouth. He passes a few tempting puddles of water but refuses to stop. If he's growing weaker, the kudu must be near collapse. So he staggers on, switching his spear from hand to hand whenever an arm grows tired.*

A mile on, he checks a bush for spoor and clumsily steps on a turtle shading itself beneath. It hisses at him. As he catches his breath, a foul odor tickles his nostrils. Curious, he follows it along a path into a nearby cluster of bushes, where he finds a dead ostrich on the ground. He's not

---

* Kayate could never outsprint an antelope, but they're poor long-distance runners, in part because they can't sweat to dissipate body heat. So in desperate straits, human beings—who are excellent long-distance plodders—can simply pursue antelope until they collapse from heat exhaustion. Injured antelope are especially susceptible, since they're simultaneously bleeding out.

Biologist Louis Liebenberg once participated in a "persistence hunt" like this in the Kalahari. He described it as a "trance-like state": "I could not only see how the kudu was leaping from one set of tracks to the next, but in my body I could actually feel how the kudu was moving. In a sense it felt as if I myself actually became the kudu, as if I myself was leaping from one set of tracks to the next." But as he grew weaker, the heat clobbered him: "At one point a cold shiver went through my whole body and...I realized that I was dragging my feet in the sand. Sometimes my legs buckled under me and I would stumble over branches...My mind was simply dragging my body along." When the antelope finally crumpled and the hunt ended, Liebenberg realized he'd stopped sweating—a dangerous state. The hunting party had to dispatch someone to bring emergency water, while Liebenberg rested under a shady tree, trying desperately to cool his body. He barely survived the ordeal.

above scavenging, but this ostrich has been picked over already—wild dogs, by the look of it—and the putrid smell of the remaining flesh puts him off. He turns back to keep pursuing the kudu.

But he struggles to get moving again. His feet are dragging in the dust, and his thighs seem to weigh more than the rest of his body combined. Then a cramp seizes his calf, sharp and shrill, slowing him to a hobble. He's teetering on the verge of dehydration. Even if he can run down the kudu in this state, what then? Unless it dies right next to a waterhole, he himself might collapse alongside the beast. He'll need to take some nourishment along.

He wobbles over to an exposed tendril and begins pounding the ground with the butt of his spear. Then he drops to his knees and paws the earth. Eighteen inches down he finds a *xwa*, or water root, a dirt-encrusted tuber. It's the size of a child's head, and has enough pulpy liquid inside to fill half his stomach. He'll carry it with him, and wait to devour it the instant the kudu dies.

But as he stares at the water root, everything else around him—earth, sky, trees—falls away. He licks his scaly lips and tells himself no—he can't give the kudu strength.

He also can't tear his eyes away. And before he knows it, he's fishing a chert knife from his pouch and making a shaky cut through the rind. He carves out a chunk of sticky pulp. He'll only take a few mouthfuls, just to get running again. No more.

But as soon as the first drops of sweet, starchy milk hit his tongue, his resolve collapses. Every bite is delicious, and deeply shameful. He's ruining the hunt, giving the kudu strength. But he can't stop himself. He swallows faster and faster, licking his fingers clean, then tosses the empty rind aside, grabs his spear, and digs up another.

After the third xwa, his head feels right; he can resume tracking. But as his strength trickles back, so too does the enormity of what he's done. He fears there's no point in tracking now: he's almost certainly jinxed the hunt. On the other hand, the thought of returning to the cave without food seems disgraceful. He forces himself to plod on.

He's soon glad he did. At the next tree in the chain, a new set of

## Africa — 75,000 Years Ago

prints grabs his attention — elephant tracks. Elephant herds are common here, but this looks like a lone one. There's something odd about the tracks, too. He gets down on his knees, to catch the sunlight from a better angle.

The view confirms his suspicion. Three of the prints are well formed and deep, but the back right one is curiously shallow. This elephant is limping, badly.

Compared to kudu, elephants make obvious tracks, and as Kayate's eyes follow this set into the distance, his mind begins whirring, trying to get inside the beast's head. Why would it go that direction? The only thing over there is a fetid watering hole, where a few wildebeests died recently and fouled the water. No animal in its right mind would drink there. So why would the elephant? It's clearly hurt, and even beyond the drought, it must be desperate for water. Only one thing could provoke such a mad thirst: the elephant is dying.

Kayate begins jogging alongside the prints, searching for any sign that they're not fresh; he finds nothing. A quarter mile on, his eyes go wide with excitement. There's a puddle of urine, so fresh there's still foam on top. It can't be more than a half hour old.

Kayate now has a decision to make — resume the search for the kudu, or gamble on the dying elephant. But the more he thinks about it, the more it seems like no gamble at all. The kudu is barely bleeding anymore, and will only grow stronger after the xwa. The elephant, meanwhile, is on its last legs. He also knows a shortcut to the watering hole — a cliff he can climb down that the elephant cannot. He can be there in minutes.

For what he hopes is the last time today, Kayate turns, fixes his position and takes off.

Archaeologists have long known that human beings ate pachyderms once, based on cut marks on elephant and mammoth bones. But the dig sites where the bones emerged were messy and hard to interpret, because scavengers and geological processes often scattered the bones over time. Moreover, soft flesh decays quickly in the wild, erasing clues about how events unfolded. As a result, many basic questions about big-game butchering remained unanswered. What tools did people use? How long did processing the meat take? How hard was the work? No one knew—at least not until the Ginsberg affair, one of the first historical forays into experimental archaeology, and certainly the most madcap.

In the 1970s, Ginsberg the elephant lived at the Franklin Zoo in Boston. She was a featured attraction there, having starred a decade earlier in a John Wayne rom-com (really) called *Hatari!*, set on an African savanna. Sadly, in December 1977, at age twenty-three, Ginsberg fell and broke her leg and died from a blood clot.

The zoo's loss was archaeology's gain, however. A scientist at the Smithsonian Institution heard about Ginsberg's death on the radio, and alerted some colleagues who were eager to study elephant butchering. Several hurried phone calls to Boston followed. Bizarrely, the zoo had already promised the head to Harvard, and had taken the viscera for its own research, but the Smithsonian got the remaining four thousand pounds.

The catch was that the zoo didn't deliver; the archaeologists would have to pick up the carcass themselves. At most places, the plan would have died there. But the Smithsonian crew simply walked down the hall to the cetacean curator and borrowed the flatbed truck he used to haul whales around. A junior archaeologist was dispatched to Boston, where he arrived in an icy rainstorm. He somehow managed to heave the carcass onto the truck, then bravely got on the interstate and started plowing south, his windshield wipers churning furiously. (Imagine being a driver that day. You're just trying to keep your little Gremlin or VW Bug between the yellow lines and not crash. All of a sudden a huge rumbling flatbed roars past with a headless elephant in back, its tail flapping in the wind. A story, perhaps apocryphal, says that a highway patrolman finally pulled the archaeologist over and asked what the hell was going on. The archaeologist shrugged and said, "Roadkill.")

The experiments to butcher Ginsberg were an odd mix of high-tech and primitive. To match the tools of Kayate's time, the archaeologists knapped sixty-five flint and obsidian choppers and scrapers. They also recruited a NASA engineer to attach electronic cables to each tool to measure the force and angle of every cut. They fed the data to a rewired polygraph machine, which scribbled everything down.

In all, it took three archaeologists eighteen hours to skin and deflesh Ginsberg—a serious investment of time that left them grumpy and sore. That said, Stone Age butchers could probably do the job faster. The archaeologists had zero experience dressing elephants, and spent the first few hours fumbling about. They also wanted to save the elephant bones for study, so they'd meticulously cut away every bit of cartilage and tendon, whereas ancient butchers wouldn't have bothered. In all, they liberated a full ton of surprisingly lean red meat. They considered eating it, until they remembered that Ginsberg had been tranquilized before her death; they feared they'd croak themselves.

Ginsberg kicked off something of a craze for elephant butchery over the next decade, as several other teams ran their own experiments, using other zoo and circus specimens. The collective results of this research were eye-opening for archaeologists. The first and most obvious thing they noticed was that elephant hide, which can be an inch thick, is an absolute bitch to cut. Even the sharpest stone knives made little headway, and sawing proved only marginally better. In either case, the tools disintegrated in archaeologists' hands, pieces flaking off left and right. One team also learned the hard way that you need to strip the whole hide off the animal right away, to let body heat escape. Otherwise, putrid gases build up inside the carcass and spoil the meat.

Some of the work got downright macabre. It turned out that cutting the soft red flesh *beneath* the hide was fairly easy—so easy, one team reasoned, that perhaps Stone Age people didn't waste good flint tools on the job; blades of bone might suffice. So they used a boulder to crack their elephant's humerus into shards and had at her with those. It sounds like the premise for a horror flick—butchering an animal with its own bones—but it worked well. One grizzled old archaeologist also decided to use Ginsberg's tendons as ropes, to bind stone points to wooden spears. To separate the tendon fibers and soften them up, he popped them in his mouth to chew. One observer recalled watching him gnaw away with "an odd string, still meaty, hanging out of the side of his mouth like a wisp of cold spaghetti."

Most archaeologists in these experiments sliced into the elephant's haunches first, probably influenced by the modern bias for premium cuts of beef. Kayate, in contrast, would have known from his elders to start with the choicest bit of an elephant—the trunk. An elephant trunk contains 40,000 supple muscles (the entire human body contains 600) and there's no bone to work around. Still, Kayate himself had never butchered an elephant before—they're rare finds. And he's about to learn that, even beyond the hard work involved, it's a treacherous operation…

Climbing down the cliff, spear in hand, Kayate can smell the fetid watering hole. Thousands upon thousands of flies are buzzing about, some so aggressive they crawl into his mouth and nose.

Worryingly, he doesn't see the elephant at the water. But when he jogs around the edge of the outcropping, his heart soars to see it collapsed in a heap on the grass. It's an old bull with a deep gash on its leg, probably from a younger rival's tusk. As elephants do, it's plastered the wound with mud—but infection set in anyway. Kayate doesn't waste time ruing its fate, though. He's got to butcher it quickly before any scavengers arrive.

Laying his spear aside, Kayate straddles the trunk and selects several tools from his pouch; he plans to remove a quarter of it. Even beyond his fatigue, it's beastly work. The blood and greasy flesh leave his hands slick, even after he dries them with grass, and the heavy folds of hide keep flopping down where he's trying to cut. He quickly dulls several blades.

After a half hour, though, he's nearly detached the shank, a hundred pounds of trunk meat. He can take some home tonight to roast, and hang the rest in a tree for safekeeping, to haul back later and smoke into biltong jerky. The thought of his first bite makes his mouth water, the

## Africa — 75,000 Years Ago

rich juices running down his chin. But even more vividly than the meat, he imagines his reception upon returning to the cave: Namkabe weeping in joy. Xate fuming in jealousy. His little niece happily suckling milk. Just a few minutes more, and he'll be a hero.

A sharp bark interrupts his fantasy. A dozen more follow, and a spasm of cold dread shoots down his spine. He snaps his head around to see a pack of wild dogs glowering at him.

When they charge, Kayate does his best. He screams and swings his spear to drive them off. But the curs dart past him on all sides, quick as demons. It's like trying to stab the wind. One finally plows into him and knocks him flat.

He hurries to his feet, but they're already devouring the piece of trunk he'd cut, their snouts bright red. Others have torn through the thin skin around the belly to snack on the inner organs. One, obscenely, has its head up the elephant's anus, munching the soft intestines.

With his family's food disappearing into their maws, Kayate grows crazed with desperation. He wrestles one dog off the trunk, grabs his spear again, and darts forward. But the dogs easily sidestep him, and after one errant jab, the spearpoint snaps on the grizzled elephant hide.

Kayate is weaponless now — and the dogs sense it. Several turn to snap at him. Desperate as he is, he knows he won't bring back any food if he's dead. He has no choice but retreat.

After backing away, Kayate prays for the dogs to hurry and finish. But they seem insatiable. And before long, he notices shadows crisscrossing the carcass. Vultures. It's a bad sign — they'll attract other animals.

Sure enough, within ten minutes, he hears the mocking laughter of hyenas. They'll likely drive the dogs off, but that's hardly an improvement. And the only things that drive hyenas off are lions. He's forced to face the truth: he's lost the elephant.

The thought crushes him. He had food — a hundred pounds' worth! — in his hands. His fingers are still wrinkled with its blood. But his chance has slipped away. The kudu is certainly gone by now, too. There's little to do except trudge back home in disgrace.

He's never faced such a long walk in his life. He scuffs along, his feet barely rising, his broken spear dragging behind him. Thoughts of every stupid mistake he's made today pound inside his skull. For all his supposed cleverness, he's acted like an outright fool.

In thinking over his path home, he remembers the turtle in the bush he passed, just before eating the water roots. He decides to grab it. At least his sister can have a little food to make milk for the baby, provided he can hide it from Xate.

Forty sweaty minutes later, he's retraced his steps. Thankfully, the turtle hasn't strayed. Kayate hoists it up by its shell, watching its leathery legs and head disappear inside.

Standing there, he catches another whiff of the rotting ostrich carcass in the cluster of bushes. He's desperate enough to reinvestigate. He lays the turtle upside-down — it can't right itself — and when he reaches the dead bird, he peels back a few inches of skin. The muscle beneath should be dark red, but this meat looks gray, and it's as much liquid as solid. The pinch he takes between his fingers is slimy and begins to drip. He forces himself to try some anyway.

One chew, and he spits it out. It's unspeakably vile; he actually scrapes his tongue with his fingers, but a film remains. He gags and spits again, and before he can stop it, the remaining water root in his stomach erupts out of his mouth. He's left kneeling in muddy vomit, ruing yet another mistake.

But while staring down into the dirt, he notices something. Ostriches take "baths" by rolling around in dust, and the patch of dirt ahead of him shows signs of a recent bath. Which is strange, since the ostrich he saw has clearly been dead for days. Is there another nearby? Kayate scans around, and spies something that confirms his hunch — a black feather snagged in a bush.

He deduces what must have happened. Ostriches are territorial, especially when defending a nest. They're quite dangerous, too — a single kick can stun and kill a lion. The dead one must have wandered through recently and lost a fight. But the fresh bath shows that the other one — and its nest — are near.

## Africa — 75,000 Years Ago

With renewed energy, Kayate rises to his feet and hurries off to grab the turtle. He's suddenly got a new plan.

To Kayate's people, an ostrich egg represented both a quick meal and a valuable means of storing water. To understand more about the eggs, I ordered one online from a farm in New Mexico for $128 and set about processing it. It arrives in a cardboard box swaddled in diapers with pictures of baby monkeys on them—a buff-colored sphere with a dimpled surface like a small moon. It has the volume of two dozen chicken eggs and is dense like a cannonball, although I can feel the fat yolk sloshing inside.

A decorated ostrich-egg canteen. (Credit: Lam Museum of Anthropology, Wake Forest University.)

The accompanying instructions (it needs instructions) suggest opening the egg with a hammer and nail, but I do it Kayate's way. I dig out a chert blade from Kent State, plop the egg in my lap, and begin scraping a tic-tac-toe diagram on it, from which I'll remove the inch-wide center square.

This proves a gigantic pain in the ass—ostrich eggshells are *thick*. For the better part of an hour I cut, saw, scrape, gouge, and furrow into the shell with one snaggletooth edge. Nothing really works. I'm left marveling at what tough little buggers ostrich chicks must be, to break through this prison when they're just weeks old.

Finally, finally, after some intensive chipping, one of the tic-tac-toe lines breaks through. Clear foam burbles up from inside the shell, as if under pressure. I dab a bit onto my tongue; it tastes vaguely of minerals and meat. Twenty minutes later, I've scored the four lines deeply enough to start (gently) smashing the center box with the heel of the scraper.

When struck, the shell doesn't shatter like a chicken egg does, in a spider-web mosaic; instead, the ostrich eggshell flakes off in pieces like dried clay. I brush the debris into my Elvis ashtray from Graceland, pleased to finally have a use for it.

After pecking out the center box, I find a membrane beneath, white and crinkly like parchment. Once I pierce that, I enter the sanctum sanctorum—the inner egg. Using a flashlight, I see the whites jiggling like thick oil, and the goldenrod yolk floating inside them. I get a bowl and start pouring.

The whites flow out smoothly. The yolk does not. It too has a membrane, and while it does start to pour out of the hole, the membrane catches it after an inch, leaving it dangling like an orange-juice loogie. I began shaking the egg violently, hoping the sharp edge of the hole will cut the yolk's membrane. Finally, there's a faint slurping sound, and the yolk comes tumbling out like a waterfall of caramel, thick and creamy. I can't resist running a finger through and trying it. It tastes rich enough to make my knees bend—intense yolk flavor with hits of umami. I greedily swipe another fingerful. Unfortunately, I get so caught up with the taste that I don't see how rapidly the bowl is filling. It nearly floods over while I scramble with my yolky hand to find a free bowl. There's a lot of egg here.

Over the next week, I eat some each day as an omelet. The yolks are every bit as delicious cooked as they were raw. As for the whites, well. They lack much taste, and while I'm usually not too picky about texture, they fluff up oddly when cooked, like spongy snot. By week's end, they're making me queasy.

I have better success rinsing the eggshell out and using it as a canteen. It holds six cups of water. And while the wall's quite thick, it's porous enough to lose small amounts of moisture to evaporation. Much like sweating, this draws heat out of the water and keeps it cool even in my warm apartment. In other words, unlike metal or glass, ostrich-egg canteens are self-chilling. The shell also gives the water a chalky tang, reminiscent of country wells. All in all, between the yolks and the canteen, I can see why ancient people coveted such eggs—and why Kayate would risk going after them…

Holding his breath against the stench, Kayate unpeels the slimy ostrich skin from the carcass. He lays the deflated hide on the ground, and uses a chert blade to scrape as much of the putrid flesh off as he can. When it's finally clean, he slices two eyeholes in the chest a few inches apart.

His costume complete, he throws the skin over his shoulder, grabs the turtle—its legs dart inside again—and returns to the ostrich bath, to find the tracks leading away. He follows them five hundred paces through some more scrub brush toward a clearing. Concealing himself, he creeps forward and spies an ostrich sitting near a clutch of eggs. Kayate studies the scene, then retreats until he finds a suitably springy sapling, at which point he begins to construct a snare.

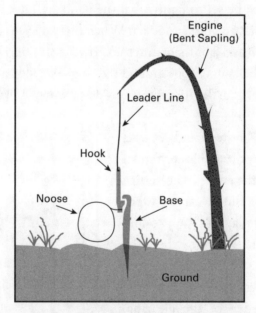

A snare to capture small game. Trappers placed bait inside the loop of cord, and animals triggered the snare by poking their heads inside and disturbing it. The cord is attached to a bent tree that springs upright and snags them. (Credit: Benjamin Hollis.)

It has several parts. First, he finds two branches on a bush and snaps them off. One branch will become a plug three inches long, the other a foot-long stake. Using a chert tool, he carves a notch near one end of the stake, then anchors it into the ground with the notch end free. The plug also gets a notch. Now he needs rope. He unwinds the rope around his waist, forms it into a loop—the snare—and secures that to the plug. To make a second piece of rope, he rips several dozen handfuls of dried grass out of the ground and rolls these on his thigh to form bundles. He then lays the bundles end to end and twists and splices them together, lengthening the rope until he's got enough to reach from his shoulder to the ground. He ties one end around the sapling and knots the other end around the plug.

Now for the most ticklish part—setting the trigger. He bends the springy sapling down, and slides the notch of the plug beneath the notch of the anchored stake, to keep the sapling bent. It's a delicate balance: the trigger can't be set so lightly that the wind activates it, but can't be set so firmly that an animal doesn't. When he's satisfied, he snaps eight twigs off a bush and jabs them into the dirt vertically, forming a ring; he then spreads the loop of rope around the ring. Finally, he baits the trap with the hapless turtle, flipping it upside-down and placing it in the center of the loop.

Normally Kayate would set up a wall of thorns or brambles to funnel game toward the snare, then return in a day or two. But he cannot afford to be patient today. He's got to lure the ostrich here, which is where the skin comes in. He picks it up from the dust and creeps back down the path to the edge of the clearing. There, he takes his broken spear and works the shaft up the neck into the head. Holding the head high, he tosses the rest of the skin over himself like a blanket. Despite his scraping the rotten flesh off, the odor

STALKING THE OSTRICH.

African hunters sometimes donned ostrich skins, to both spear the birds and steal their eggs. (From *Through the Kalahari Desert* by William Leonard Hunt.)

inside leaves him gagging. But he finds the eyeholes he cut and, with a confidence he doesn't feel, struts into the clearing, lifting his legs high in the slow, bobbing prance of an ostrich.

When Kayate first saw an elder do this many years ago, he thought it looked preposterous. How would a rotten skin on a stick fool an ostrich, especially with two human legs beneath? But the birds readily fall for the trick—as long as the person in the skin commits.[*]

As Kayate struts into the clearing, the ostrich's long-lashed eyes snap to attention, and its serpentine neck begins swinging back and forth. Kayate has seen elders roar in perfect imitation of an ostrich, but he doesn't dare risk that. Instead, he fake-charges, running several steps and throwing his elbows out to make the wings flap. Warily, the real ostrich hops up, unfurling to its full eight feet of height. Marching forward, it spreads its wings and hisses. Kayate hisses back, hoping that he seems threatening.

After a short standoff, Kayate begins to back away into the bushes, albeit slowly. He needs to agitate the ostrich enough to lure it toward the snare, but not so much that it sprints and attacks. So as he retreats, he stops every so often to hiss and flap his arms.

It's a delicate dance, but the ostrich follows Kayate's lead. Fifty yards later, he steps over the snare, relieved to see the turtle still there. He hisses one last time, to lure the ostrich forward. Then he turns and runs, as if he's decided to flee.

A moment later, he ducks behind some scrub and discards the ostrich skin, grateful he can breathe again. When he creeps back along the path, he finds the ostrich standing over the turtle, considering this potential meal. Suspicious, it prods the turtle with a talon, turning it in a small circle. Kayate watches with clenched breath. At last, one final nudge knocks the turtle against the trigger. An instant later, the bent

---

[*] People really did hunt ostriches once as described here. The trick worked because, when recognizing their fellow creatures, most animals don't focus on the gestalt, the whole; they focus on specific parts. (With ostriches, it's the bobbing head.) So as long as you imitate those parts convincingly and avoid any strange, giveaway movements or noises, some otherwise intelligent animals are surprisingly easy to fool—at least for a bit.

sapling springs upright and the loop snares the ostrich's leg. There's a strangled yelp as it's torn off its feet; its rump flies upward and its head slams into the dirt.

Kayate takes off running. Not toward the ostrich—he could never dispatch it with a broken spear. Instead he swerves around the furious bird and toward its unprotected nest.

He arrives to find a dozen eggs in a shallow depression. He skids to a stop, drops to his knees, and balances as many as he can in his arms. Each weighs a few pounds and is the size of a small melon. He can easily carry three—one for each adult in his family—but badly wants an extra for Namkabe.

While he fumbles, he hears the ostrich thrashing. He peers over the tops of the bushes to see the bird upright again, trying to tear the sapling out with its powerful leg. The roots must be barely holding.

But at last, Kayate manages to wedge all four eggs into his arms and stand up. While the bird flails and squawks, he hustles off with them knocking against his heart, a smile lighting up his face for the first time all day.

Dusk is falling by the time Kayate sees the anteater rocks in the distance. He stashes one egg in a tree hollow, so Namkabe can eat it later, then trudges toward the cliff. The oldest person he's ever seen in his life was his sixty-year-old great-grandmother, a toothless woman with deflated breasts who practically had to be carried the last mile every time they moved camp. She died years ago, but Kayate thinks of her now as he drags himself up the final few feet; his muscles feel every bit as weary as she always looked.

It's nearly pitch black inside the cave. He treads softly, but his sister pops awake anyway—her eyes wide in the dark. She's holding the limp baby to her chest. Xate lies motionless beside her, snoring.

—Did you bring back any food?, she asks.

He holds up an ostrich egg. She wrinkles her nose.

—You go after a kudu, and this is all you return with? They're probably rotten.

Kayate's heart swells to hear this. His clan has strict taboos about praising people after hunts. Instead, they disparage the meat and belittle the hunters' skill, all to prevent them from getting swollen egos that might lead to conflict or hoarding. The group always comes before the individual. And the more desperately the group needs food, the more miserable they pretend to be. Namkabe would never be so harsh if she weren't so grateful; if she were truly disappointed, she'd turn aside in silence.

Kayate wants to smile, but knows he has a role to play, too. He twists his face into a mask of shame.

—I know it's miserable, but please accept it.

—I have no choice. Now sit, tell me what happened.

This too is vital—the telling of the hunt. It fixes lore in people's minds: which tactics work, which don't. So as tired as Kayate is, he relays the whole affair. The kudu tracks, the elephant, the wild dogs, the turtle and the ostrich and the snare. He supposes he'll have to tell it to Xate again tomorrow. Which reminds him.

—How is Xate's hand?

—We closed the wound with tree sap. He should be able to hunt tomorrow, or the day after.

Kayate is relieved to hear it. He hands the eggs to Namkabe, who nestles them into the hot sand near the fire to cook. Kayate settles down on his bed of grass to wait. They're one day closer to the rains.

# South America — 7500 BC

⁓⁓

After the emergence of modern humankind in Africa, *Homo sapiens* began spreading across the globe, first into the Middle East, then into Eurasia and elsewhere. This was not a clean or linear process. We skipped over some spots, paused in others, and sometimes made initial forays into a region only to abandon it and migrate back home. Human colonization, then, was not one unbroken wave of expansion, but a series of fitful swells and retreats, especially with the last two continents to be populated, the Americas.

The ancestors of today's native North and South Americans came from Asia, probably Siberia and China. They likely walked across a land bridge between Russia and Alaska at a time when sea levels were far lower and the modern Bering Strait was dry land. The details of this crossing remain murky, but the first migrants possibly paused on the now-submerged land bridge for a few thousand years before entering the Americas and dispersing around 17,000 years ago.*

Regardless of the exact details, as soon as people started migrating down from Alaska, they moved remarkably quickly into North, Central, and South America—either by walking or by taking canoes along the Pacific coast. Even the most forbidding parts of the Americas, like

---

* Contributing to the murk, the recent discovery of fossilized footprints in New Mexico could push back the arrival of the first Americans to 23,000 years ago or even further.

the Andes highlands of the next chapter, show signs of sustained human activity by 12,000 years ago.

None of these first migrants, however, were entering empty land. Animals of all sorts roamed North and South America long before humankind existed. And some of these animals resented the intrusion of their loud, clever, and quite deadly neighbors...

Asana tosses aside her vicuña-fur blanket and steps outside the grass hut. She stoops to grab a gourd full of water and cracks her thumb through the overnight crust of ice. Between drinks, she studies the landscape around her, thinking about the day's hunt. Their camp is nestled in a sunken valley, with an escarpment to the west. To the east, the patchy dirt rangeland is dotted with thousands of scraggly bushes and a few marshy depressions. It rolls onward for an empty mile before terminating at the foot of several low, black hills, hunched like giant lumps of coal. A single darting bird breaks the stillness.

But a moment later, the wind kicks up, lifting her hair. The swirling gusts bounce between the hills, and before long, they conjure up a rare morning dust devil. It must reach ten yards high, a writhing brown snake far taller than any tree she's ever seen. It skips along toward the black hills, hopping across the dirt. People in her clan generally fear dust devils as bad omens. Asana doesn't. She loves their furious, churning energy.

When it dissipates, she sets down the gourd and kneels by the smoldering fire, brushing the sandy soil aside to reveal a cache of rocks she buried last night. They're made of chert, and still warm to the touch. Fire transforms chert, making it glossier and easier to knap into sharp spearpoints. Even more magically, it changes the stone's color, dappling the bland cream base with swirls of green, streaks of yellow, and pink and indigo spots. Asana clutches them like a gnome's treasure.

—Asana! The fire's dying!

Asana sighs. It's her aunt, sticking her round face out of their hut. Asana promises to fetch some fuel, but her aunt keeps babbling.

—If the fire dies, how will we eat? We'll starve. Or we could freeze, or a puma will come and...

Asana rolls her eyes. Both women have long black braided hair, and they're wearing tailored hides of vicuñas, small relatives of llamas. Like most people in elevated lands—they live above 13,000 feet—they're both short, an adaptation to cut down on caloric needs. But while

Asana's aunt is thirtysomething and stout, with wrinkles of worry creased into her face, the teenage Asana remains lithe and wrinkle-free.

Normally they'd be traveling with a few dozen others, but it's Asana's initiation week. To become a full-fledged hunter in her clan, she has to spend a week tracking and bringing down animals—a quota of at least three vicuñas, and ideally a bigger creature or two as well. Her tribe does most of its hunting communally, cooperating to dispatch game. But given the importance of hunting for food—there's little to forage up here—each hunter must be skilled enough to kill alone. Last month, Asana requested permission to head out and try, but she was denied. She's the only teenager in the clan—everyone else is a decade older or younger—and the elders often lump her in with the children. It enrages her; she feels herself being groomed for a menial role like her aunt.

So three nights ago, Asana made a decision: permission or not, she would head out hunting and return with proof that she could handle herself. After dusk, she waited until everyone had settled down to sleep, then sneaked off—or tried to. Her aunt caught her, and threatened to tell the elders. After an angry, whispered standoff, they compromised and set out together, to their mutual misery. Even worse, due to her aunt's slow pace and anxious demands that Asana keep close by, three empty days have passed with no kills. When the clan catches up—and it will—she'll need to have vicuñas to show for herself. Otherwise, she risks permanent servility.

Her aunt, however, has other priorities.

—The fire?

She stands there wringing her hands. Without another word, Asana stomps off for fuel.

In most of the world, "fuel" means wood, but the altiplano highlands lie above the tree line. Asana is seeking vicuña dung. She walks east, away from the escarpment, until she finds a mound of small black pellets. She scoops several handfuls into her outstretched coat, and returns to dump them on the fire. Feeling guilty—given the cold here, the fire *was* dangerously low—she makes a second trip to save her aunt the effort. It's as close as she'll come to apologizing.

Still, she doesn't have to endure her aunt's company. Asana orders her aunt to start breakfast, grabs her heat-treated chert, and retreats to knap in peace.

As soon as she begins working, Asana feels her mood lift. And while she'd really love some obsidian, she's confident her luck will turn around today with these new chert points.

After shaping a number of them, Asana moves on to hafting—binding each of the best dozen points to a wooden shaft with resin and vicuña tendons. When all twelve are finished, she checks each one's alignment by balancing it in the dirt point down and spinning it between her palms to see whether it wobbles. Eleven of the dozen look perfect, and even the misfit has its use.

With the preparations for her hunt complete, Asana has worked up an appetite. So she returns to the hut and calls to her aunt for the staple breakfast of the Andean highlands—the succulent, nutritious, and highly poisonous wild potato.

As Asana settles herself in the hut, her aunt removes a score of finger-sized potatoes from a crude earth oven. Asana points out that they're overcooked—she overcooks everything these days. Her aunt doesn't answer, just hands her some clay dipping sauce in the concave shoulder blade of a vicuña.

By the time her aunt finishes her first mushy potato, Asana has wolfed down three. There's clay sauce all over her fingers and chin. She even burns her mouth on one, but reaches for another before she's done swallowing. When her aunt scolds her to slow down, Asana protests through a full mouth that she's still hungry. She wants more.

—That's your problem. You always want more.

—What's that supposed to mean? Besides, we don't need to save any. I'll get a vicuña today.

—You said that yesterday.

The words infuriate Asana.

—What do you know about hunting? You just sit around all day.

—You think that's what I do, just sit around?

They've clashed on this point before—indeed, it's a classic quarrel, hunters versus gatherers. Gathering plants provides steady, consistent calories to fill the belly. But it's a low-prestige activity, the province of children and the feeble. Hunting, meanwhile, succeeds only intermittently; failure is the general rule. But all failure is forgiven when someone drags an animal back, and people can sink their teeth into its flesh. Hunting, then, is exalted, gathering scorned.\* In truth, societies need both hunters and gatherers to survive. But that truth doesn't prevent tension from arising. Asana holds up her feet.

—Look at the cuts on my soles. I spend all day walking and climbing. How long do you work?

—Long enough to fill your mouth.

—Vicuñas are a little harder to catch than potatoes. But I guess they're the only food you can outwaddle. How can you grow so large when you cook so badly?

Seeing her aunt's face fall, Asana regrets the barb. Not about her weight, but her cooking. Before she can take it back, her aunt hurries from the hut, her lip quivering.

Asana groans. However overprotective, her aunt means well. And if she overcooks everything, well, that's understandable given how Asana's family died of tainted food. She just wishes her aunt didn't make her so angry. She shoves another potato into her mouth, licks the clay from her fingers, and heads outside.

She blinks in the bright sun—it's a clear day, utterly cloudless. Her aunt has her back to her.

—Auntie? Auntie, look at me.

She stubbornly keeps her back turned. Asana feels her bile rising again—until her aunt lifts her arm. She's pointing at something.

Asana raises her eyes to the escarpment fifty yards away. There, descending it, is the last thing she ever expected to see. A woman from an alien tribe.

---

\* This difference is mirrored in the rites, or lack thereof, associated with each activity. Hunter-gatherers collect and eat plants without fuss—you just grab and scarf. But success in hunting is so fickle that it virtually always has elaborate rituals associated with it. You don't need the gods' help to dig up potatoes; to bring down vicuñas, you do.

# South America — 7500 BC

Potatoes were first domesticated roughly 10,000 years ago around Lake Titicaca, near the modern border between Peru and Bolivia. I visited the region to learn more about ancient potato varieties, and the first thing that struck me is their astounding array of colors. Some heirloom spuds are canary-colored, some fuchsia, some deep purple or blue. Some reveal kaleidoscope starbursts of color when sliced open. The shapes are a hoot, too: pinecones and curlicues, boomerangs and swollen raspberries, fingerlings and golf balls. In comparison, the homogeneous brown binfuls in modern grocery stores seem bland and unimaginative.

The only problem with those lovely varieties is that they're poisonous. Potatoes belong to the same botanical family as deadly nightshade and are teeming with toxins called glycoalkaloids. The potential for harm here ramps up quickly. Nibble a wild potato, and your lips and tongue will burn. Eat a few bitter mouthfuls, and you'll vomit and cramp with diarrhea. If you somehow manage to choke down more, you'll start convulsing and seeing things before you pass out and die. Potato peels are especially toxic because glycoalkaloids evolved to repel insects and fungi, and the peels are the battleground where those pests attack.

However dull, modern supermarket varieties have had the toxins bred out of them; your Thanksgiving sides won't make Aunt Betty hallucinate. (Unless they're tinted green—in which case throw them out. The emergence of that green chemical, chlorophyll, correlates strongly with an eruption of glycoalkaloids.) But for most of their history as food, potatoes contained toxins, and Andean peoples had to develop tricks to make them edible.

One trick involved freeze-drying potatoes to make *chuño*. Most chuño recipes call for soaking the potatoes in achingly cold streams, then spreading them on blankets to freeze overnight. These processes break down the potatoes' cells and allow the toxins inside to leach out. (In some recipes, people also stomp the potatoes with their feet à la French peasants making wine, although less to mash them than to work the toxic skins loose.) Once this step is complete, they dry the potatoes in the sun for a few days to produce chuño. The final product resembles dried mushroom caps — either dusty black or bone white, depending on the recipe. They keep for a decade. To eat them, you simply add some water to plump them up.

Several experimental archaeologists have investigated potato-processing, but people in the Andes still practice the art today. So rather than rely on an outside perspective, I decided to go right to the source on my trip to Peru. And call me a geek, but I'm excited to try chuño. Freeze-dried food! It seems like an incredible fusion of past and future — ancient Andean grub meets astronaut ice cream.

Alas. I taste my first chuño at a rough-hewn table in a rustic Peruvian village. I get a plain bowl first. The chunks are mealy and dry, lacking the creaminess of a fresh-baked tater. So I order chuño soup, thinking they'd be moister. No: they still crumble in my mouth like soggy bran. As for the taste, "freezer burned" sums it up well. I can detect hints of potato in each bite, much like you can squint at a mummy and sense a real live person somewhere beneath the shriveled skin. But it's a dim connection at best. I walk away from chuño disappointed.

Another ancient method of detoxifying potatoes involves geophagy, or eating them with clay sauce. The chemistry here is grade-school stuff: opposites attract. Glycoalkaloid toxins are positively charged, while certain molecules in clay are negatively charged, so they stick together. Your body can't digest clay, and as the clay molecules chug through your intestines, they drag the toxins along. No one knows how people first got the idea to mix potatoes and clay, but parrots and vicuñas in Peru also eat clay before munching on toxic plants, so perhaps Asana and company copied them.

The potato-clay combo is called *chaco* or *pasa*. I get to make some at a tourist village near Lake Titicaca, playing sous chef to four stout grandmothers wearing vibrant red skirts and straw hats. Our countertop is an uneven stone wall in the bright sunshine outside their home — waist-high for them, mid-thigh for me. They harvested the clay from nearby riverbeds, and present the raw pieces to me in a wooden bowl — gray flakes the thickness of nickels. The grandmothers forbid me from eating the clay dry, but I nibble a bit on the sly. You should always listen to grandmothers. It tastes of dirt, and gets so stuck to my molars that I have to pry it off with my fingernail. Classy.

To make clay sauce, we have to mix the flakes with water. It sounds easy: I figure we just sprinkle the water into the bowl, and that a few brisk what-have-yous with a spoon will dissolve it all. Nope. I spend fifteen minutes grinding away with the spoon in the noonday sun. My forearm starts cramping, but the bottom of the bowl is still thick with inedible sludge. I've also managed to slop it all over my pants. One of the grandmothers, Pilar, finally takes pity and finishes for me.

Our sauce ready, Pilar grabs some freshly cooked potatoes for us to skin. She's boiled them over a peaty dung fire, and seems not to mind that they're still roughly 211°F. I, meanwhile, am playing literal hot potato, and scald my fingers several times running. It's at this point that I admit a small crisis of confidence: I'm singeing my hands to prepare a toxic food that I have to dip in dirt sauce simply to avoid severe upchucking and/or diarrhea. Bon appétit.

And yet. When all is said and done—when I've blown one potato chunk down to a humane temperature, slopped some wet clay on, and popped it into my mouth—it's kind of yummy. The flavor is a nice balance of silken and salty, similar to peanut sauce, except with earthy overtones. Best of all, you can eat regular old fluffy potatoes with the clay sauce; no need to freeze-dry them into mummy food. Despite my initial prejudices, I'll take chaco over chuño every day of the week and twice on Sundays.

Beyond potatoes, I sample several other traditional foods in Peru, including a variety of meats—which unsettles me, given that I've been vegetarian for twenty-plus years. But Asana's people didn't have the luxury of eschewing meat, and you can't understand an era anyway unless you grasp its food. So at a restaurant one night, I feel compelled to try a few dishes.

I start with alpaca, a llama relative that's as close as you can get nowadays to eating vicuñas, a protected species. It tastes like especially juicy steak, and I can see why Asana loved it. I also sample cuy, a guinea pig that comes served like a skinned woodchuck. I quickly learn why Asana despised it. Cuy tastes like rabbit mixed with fried chicken; globules of fat squish between my teeth with each bite. It also has an ungodly number of small, hard bones inside, some of which look alarmingly human, like doll bones. Even native Peruvians I meet, people who gobble down freeze-dried chuño like delectable pastries, wrinkle their noses at cuy. I have to agree.

A depiction of a nine-thousand-year-old female hunter from the Andes. She's going after a vicuña with an atlatl. (Credit: Matthew Verdolivo. Copyright 2020, the regents of the University of California, Davis. All rights reserved. Used with permission.)

Later, I get to observe cuy in the wild, although mostly just their plump backsides as they scramble beneath bushes. Wild vicuñas, in contrast, are truly majestic. Vicuñas, llamas, and alpacas are all types of camelids. They have huge dopey eyes and fluttery eyelashes, and they hinge their jaws sideways when they chew. Most vicuñas are tawny on top with white faces and long white chest hair that flits in the wind like a bad comb-over. They're best described as *svelte*, with taut bodies and serpentine necks. I spend one lovely morning watching thousands of them canter around a highland plain. They don't run so much as lope, their bodies seesawing back and forth like rocking horses. Afterward, on the drive back, I'm startled to see a cow on the side of the road. It looks so stodgy in comparison, so grotesquely swole.

Even vicuña poop—a popular fuel source—proves intriguing. Given how dry the Andean highlands are, vicuña dung quickly solidifies into black pellets the size of marbles. Upon seeing some for the first time, I nudge a few with my boot toe, surprised at how hard they feel. Then, figuring I have a journalistic duty to put myself in Asana's place, I kneel down and—God help me—rub one between my fingers. It has a pumice-y texture, like a charcoal briquet. My fingers don't smell afterward (I check several times), but I'm nevertheless grateful when my bemused guide offers a squirt of sanitizer.

The highlands where vicuñas live—a region called the altiplano—resemble the grass scrublands of the American West: open, treeless, dusty, beige, with lizards darting underfoot and skeletons baking in the sun. But the elevation makes things even harsher. Temperatures drop quickly as you rise in the Andes, four degrees every thousand feet. The air becomes chokingly thin, too. (Halfway through my drive from the Peruvian coast to the altiplano, the driver stopped to let air out of his tires, to prevent them from bursting.) Overall, the altiplano has 40 percent less air than the seaside, and the gasping lack of oxygen can induce a suite of hangover-like symptoms known as altitude sickness—headache, nausea, malaise.

But if the altiplano looks familiar to anyone who's ever visited the West, another highland landscape, the *bofedales*, would look alien to anyone not from Peru—indeed, alien to anyone not named Dr. Seuss. These spring-fed wetlands seem like the natural habitat of the Cat in the Hat or the Lorax—they have the same bold Seussian colors, the same bulging curves, the same charming strangeness. They mostly consist of thousands of tiny, puffy islands crowded together, with sinuous ribbons of streams running between. Each island supports foot-tall tufts of surprisingly poky *ichu* grass; when I carelessly jab a blade with my finger, it draws blood. The islands also support dark-green cushion plants. They look like giant gumdrops of moss, and I expect them to be, well, cushiony. They're not. I press my hand into one, anticipating a spring. It feels like Astroturf on rock, coarse and unforgiving. They have a vaguely piney smell.

Despite their wildly different appearances, one thing the altiplano and bofedales share is emptiness: my guide and I drive across them for hours, and see no signs of human life. But that was less true in the past, when clans of hunter-gatherers regularly ranged across these landscapes in search of game and treasures like obsidian. In addition, the highlands attracted lowland traders hauling up wood and other scarce resources. As a result of these trade networks, Asana has certainly seen foreigners before, some from great distances. But a lone foreigner—especially a woman—is a rare and suspicious thing indeed…

The woman, perhaps twenty, is descending the escarpment with her back to Asana. And given her direction of approach, into the sun, she likely didn't notice the grass hut camouflaged against the terrain. Still, as she descends, she never turns around to heed Asana's shouts—first in wary greeting, then with increasing aggression. Why is she brazenly ignoring her?

As Asana creeps closer, she can make out the intruder's clothing—a breechcloth and loose cord poncho, far inferior to the tailored hides necessary to keep warm on the altiplano. Her legs look pale, despite being streaked with decorative ochre. All the while Asana keeps calling, and she keeps boldly descending.

Asana finally hurls a pebble as a warning. It hits its mark above the woman's head, and dust sprinkles onto her face; she stares upward in confusion. She then hops down the last few feet, turns around, and nearly jumps out of her skin upon seeing Asana.

Like Asana, she has long, dark braids and brown eyes. Her forehead slopes back steeply, and she has a crooked nose. Her neck is ringed with a seashell necklace, and there's a satchel slung over her shoulder.

It's not until she opens her mouth that everything falls into place. The skimpy clothing and seashell necklace mark her as being from the coast, where rumor has it the people lead a bizarre lifestyle. Rather than

## South America — 7500 BC

shift camp every few weeks, they often squat in one spot for months, even a year—an all but sedentary life. Imagine living week after week amid your own filth! And rather than hunt for game, they slurp down shellfish they gather from the ocean. Where's the challenge in that? A whole culture living like Asana's aunt.

Asana has heard more, too. The fierce ocean surf limits the size of shellfish along the coastline. This forces people to dive into deeper, more frigid water to gather them, and the cold somehow damages their ears. Which means this stranger who ignored Asana's shouts is more or less deaf.\* When she opens her mouth, her voice comes out flat and toneless.

She calls herself Moja and says she can read lips. She's come to trade; specifically, she wants stone hunting points. Asana asks if she has any obsidian. Her clan always needs some, and obtaining it would impress the elders. But Moja sighs.

—Everyone always asks that.

—What do you have, then?

Moja digs into her satchel and holds up a cuy guinea pig carcass by its legs. She assures Asana it's fresh.

Asana laughs. Although she'd spear a cuy if desperate, she barely considers them food. She tells Moja to stop wasting her time.

Moja's crooked nostrils flare. But she tucks the cuy back into the satchel and draws out a small pouch.

This gets Asana's attention. Her eyes gleam to see the emerald green of coca leaves.

—How many stone points for the pouch?

—All fingers, Moja says. Ten.

—I'll give you one hand minus one.

Moja swallows.

—One hand, plus one.

Given how quickly she's dropping her price, Asana senses she's

---

\* This disorder is sometimes called "surfer's ear." For unknown reasons, repeated exposure to cold water spurs the growth of bony knobs in the ear canal, blocking it off and blocking out sound. Archaeologists have found several skulls with narrowed ear canals from the first quasi-sedentary coastal divers in Peru.

desperate. And when Asana realizes that, she suddenly wants more. She offers one hand, five points, but also demands Moja's seashell necklace. She's learned from traders that if you tuck a shell into your cheek, the minerals inside react with coca leaves to draw out more juice and provide a stronger buzz. Asana's mouth waters thinking about it.

Moja fingers her necklace and says it's not for sale. Asana shrugs.

—Then we don't have a deal.

—You're already getting a full bag. No necklace.

—Then no points.

The two women stare at each other. But Moja ducks her eyes first, and begins untying her necklace.

Giddy, Asana practically skips back to the grass hut. Predictably, her aunt has a dozen worrywart questions (Who is that? Is she cheating you?), but Asana ignores her. The coca will impress her elders—a bonus on top of the vicuñas she'll get today. She picks out five stone points left over from this morning—not her best, not her worst—and jogs back.

After they make the trade, Asana tries to chat—she rarely sees people her age—but Moja turns and hurries north. Asana walks away miffed. Coastal people are so strange.

Back at the hut, her aunt is desperate to know what happened. Asana hands her the pouch in response—thoroughly enjoying her gape of surprise—and ducks inside to gather her gear. It's time to go hunting.

A moment later, her aunt screeches.

—Asana, what did you do?

Confused, Asana pokes her head out. Her aunt is clutching a handful of coca leaves from the top of the pouch. But when she turns it over, nothing but dead grass drops out.

—Did she cheat you?

Asana's cheeks burn—it's a child's mistake. Her eyes dart to the horizon, and she takes off after Moja.

Asana screams her name, forgetting that she can't hear. But Moja eventually glances back, and jumps in fright—then takes off running.

It's a pitiful chase. The altiplano has far thinner air than Moja's seaside home, and Asana can see her sucking wind, her chest heaving.

When Asana closes the distance, Moja tosses her satchel aside and gamely turns to fight. But she's too short of breath. Asana rushes her, flinging her to the ground and jumping atop her. Within seconds, she pins Moja's legs, then rears up and raises her fist to strike.

Only to stop cold. Moja's poncho sleeve has slipped down, and she's shielding her face with a bare forearm. Asana is shocked to see how crooked it is—bent at a sickening angle, as if she had another joint there. There are sores all along it, too, red and weeping pus. Asana's fist drops to her side.

Moja wriggles free and crab-scuttles backward. As they lie there panting, Asana asks what happened to her arm. Moja bites her lip.

—It's no concern of yours.

—You're alone. Are you fleeing someone?

—I could ask the same of you. Where's the rest of your clan beyond your mother?

—That's my aunt. My mother...

Asana swallows. She's never said this aloud before—never had to. Everyone in her life simply *knows* what happened. But she explains, haltingly. Her parents were the tribe's top hunters, a dynamic team. Her feckless aunt cooked. But one night she didn't cook the food properly—the vicuña was too raw, the potatoes too toxic. Her parents spent several days violently ill, then both died. Along with, Asana doesn't add, her sister—who was roughly Moja's age.

Moja says she's sorry to hear this. But her pity makes Asana restless. She points at Moja's forearm.

—So what happened?

—My husband broke my arm three times last year. My nose, too. So I stabbed him in his sleep.

Asana expected as much—it's another rumor about coastal people.*

---

* Archaeologists have found signs of extreme violence written onto the bones of lowlanders in ancient Peru: dented skulls, spines with embedded stone points, female forearms shattered during presumed beatings. It's not simple cause-and-effect, but sedentary living allows people to accumulate material goods, which can lead to jealousy, power imbalances, and violence. In Moja's time, these processes are just starting, and will accelerate over the next few millennia.

Looking closer, she can see scars on Mojo's legs, too, half-masked by the ochre paint. She's staring so hard, in fact, that it takes her a moment to realize that Moja is picking through the goods from the satchel that spilled out. A moment later, she reaches out and drops the five stone points into Asana's palm.

—Can I leave now?

Asana feels her anger deflating. She tells Moja to stand up and follow her. Moja looks wary.

—I gave your points back.

—Do you want to eat today?

—I have cuy.

—One cuy. You'll be starving again in an hour. Come hunting with me.

Asana can see how tempted Moja is—her eyes flash with hunger. Asana tells her to wait and takes off jogging.

Back at the hut, her aunt now has a thousand questions, but Asana shushes her. She grabs her leather hip pouch and fills it with hide-scrapers, blades, ochre, and extra resin. She also grabs her best bone atlatl and her newly hafted darts—plus the scattered coca. Her aunt makes one last plea.

—Can you trust her?

—I know what I'm doing, Asana lies. She hurries off.

But when she jogs back to the scene of the fight, Moja has vanished. Asana scans left and right. The only thing remaining is some trampled grass, and the dead, limp cuy.

Unexpectedly hurt, Asana turns and marches off alone.

To placate her aunt, Asana has hewed close to camp the previous few days while hunting. But with not a single kill so far, she's decided to go ranging today; it's her only hope of landing three vicuñas by week's end. She takes her bearings every time she turns, sometimes with snow-peaked mountains, sometimes with glaciers or rock spills. After an hour, she pulls two coca leaves from her pouch and wads them into her cheek. She soon has a mild buzz that warms her down to her bare feet.

Along the way, Asana sniffs something interesting on the breeze—an unusual dung, wetter and fouler than vicuña droppings. Meat-eater dung. She locates the scat: two lumpen masses the size of her head, quite fresh. She scans the horizon until she spots the source: a pair of giant sloths lumbering along, each over a dozen feet tall.

One of Asana's most vivid memories of her parents involved them bringing home a giant sloth paw they scavenged. Its immense size seared itself into her mind. She could barely wrap both hands around a single claw—and it had five on each forepaw, as sharp as hawk talons. Ever since then, she's been inflamed with the fantasy of hunting one.

She's tempted to divert her path. Sloths are the biggest of big game, and taking one down would all but guarantee her initiation as a hunter. But it's too risky: she simply cannot go home empty-handed. Besides, she'd need obsidian points for that—as well as a second hunter for support. So as badly as she wants to pursue the sloths, she walks on.

She soon comes across a promising hunting ground—a scrub plain full of vicuñas, with a thin, shallow canyon snaking through. The canyon is perfectly blinded, and she picks her way along the silty, rock-strewn bottom, occasionally peeking above the rim to check her position.

She finds the animals a bit too far away, so she deploys a trick. She takes her atlatl—a launching stick—and loads it with the semi-crooked dart from this morning's hafting session. Then, from the canyon bottom, she heaves the dart at a lofty angle. It clatters down behind some

The atlatl, or throwing-stick, was perhaps the most widespread hunting tool in prehistory. (Credit: Sebastião da Silva Vieira, Wikimedia Commons.)

vicuñas grazing fifty yards away. They flinch and snap erect, their necks periscoping for danger. Several lope away from the noise—and right toward Asana in the canyon.

They stop within forty paces. When they relax enough to resume grazing, she creeps up to the canyon rim, readies another dart in the atlatl, and springs up to let fly.

Although powerful and accurate, atlatls require a big, obvious throwing movement, which can spook animals. Sure enough, the vicuñas scatter when she hurls the dart. But Asana has good aim, and twenty yards is an easy toss. The dart punches into the rib cage of one beast. It staggers two steps and collapses sideways.

By this point Asana has already loaded a second dart. Unfortunately, the vicuñas are now stampeding, and hitting an animal at full gallop is a much tougher task. But she leads her target perfectly and snipes it in the back. It falls and tries to rise, but its legs buckle a moment later. It sags earthward, its chest still heaving.

Asana throws her arms up and screams in triumph. Two throws, two kills. That's virtually unheard of. She doubts the rest of the tribe will even believe her.

Now to feast. She spits out the coca leaves, then gathers some dung and grass. With the drill-stick in her hunting kit, she starts a small fire. After the long walk, she's ravenous, and while the flames build, she indulges in a special treat. Holding a vicuña head between her knees, she pops its eyeballs out with a knife and peels off the delicious gobs of fat behind them. Using an atlatl dart as a spit, she roasts the fat until it's dripping like resin, then pops it in her mouth. It melts deliciously down her throat.

She goes after the succulent organs next, splitting the belly and roasting a slice each of liver and kidney. Her aunt always overcooks organ meat, but this time she can enjoy them rare. Then she hacks off a strip of vicuña steak, juicy and pink and barely chewy. After three days of clay and potatoes, it's one of the best meals of her life—in part because she secured it herself.

By meal's end the coca stimulant has worn off. And with her belly full, Asana feels drowsy. Aside from the occasional puma or giant sloth, the altiplano lacks big predators and scavengers, so she can leave the carcasses lying around without much fear. She's soon fast asleep.

But if the altiplano lacks predators and scavengers, it doesn't lack danger or surprises. When Asana stirs awake a half hour later, she hears the familiar sound of flesh being butchered.

Her eyes flutter open. Hovering over the dead vicuña is a figure with dark braids, dark eyes, and a familiar crooked nose.

Beyond the potatoes and vicuñas, the unquestioned highlight of my trip to Peru is a visit to an ancient gravesite—one of the most paradigm-shattering graves in the history of archaeology. Why? The idea of an ancient female hunter like Asana might strike some people as strange. After all, if laypeople "know" anything about archaic tribes it's that guys hunted for food and ladies foraged. Man the Hunter, and all. But as that grave in Peru shows, that tidy division of labor is wrong.

One day in 2013, a farmer on the Peruvian altiplano named Albino Quispe was plowing his field when he accidentally turned up some old stone tools, including stone points for weapons. Luckily, Quispe had worked on archaeological digs before and reached out to an archaeologist at the University of California, Davis, Randy Haas, who's since moved to the University of Wyoming. Haas swung by to take a look. The field—now called the Wilamaya Patjxa archaeological site ("will-a-MYE-uh PAH-ta")—turned out to be a bonanza. Haas and his team eventually excavated twenty thousand artifacts and six bodies—one of which dated back nine thousand years.

Haas is a slender fellow of medium height with slightly bulging eyes that give him an intense look. And it's not just his eyes—he's intense about many things. If he's looking for a bone sample or stone point, the task absorbs his attention entirely; questions to him in such moments go unanswered, even unheard. Before our first interview, he interrogated me warily about this book's topic for six minutes before I got to ask anything back. But by the time I visit Wilamaya Patjxa with him in Peru, we've developed a nice rapport, and we warm to each other immensely during my trip.

Wilamaya Patjxa is hillier than I expect, rolling fields of soft, powdery dirt cut with furrows of potatoes, barley, and quinoa. A red tractor with a wobbly wheel sputters by, and several ominous dust devils kick up. At one point, a local farmer wanders up and lets me sample his stash of coca. The papery leaves poke my cheeks, and taste sharp and astringent, like raw tea.

One discovery from Wilamaya Patjxa—a scattering of bones and teeth a yard beneath the topsoil—especially excited Haas because it included a complete hunter's kit: "all the tools necessary to kill and process an animal," he explains, including chert points, stone scrapers and knives, and red ochre for tanning hides. The tools were "neatly stacked in a bundle at this person's hip," he adds, presumably still lying where a leather holding pouch rotted away around them.

Inflamed by the discovery, Haas fantasized that his team had unearthed a "great chief." But things took a swerve when he consulted with a former professor of his, University of Arizona archaeologist Jim Watson. Watson specializes in reading human bones, and Haas asked him to check the big chief's remains for signs of disease or other subtleties. Instead, Watson exploded Haas's fantasy.

The issue was the big chief's femurs, the thigh bones. I see them in Peru, laid out on clean white paper in a lab. They look like cracked-open sticks caked with dirt and roots, so fragile that even the expert Watson seems reluctant to handle them. But the thing that stands out most is how slender they are—barely wider than my middle finger, and I hardly have meaty plumber hands. They look like children's bones, and might have been mistaken for such except for one thing: The femurs were found alongside mostly mature adult teeth. And that combination of tiny bones and grown-up teeth "really gave me pause," recalls Watson. "There's just no way that a nearly adult male would have such slender femora." Which meant, Watson realized, that the "big chief hunter" was actually a hunt*ress*.*

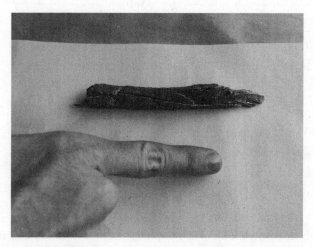

A slender femur fragment from an ancient Andean hunter. (Copyright: Sam Kean.)

Haas admitted to being startled at this news; he, too, had absorbed the pervasive Man the Hunter concept in grad school. Indeed, he wondered whether his huntress was merely an exception to the rule. "Maybe she was the original gender-bender," he mused. "Or maybe this was an unusual culture that bucked the trend." After all, the altiplano is cold and desolate. Securing precious fats and proteins was hard, and maybe clans couldn't survive there when half the population was barred from hunting.

Still, science doesn't stand on musings and maybes. So in his intense way, Haas looked into the matter. While his colleagues begged him to hurry up and publish the discovery of the huntress—this was quite a scoop—Haas ignored them and delayed for an entire year while he

---

* Statistically, the size and shape of men's and women's bones overlap, so you cannot look at a single bone from a single skeleton and declare that such-and-such was definitely male or female. But Haas and Watson confirmed their initial assessment with another test: On a molecular level, men and women produce different versions of certain proteins in our enamel. They submitted one of the Wilamaya Patjxa hunter's teeth to this protein analysis, and it revealed that she was indeed female.

scoured the archaeological literature, looking for other cases of ancient Americans buried with hunting tools. He eventually found twenty-seven instances. In eleven of those twenty-seven cases, ranging from Alaska to Argentina, the hunters were women. That's not a 50/50 split, but it's vastly closer to 50 percent than any archaeologist would have ventured before.* The bottom line is, women hunted way back when, probably a lot. Haas says, "As the numbers were coming in, I was like, *Holy shit*." He pauses. "Don't quote me on that. Oh, well, I guess you can, I don't care." As he knows, the discovery of female hunters was some pretty important shit.

The clan of the Wilamaya Patjxa huntress made stone points with a characteristic shape—long pentagrams with pointy "ears" sticking out. One point I examine in Peru looks especially fetching, a creamy pink rock with dark polka-dots. But points are only half the game. To kill a vicuña, or any sizable animal, you have to haft the point to a wooden shaft.

Hafting isn't nearly as sexy as knapping, but it's every bit as important for an effective weapon. It also marks a key step in human technological development—combining one tool with another to make a superior joint tool. Metin Eren at Kent State walks me through the steps of hafting. First we saw a dowel of poplar wood in half with a chert blade. (Eren suggests using a hacksaw initially, but we can't locate one; it's easier to find stone tools than metal ones in his lab.) The chert blade looks like a clam, and I grip it in my palm as I saw. Poplar's a soft wood,

---

* Inevitably, some archaeologists challenged Haas's conclusions. For instance, two of the females buried with hunting tools in his literature review were infants, and it's reasonable to ask whether those cases should count, given that infants don't hunt. Other critics suggested that the women had simply been buried with tools for symbolic reasons that we cannot hope to parse thousands of years later.

Haas doesn't buy that last criticism. Despite how much burial practices differ across the globe, he says there is one consistent rule: that "the objects that accompany people in death tend to be those that accompanied them in life." In other words, you wouldn't bury a woman with hunting tools unless she hunted with them.

but I'm shocked how quickly the chert chews through it. A hacksaw with a handle would have been more comfortable, but only marginally faster.

Next, we carve a slot into the end of the dowel and jam the point in. Now for the adhesive. In the ancient world, people made adhesives from sticky resins they scraped off trees; Asana's clan likely gathered resin from either cushion plants or *yareta* shrubs—bulbous yellow-green mounds that look like melting broccoli. People likely stored the resin in balls on sticks, like brown lollipops, and might have added dung to make the resin smoother or beeswax to prevent it from becoming brittle upon drying. In Eren's lab we use thermoplastic, a synthetic compound with properties similar to resin. It comes in translucent amber blocks that look like fat magic markers. We heat our blocks over a flame until they begin dribbling like honey, then lace the liquid around and around our points. It looks like confectionary, and I'm perversely tempted to lick it, although I'm glad I don't: while working on his spear, Eren gets distracted for a moment and burns his finger badly. He explodes into curses, and his shaft and point go flying. Add another injury to the green blob-man.*

When the thermoplastic honey cools, Eren and I bind the shafts and points together. To do this, hunter-gatherers used animal tendons that they perhaps pounded with rocks to tease out individual fibers. As a bonus, tendon contracts as it dries, so if you apply it wet, it hugs a stone point tighter upon drying. In Eren's lab we use strings of synthetic sinew (waxy twine), winding it around and around the point in tidy, parallel rows. For extra sturdiness, we dribble one more layer of thermoplastic honey over this, letting it soak into every crevice. And, voilà—a hafted spear. I assume we're finished at this point, but Eren grins and begins rotating his wooden

---

* When we retrieve Eren's point, a fat chip is missing from one side. But as Eren notes, the flaw makes the point look more authentic, more like a true prehistoric artifact. In fact, he says that people who churn out fake arrowheads and scam people online will deliberately gouge chips from them to boost their value. Eren mentions other counterfeiting tricks, too, like freezing points in ice to pry off the "lips," the subtle edges that are a giveaway of newly knapped tools, or smearing animal fat on points and burying them in dirt to give them the proper patina.

shaft in the fire, scorching black snake-bands into it to make it look more badass. "You can't underestimate the cool factor," he says.

To make shafts for their spears, Asana's clan would have traded with lowlanders for branches or perhaps sought out the wood of the *queñual*, about the only tree that grows in the highlands. But using the latter would have been difficult. Queñuals are gorgeous in a Gothic sort of way, twisted and craggy with shiny red-brown bark the color of a cockroach. Unfortunately, that craggy aesthetic makes the queñual a lousy source for shafts, which have to be straight.

Honestly, though, it's unfair to single out the queñual for having branches that make less-than-perfect spears and darts. However ubiquitous wood is in some parts of the world, much of that wood is crooked, stunted, knotty, or otherwise useless for weapons. There are tricks to fix such problems (e.g., heating wood over coals or steam and bending it) but they're slow and laborious and can damage the timber.

Given all these hassles with wood, Eren says something that completely flips my perspective on ancient tools. Even with poor rocks, a master knapper like him can knock out several stone points in a few minutes; as a result, points are fairly dispensable and easy to replace if they break. The truly valuable bit of a spear, then, is the wooden shaft, which takes vastly longer to find and prepare. Shafts would be especially precious in places like the altiplano that lacked trees. Yet because wood disintegrates over time and disappears from dig sites, archaeologists tend to neglect it. "Maybe the stuff we don't find"—like wood—"is the stuff they cared about," Eren muses. "Maybe instead of the Stone Age, we should call it the Wood Age." It's a bracing insight—and one you wouldn't get from textbook archaeology. You have to actually build tools from wood and stone to grasp that truth.

At this point, I'd like to clarify something. So far, I've been using the terms *spear* and *dart* a bit loosely. In truth, a spear is similar to what Kayate used in Africa—a stone-tipped javelin hurled by hand. Asana's darts were not thrown by hand, at least not directly. Instead she launched her darts with a throwing-stick called an *atlatl* ("at-LAT-ul").

The atlatl is a two-foot-long stick with a hook or spur on one side. The darts are wooden shafts with a stone point hafted to one end and a concave cup carved into the other. (Asana prefers shorter darts, a yard or so long, since they can double as spears in a pinch.) To load an atlatl, you hold it at shoulder-height, parallel to the ground, and fit the cup end of the dart into the spur-hook. To fire the dart, you step forward and snap the atlatl down with your wrist. Imagine flicking paint off a paintbrush—same motion. Overall, your thighs and core generate the power, which gets channeled into the dart via the arm and snapped wrist.

To the uninitiated, the atlatl probably seems baroque. Why not just hurl a spear, instead of using a stick to fling it? A detailed answer would require a long digression into the physics of levers and rotational velocity, but the basic idea is this: the longer your arm, the faster you can throw something. (Think of those long plastic ball-throwers for playing with dogs.) Atlatls effectively lengthen your arm by a foot or two and therefore provide a huge speed boost: experts can fling the darts 80 mph, while spears alone top out around 50 mph. Speed is a major factor in a weapon's penetration ability and knockdown power, so atlatl darts are pretty darn deadly.

Despite their obscurity nowadays, especially compared to spears or arrows, atlatls were probably the most widely used hunting weapon in prehistory, in climates from the tropics to the poles. As a result, there's a serious contingent of atlatl enthusiasts within experimental archaeology.

I get to try atlatls myself during an undergraduate class that Metin Eren teaches. We meet at a frisbee golf course near campus, where he unloads several dozen atlatl darts from his pickup. Each is around six feet long and a bit less than an inch thick, and they're fletched with fake feathers—neon green, black-and-blue, crimson. I'm surprised how bendy they are, quite flexible. Atlatlists debate why that flexibility matters, but rigid darts simply don't fly as straight or true; darts need spring.

The atlatls Eren hands out are pretty basic—wooden sticks with hooks. Most people throughout history used something similar, but the inhabitants of the treeless altiplano would have saved their wood for darts and made atlatls from the leg bones of vicuñas. (I actually stumbled across one such bone on a walk in Peru. It was bleached white, and differed from traditional atlatls in that it had a kink in it. But when I whipped it downward with my wrist, it felt perfect.)

For his class, Eren sets up targets for us to throw at, roughly 25 yards away. The first two are black foam cubes, but everyone ignores those as boring. The sexier targets are the foam animals. One is a ⅓-size caribou fresh out of the box from Target. It has snap-on antlers as well as, oddly, two plastic tubes snaking out of its torso—perhaps, someone suggests, to pump fake blood inside for gory special effects. The second animal target is the Kent State lab's mascot, the much-abused Bambo, who's been so riddled with atlatl holes over the years that all four legs have crumbled off or been amputated. Eren plops it on the grass anyway, like a deer at rest.

With the targets arrayed, Eren lines the class up and lays out the rules. ("Rule number one: no battle cries.") He also punctures our expectations, explaining that we probably won't be very good at this: Despite how simple atlatls look, hunting with them is tricky. Indeed, they prove to be even more maddening than he lets on.

On the one hand, the darts are easy to throw—and they go fast. When you step and flick your wrist, they bend hypnotically for a fraction of second before snapping straight again and screaming off. *Whoosh*. The darts fly far as well. I could fling one most of a football field without much effort. (The world record for an atlatl toss is 282 yards; the javelin world record is 115.) In Eren's class, darts are soon whizzing everywhere up and down the line of students—*whoosh, whoosh, whoosh*. It's incredibly fun. Eren even raises the stakes by promising extra credit points every time a student hits a target.

Unfortunately, hits are few and far between. However easy to fling, the darts are nearly impossible for beginners like us to aim. While Bambo and the mini-caribou bask in the sun, we miss them in every possible dimension—too high, too low, too left, too right. With the darts stuck in the dirt all around them, they look like smug little ticks on a porcupine. Eren roars with laughter. "If we were hunting slugs, we'd be eating like kings," he teases.

Still, we're just amateurs. Expert atlatlists can consistently spear a target at forty yards. And even among us tyros, some stand out as better than others—especially some young women. In fact, on a societal level, that's one big advantage of atlatls compared to other weapons: they level the playing field between men and women and expand the pool of potential hunters. That's because atlatls are finesse weapons. Strength and power matter less than form and speed, and men and women can throw them equally well.

Indeed, in the Kent State class, the ladies kick the lads' butts. There are roughly equal numbers of men and women throwing, but the first four "kill shots" to *thunk* the deer are all thrown by women, as are nine of the dozen kill shots overall. One wavy-haired woman in jeans and a navy T-shirt nails the caribou three times—once in the throat—and jumps up and down squealing after each strike—more extra credit. Armchair archaeologists who've long discounted women's ability to hunt would have much to ponder watching the guys flail.

Including me. I nick the caribou's kneecap once, which I tell myself would have hobbled it. Frustratingly, though, a half dozen other shots land within a yard or two without connecting—and when it comes to weapons, *close* only counts with hand grenades. Other tossers have an even more aggravating afternoon. One graduate student named Dan gets within inches of a hit a dozen different times, practically tickling the caribou's chinny-chin-chin. And with each near-miss, he grows more determined to score a kill shot. *One more toss,* he keeps telling himself. *Okay, just one more. I can't quit now.* Ultimately Dan forces himself to stop. "I feel like that's how you get addicted to gambling," he says. I know what he means.

Of course, for Asana's people, hunting wasn't a game. An unsuccessful hunt could mean death—either from starvation if an animal gets away, or from the counterattack of a wounded beast if the hunt goes wrong…

Asana and Moja both jump at the same moment, like body and shadow. Despite her fright, Asana can't help but notice how much of the vicuña Moja has eaten already. She nearly laughs at her bulging cheeks.

—Why did you hide before? *Where* did you hide?

Moja swallows half the food in her mouth. It looks painful.

—There was a rotten yareta bush. I burrowed in.

Asana nods. Yaretas can reach several feet across and have thick canopies. They're ideal hiding spots. She sees Moja's hair is now streaked with resin.

—But why?

Moja shrugs. Asana asks again, to no avail; Moja plays up her deafness and pretends not to understand. Asana finally loses patience.

—First you cheat me, then you sneak up on me while I'm asleep. Give me one reason I shouldn't spear you.

Moja reaches into her satchel. She removes a piece of hide and unfolds it to reveal several glassy black flakes. Asana gapes: it looks like obsidian. Then she narrows her eyes.

—Are you cheating me again?

Moja picks up a flake, kneels next to the vicuña, and grips a handful of fur. Vicuña fur often has a weathered, dirty crust on the outside and can be tough to shear. But even with her crude blade, Moja slices right through. She peels off a tawny mat that looks almost fatty in its consistency and tosses it at Asana's feet. The flake sure looks like obsidian to Asana—it cuts so effortlessly.

—You said you didn't have any.

—No, I said people always ask about it.

—Where did you find those?

—There's a mountain pass. I can show you, but…

—What do you want? More stone points?

Moja gazes hard at her.

—Do your people ever take in outsiders?

Asana frowns, surprised at the question. People in her clan often marry into different groups, or quarrel within their own group and fission off; she's fantasized about leaving her aunt many times. But those

## SOUTH AMERICA — 7500 BC

swaps all took place among people already from the highlands. She's never heard of, never considered, taking in a true outsider.

—You've surely met other people. Why are you asking me?

Moja squirms, her fingers gripping her crooked forearm.

—Because you asked about my arm. And I thought with your family gone...

Asana feels a pang. Moja is clearly clever, clearly resourceful—she'd be a valuable addition. And Asana's always longed to have someone her own age around. But taking someone in without permission from the clan would be a grave offense. She simply cannot say yes without consulting her people.

Moja studies her face, anxious. She must sense Asana's hesitation.

—I'll pull my own weight—you'll see. You can teach me to hunt?

At that, another thought flits through Asana's mind. She suddenly wants more.

—Yes. In fact, we'll start today. Have you ever seen a giant sloth?

After shaping Moja's obsidian into points, Asana slices a foreleg bone out of the vicuña to make a second atlatl, scraping it clean of gristle and hardening it over the fire. When it's dry, she sets up the vicuña carcasses as targets and teaches Moja how to throw darts. They'll have a better shot of taking down a sloth if they can hit it from multiple angles.

Moja's not terrible, but her crooked forearm causes her tosses to hook. One errant throw even snaps a precious wooden shaft; Asana winces. Nevertheless, she decides her pupil's good enough after a few near hits. A sloth's a much fatter target than a vicuña anyway. She also teaches her how to thrust and stab with the darts at close range, the killing blows.

The lesson complete, Asana gathers all her atlatl darts, plops down near the fire, and begins melting the adhesive on them. She shows Moja how to unwind the sinew, then re-hafts them with obsidian points using extra resin from her toolkit. Obsidian's much sharper than chert, and has a better chance of wounding the sloth.

When the shafts are ready, they load up and backtrack for an hour to the pile of dung. Then the two women turn to pursue the scuffling sloth tracks into the distance, Asana leading the way. As they walk, Moja asks her several nervous questions about hunting sloths. Asana explains all she knows, then reassures her.

—You'll do fine. Just flick your wrist like I taught you. If we take one down, I guarantee my clan will welcome you.

Asana gives her coca leaves to quell her nerves. Asana takes a plug herself for the same reason. Because the truth is, despite everything she's been telling Moja, she's never hunted a sloth in her life.

As they follow the tracks up a ridge, Asana hears the crickety squeak of the wallata bird, the Andean goose. Wallatas always seek water, and when the two women crest the ridge, Asana sees a new landscape spread before them—the spring-fed wetlands known as bofedales, with thousands of puffy islands crowded together. Sloths are too massive to navigate these tiny islands, so they likely circled around. Asana decides to cut through to make up time, hopping from tuft to tuft.

The day is disappearing fast. When scanning the horizon for the sloths, she has to shield her eyes from the setting sun. Finally, she spots them. They have indeed circled around the bofedales, to some muddy flats on the outskirts. They're both knuckle-walking along, oblivious to any danger—there's no animal even remotely their size for a thousand miles.

There seems to be a male and a female—one's a few feet taller. As they draw near, Asana tests Moja.

—Which one should we go after?

—Neither?

—Very funny. Which one? And why?

—The male. The female might be pregnant. Look at her belly.

Asana shakes her head at this silly sentimentalizing. Hunters are ruthlessly practical: you always go after the smaller, less dangerous beast. Moja doesn't look pleased, but she'll have to learn to be tough.

Unlike with vicuñas, there's little need for stealth here; sloths can't dash off. When the two women arrive within fifty yards, the sloths

South America — 7500 BC

Giant South American sloths went extinct several thousand years ago. The largest stood twenty feet tall. Notice the fearsome claws. (Copyright: Associated Press.)

catch a whiff of them and crank their giant heads. They bellow, and Moja flinches. Asana nearly does, too, but keeps marching.

At twenty yards, the two hunters fan out. Asana finds herself staring into the female's wary eyes. Both hunters load their atlatls, and with a nod from Asana, they let fly.

Both miss. Moja's shot hooks, while Asana's sails high.

Asana swears to herself. Again, she orders.

But Moja has lost heart. The male is heaving and snorting now, and she backs away. Truth be told, Asana's nervous, too. But she takes a few deep breaths, then hurls her dart.

She can tell when it leaves her fingers that it's a perfect shot—straight and true. It slams into the female's rib cage—*thwack*—and the shaggy beast howls in rage.

Then, with a flick of its arm, it swats the dart out and starts to knuckle-walk away.

Asana stares, stunned. A shot that hard would have run a vicuña through. But the sloth brushed it aside like a twig in its fur.

When the sloths are a safe distance off, she creeps up to retrieve her atlatl dart. She's hoping to find a problem with the hafting. Maybe it hadn't dried yet and the tip slipped when it struck, blunting the impact.

But no. The obsidian looks secure. Somehow, though, there's barely any blood on it. It couldn't have penetrated more than an inch.

—Now what? asks Moja from behind her.

Asana scans the landscape, trying to think. What indeed. Then she spots something, a particularly muddy stretch of the wetland. She grabs Moja's hand.

—Come on. We don't have much daylight left.

There's no denying that atlatls killed modestly sized animals quite well. Their efficacy with big game is another question. And the answer has big implications for one of the greatest mysteries of the human past: the sad, sudden disappearance of the megafauna.

## South America — 7500 BC

Megafauna—large animals—once thrived across the world, including giant sloths in the Americas. No one quite knows when giant sloths died out, since they seem to have survived in refugia here and there, but butchering marks on ancient bones reveal that they overlapped with the first Americans, and some scientists believe they survived into Asana's day. Some species weighed four tons and stood twenty feet tall when they rose on their hind legs. They had shaggy gray-white coats, and were almost certainly less lethargic than their modern counterparts. (Taxonomically, sloths are related to anteaters, which aren't exactly speedy but aren't molasses-slow, either.) And unlike tree sloths, giant sloths were not above scavenging—or perhaps killing—for meals.

But giant sloths were hardly alone back then—beasts of mythological size thrived on every continent. Australia had bear-sized wombats. Eurasia had giant woolly mammoths and elk with ten-foot antlers. South America had armadillos the size of automobiles. North America had saber-tooth tigers and dire wolves and beavers so large you could have ridden them like stumpy burros.

Then, most everything disappeared. Every continent suffered losses,* but South America got absolutely clobbered, losing 83 percent of its megafauna overall, as well as every species that topped 700 pounds. Collectively, these extinctions undid 65 million years of steady evolutionary gains by mammals across the globe. Before that point, mammals were tiny rodent things scurrying around the underbrush—hardly impressive. Then the dinosaurs got wiped out. New ecological niches opened up, and mammals got steadily bigger, millennium after millennium, until they swelled to an average size of a thousand pounds. The great megafauna extinction, however, threw everything into reverse: today, the average mammal weighs a few dozen pounds.

The big question is why. What killed off these magnificent beasts?

---

* Africa suffered much less than other continents, which explains why giant animals still exist there today. Big animals in Africa evolved alongside humans and therefore learned to grow wary of them. Outside of Africa, big animals were probably naive—easy pickings for smart primates with atlatls and advanced hunting strategies.

The two most popular theories pin the blame on climate change or overhunting. Each has its passionate—even zealous—adherents.

Team climate holds that drastic changes in temperature and precipitation disrupted key habitats long ago, and that larger animals simply couldn't adapt or find enough food. Meanwhile, overhunting adherents point to some awfully suspicious timing. Weather patterns have changed in the past, and big beasts pulled through before. So is it really a coincidence that shortly after human beings arrived in certain places, the megafauna disappeared as well?

Experiments could help resolve this dispute. The last chapter mentioned several cases where archaeologists carved up zoo and circus elephants with stone tools. During the course of this work, some researchers also hurled atlatl darts and spears at the carcasses to test their effectiveness. In particular, they wanted to know whether human beings could hurl these ancient weapons with enough force to penetrate the hides of truly mega megafauna. After all, elephants have skin and fat layers up to two inches thick. Mammoths were even better armored, with up to four inches of skin and fat, plus dense woolen undercoats of fur up to six inches deep—all of which you'd need to pierce to kill them. Could noodle-armed humans have accomplished this?

Some archaeologists walked away from the experiments optimistic. One reported, "Projectile points used with either atlatl and dart or thrusting spear will...inflict lethal wounds on African elephants of all ages and of both sexes." Other archaeologists remain skeptical, pointing out that some of supposedly successful trials involved the carcasses of juveniles, who have thinner skin and less fat and muscle. In another successful trial, the archaeologists threw their weapons from just seven yards away—awfully close unless the elephant was trapped or injured. Moreover, a successful hunt would seemingly require high levels of skill: the penetrating power of darts and spears in the trials dropped quickly unless they struck the hide at 90 degrees. Plus, the most vital organs (heart, lungs) are lodged inside the chest, and slotting a dart between the stout "picket fence" of an elephant rib cage would be tricky.

To be fair, other spots on an elephant's body are more forgiving. A few vulnerable organs sit right behind the thin skin of the abdomen, and sure enough, several belly thrusts in the experiments did reach the necessary foot or so depth for a kill. Still, these trials took place under much more favorable conditions than you'd ever encounter in real life. In some cases the elephants were slumped on their sides, and in all cases they were stationary. A live, upright (and possibly charging) elephant makes for a much tougher shot—unless you were foolhardy enough to dash between its legs and risk getting crushed. (And remember, elephants have little hair and not much fat; blubbery, hairy mammoths would be even harder to kill.)

All in all, experimental archaeology casts doubt on the idea that human beings regularly took down animals as thick and sturdy as elephants. (Even in modern times, elephants who've died of natural causes have been discovered with hunters' bullets lodged inside them.) Like Kayate last chapter, any meal of pachyderm meat probably involved scavenging. Score a point for the climate theory of megafauna decline.

Still, not all beasts are as big and gnarly as elephants. Did we perhaps hunt some of the minier megafauna? With giant sloths, the answer seems to be yes.

Giant sloths lived in both North and South America, and in 2018, archaeologists working in New Mexico announced an astounding discovery: an interlocking set of ancient human footprints and ancient sloth footprints, with the humans apparently tracking the sloth. The initial sloth prints showed it ambling along on all fours, but as soon as the humans drew near, it rose up on two feet. The prints then show the sloth dancing around in what the scientists called a "flailing circle," with occasional crashes back down onto all fours—presumably after lunging. Unless you think the humans were tickling the big fella, it's hard to interpret these prints as anything but a hunting scene—with the humans taking swipes and the sloth rearing up to defend itself with its vicious talons.

In all, humans probably did attempt to hunt sub-pachyderm beasts. But there's a healthy gap between attempting and succeeding. Based on fossils, paleontologists know that sloths were densely muscled, and almost certainly had shaggy coats to boot—not mammoth-thick, but enough of a chain mail to blunt or turn aside even the sharpest points. All of which Asana learned the hard way on the altiplano...

Stories about hunting make up a large fraction of the lore of Asana's clan, and several of the tales involve driving game into swamps or mudflats to trap them. A variation on that plan seems like Asana and Moja's best hope now: forcing the male and female apart, immobilizing them in muck, then attacking.

First they need to separate the sloths. Asana grabs a rock and smashes through the rind of a dead cushion plant. Gripping a branch inside, she wrenches it back and forth until it snaps free, then repeats the process with several other branches. Satisfied, she starts a quick grass fire and lights two of them for now, one for her and one for Moja. Cushion plants are full of resin, so they burn bright and produce loads of choking smoke—perfect for scaring animals.

Indeed, if the sloths were scared before, they're terrified now as Asana and Mojo approach with torches. Both women—Moja with more hesitation—begin darting forward, shouting and waving the flaming brands. Sloths have poor eyesight and the smoke limits their vision even more. Just as Asana hoped, the male and female get separated in the chaos.

She and Moja drive the male into the boggy mud. As it flails and stumbles, its white fur becomes smeared with black gunk. Asana wishes the muck were deeper—its legs are mired only to its belly—but this will have to do. She and Moja retreat a bit, light more brands, and drop them at the mudflat's edge to pin the male in. Then they turn for the female, who's lumbered away from the thickest smoke. Moja sees her belly and hesitates, but Asana grabs her arm and yanks her forward. They have no time to waste.

Initially, Asana planned to drive the female into the mud, too, but she figures now it's best to dispatch it quickly. She and Moja grab their atlatl darts for spears and stalk toward it. It rises on two feet and snarls. The sight makes Asana's spine tingle. It can now swipe its claws like scythes. But the posture exposes her soft belly, too.

Asana explains the plan to Moja in a mixture of shouts and gestures. They'll need to spread apart and circle the sloth in tandem — first one way, then another, changing directions rapidly. This will keep the sloth off-balance, allowing them to dart in and jab the beast in the stomach.

Given Moja's poor hearing, Asana makes hand signals whenever they need to reverse direction. For all her timidity before, Moja plays her role bravely now — they move in perfect synch, as if choreographed. The sloth soon seems dizzy, swaying whenever it rises up. She bellows once or twice, and her male companion answers from the distant swamp.

The only trouble is the sun. It's touching the horizon now, and it keeps blinding Asana. She misses two good opportunities to dart forward. So she creeps closer, determined not to let another chance slip by.

It soon arrives. They're moving to Asana's left. The exhausted beast pushes onto its hind legs and bellows again. The male answers, surprisingly loud this time. Moja screams, too — a perfect distraction. The female turns away to look. And miraculously, right at that moment, a cloud or something shades the sun. Asana's vision clears, and before she loses another second, she raises the dart in her fist and plants her foot to spring forward.

But after one step, something snags her mind. There hasn't been a cloud in the sky all day, much less one thick enough to block out sunlight. Something about the male's last call unsettles her as well. It was so loud. She tears her eyes away from the female sloth — and is left gaping in horror.

The male sloth, covered in dreck, has escaped the swamp. It's standing on its hind legs, its huge size eclipsing the sky. More worrisome, it's standing just five yards from Moja, who tries to backpedal but stumbles and spills onto the ground.

The sloth drops to all fours to pursue her. As he sinks, the sun re-blinds Asana. But she doesn't hesitate. She darts to her right, arcing just beyond the female's reach. Then she pivots toward the male.

Luckily, the face-off with Moja has distracted it. Asana streaks in, lunges—and buries her dart deep in the male's neck.

It's such a perfect strike that she can only step back, stunned, and listen to the sloth gurgle. When it crashes to the ground, the dart is driven in further. After all the fear and drama of the hunt, one clear thought rings through Asana's mind. *I did it—we did. We took down a giant sloth.*

It's the last thing she ever thinks. As Asana stares at the gurgling male, the female rumbles up behind her, its gait quickened by desperation. Moja screams again, but by the time Asana turns, she can only raise her forearm in feeble defense—an echo of what Moja endured so many times back home. As the beast connects with Asana's skull, everything goes black.

*Days later, scouts from Asana's clan find her. They circle the site, trying to make sense of the scene. Asana's body is lying face-down, alone, her skull cratered and forearm cracked in half. There's also a sloth carcass, partly eaten, along with some dull obsidian blades. A light coating of frost covers both corpses. What had Asana been doing? They can't guess. They hoist the tiny figure over their shoulders to bring her back to her aunt, who's been hysterical with grief. They will bury their huntress tonight.*

# Turkey — 6500s BC

~~~

There was no such thing as the Neolithic Revolution, and this is a chapter about it.

The Neolithic Revolution was the biggest transition in human history—when small, mobile clans of hunter-gatherers gave up wandering to settle down and live in one spot permanently, often to farm. But the word *revolution* implies a haste and purposefulness that's misleading. In reality, the shift to sedentary living happened at different times and different tempos in different corners of the world, often over millennia, with usually only a dim awareness during any one person's lifetime that change was afoot.

Sedentary living also gave rise to cities, and the establishment of the first cities, which appeared in southwest Asia, was a similarly fitful process. Archaeologists still debate how and why the transition took place. Some argue that tiny farming villages simply swelled in size, generation by generation, until they reached a critical mass of people. Others have found evidence that sedentism predated farming. Proponents of this theory argue that small clans of hunter-gatherers, despite wandering for most of the year, would get together with other clans during sacred times for festivals, feasts, and religious bacchanalia. Gradually, some clans settled down near the festival sites, sleeping in camps that morphed into permanent, purpose-built dwellings. Perhaps both processes occurred in different places.

Regardless of the specifics, the rise of cities led to radically different lifestyles. In the mobile, hunter-gatherer days, people wandered from shelter to shelter and carried only necessities and a few precious trinkets.

Now, for the first time ever, people had not just shelters but *homes*, and they began filling those homes with all sorts of stuff, necessary or not.

This chapter examines the emotional and psychological fallout of the shift to cities, in particular at Çatalhöyük in central Turkey, arguably the first true city in history. Çatalhöyük ("cha-tahl-HEW-ook," meaning "forked mound") was founded around 7500 BC. Because people there built new homes on top of old ones, a mound 70 feet high eventually rose, with more than a dozen occupational layers spread over 34 acres. Most people around the world then would never have seen more than a few hundred individuals in one place their entire lives. Çatalhöyük boasted of up to eight thousand.

The most striking feature of Çatalhöyük was its architecture. Cities nowadays have structures of all different sizes and purposes—houses, shops, temples, palaces, government offices. Not Çatalhöyük. There were no separate spaces for living, working, worshipping, or anything else—that idea hadn't been invented yet. Instead, every building was part home, part workshop, part church. Furthermore, while these buildings varied somewhat in size, most were roughly four yards by seven yards and had the same basic layout. There were no mansions, no hovels, no obvious material differences between rich and poor. Every household made its own goods, too—knives, pots, floor mats, flour, leather clothing. As far as archaeologists can tell, the city was completely, perhaps ruthlessly, egalitarian.

Even more striking, there were no streets in Çatalhöyük, nor any parks or public spaces. In many places, houses were packed together like molars, with just inches between them. (Empty lots and abandoned homes served as latrines and trash dumps.) There were no doors or windows on homes, either, just a single hole in the roof that served as a combination chimney, skylight, and access hatch. Ventilation would have been poor at best, and the stench of cooking smoke would have permeated everything. People entered and exited these houses via ladders, and got from place to place by walking across their neighbors' roofs. Because people built and rebuilt homes on top of each other at different rates, the houses existed on many different vertical levels, like a Neolithic M. C. Escher sketch.

For roughly the first millennium of its history, Çatalhöyük was conservative in the extreme: in layer after layer, virtually nothing changed in the layout or decorations of the homes. It was the same ovens, the same bed platforms, the same religious shrines generation after generation. Around 6500 BC, however, the social order cracked. Habits and customs changed. Homes got bigger, with more space between them. Households grew more independent. Then, around 6000 BC, the original Çatalhöyük gave birth to a splinter city on a hillock 200 yards west, just across an ancient river. No one knows why the second settlement sprang up, but the people there largely abandoned the rigid equality that had defined Çatalhöyük for centuries. The rift no doubt tore the community apart, pitting friend against friend, sibling against sibling. But those clinging to the old ways would not have given up without a fight...

It takes a long minute for the smoke to shake Darga awake. After an exhausting night of tanning hides, she'd gotten a little too cozy wrapped in her buckskin blanket and fallen asleep, slumped against the wall of her bed platform. Forty years at Çatalhöyük—inhaling smoke every night, every meal, every hacking breath—has left her more or less immune to the smell anyway. By the time her eyes snap open, thick billows have filled her home. She can just make out the glow of the earth oven—as well as her brother's leather burial shroud being consumed by flames above it.

Her bad hip groans as she jumps up and tears the shroud off the wooden smoking rack. She beats the flames with her hands. Some die down, but others slide around like something liquid, igniting new pockets of fat in the hide. Her leather tunic catches fire, too, scorching the left side. Desperate, she finally tips a jar of water onto the skin, spilling the grains she was soaking for breakfast. The flames vanish with a hiss of protest.

She lies there panting and coughing, clutching the ruined garment. How could she have been stupid enough to fall asleep? Her brother's burial has been delayed long enough, and without a shroud—without a second caul to be reborn—it will be delayed even more. Stupid, stupid, stupid.

Before she starts to cry, she prays to calm herself. Then she wrenches upward, wincing, and drags the ruined shroud up the ladder to dispose of it. It's dawn outside, and as the frogs in the nearby swamp quiet down, she can see Çatalhöyük spread before her: a massive, uneven checkerboard of boxy houses with roofs at different heights. Just past them is a hillock 200 yards west. Her tired eyes can just make out the ruins of an abandoned home. There's a ring of mud bricks, now toppled, and a grinding stone for grain, now smashed. Her brother's bones, thankfully, she can't pick out.

The squawk of vultures startles her. She no longer views their presence as a menace, but even the memory of that time nearly makes her vomit.

—Darga?

She looks over to see her young cousin Binya, nervous as a sparrow, emerging from the skylight next door. She says she heard a commotion. Darga sighs, and holds up the blackened shroud. Binya looks heartbroken, and hurries over, stepping down onto Darga's roof. Binya's fussiness often annoys Darga, but she's grateful for the sympathy now.

—What's all the noise?

She's less grateful to hear this voice. It's Binya's thuggish husband, Çekiç. He'll be furious to know that Darga almost started a fire again; she hurries to stuff the blackened shroud into the gap between the homes.

Çekiç emerges and looks around, as if he can sniff trouble in the air. Çekiç means *hammer* in their language. Darga's late brother gave him the nickname as a sneering insult, but he gleefully adopted it. The name refers to the giant stone sledge he carries around with him. He's the public enforcer of morals, and treats the hammer as a symbol of his office.

Right now he's not carrying his hammer, however, just a bundle of straw with his morning shit inside. Instead of carrying it to the communal latrine a few houses over, like any decent person would, he wads the straw up and jams it into the gap between the homes. In doing so, he notices the scorched shroud, which he pulls out over Darga's protests.

—If you burn down our new home, old woman, I swear...

—You'll what? Murder my brother again?

Çekiç actually smiles beneath his beard.

—Your brother chose his fate. I think his spirit wandering for a year of years is a suitable punishment for the chaos he wrought. Unless you think he deserves more?

It's her people's most sacred belief regarding death. In any permanent settlement, bodies require quick disposal for hygienic reasons. Over time this rule became codified, then theologized. Now Darga and everyone else there believes, unshakably, that a quick burial is a moral obligation. Families have three days of grace to say prayers and let the stiffness of death fade. But every day you delay burial after that, a person's spirit must wander the Earth for an entire year before finding rest in the underworld—exposed to rain and snow, buffeted by winds, unable to eat or drink anything. With the yearlong ban on her brother's burial, he's already doomed to several centuries of this—a year of years. Rather than risk more, Darga holds her tongue.

They're interrupted by the arrival of Çekiç's minions—his deputy enforcers of public morals, two long-haired twins named Sersem and Sirkin. They insist on speaking with Çekiç alone, which is fine with Darga. She likes them even less than Çekiç. She tells Binya she'll return later with the offerings and takes her leave.

Back home, Darga adds reeds and sticks to the earth oven and lays three goat bones on top to cook the marrow. The smoke produced now is tame compared to before, but she starts coughing again, and spits gray phlegm into the straw in the corner. Then she scrapes together the pale yellow emmer grains she spilled earlier, rinses them in a pot of water, and samples them; they taste fine still—nutty, with a spongy texture. She drops them into a gourd bowl and sprinkles in roasted einkorn grains for a texture contrast—little nuggets that crunch like pepper pods between her teeth. Now for the marrow. After ten minutes, the bones have changed color from white and pink to tan-yellow; the ligaments look like singed wicks. Using two sticks that she hafted into tongs, she drags them off the fire onto a rock anvil and bashes them with a hammerstone. After a few whacks, each one makes a satisfying crunch. She pours the oily marrow onto her emmer, then digs out the stringy bits inside the shaft with a bone needle. After a sprinkle of salt, breakfast is ready.

Darga never eats in her home's dirty half, where the oven and straw toilet are. Instead, she wipes her feet with a leather rag (most people don't bother nowadays; she finds this scandalous), then grabs her bowl and steps over a red ochre line onto the raised surface of the home's clean half. This section contains her bed platform, blankets, clothes, and tools, plus several bucrania on the wall—bull skulls covered in plaster and mounted in a column like a totem pole. The plastered floor has the firm sponginess of hard cheese, and unlike the smoke-stained walls near the oven, the walls here are painted white and covered with ochre murals—geometric figures, deer, her and her brother's handprints. She'll have to paint over them tomorrow for her brother's burial, which depresses her. But the rite demands it.

Residents of Çatalhöyük entered and exited homes through a combination skylight and chimney, which they reached via ladder. Wall decorations included plastered bull skulls. (Credit: Elelicht, Wikimedia Commons.)

If only her brother hadn't been so stubborn about the millstone.

Their people grind grain on massive slabs of stone, which have to be lugged down from the mountains twenty miles distant. Given the huge investment of time and labor—a round-trip haul can take weeks—millstones represent wealth, and therefore danger: wealth can be abused to manipulate people or accumulate power. As public enforcer of morals, one of Çekiç's main jobs is to smash the stones with his hammer when the owner dies, rendering them useless. That way parents cannot pass wealth to their children. It's publicly sanctioned vandalism, to keep everyone equal.

Her brother Alev despised this law. He argued that it made no sense to destroy the most valuable things in town. And when their father died last year, he claimed the old man's millstone as his. A huge public dispute followed, escalating day by day until Alev took to sleeping on the millstone at night, so Çekiç and his minions couldn't get at it.

Everything came to a head last summer. Alev managed to sneak the millstone out of his and Darga's home and carry it to the hillock across the river. Darga woke up to find every person in town on the roofs, all pointing in that direction. Many turned a hard eye on her, saying she must have helped him — but she hadn't. In fact, her brother's behavior outraged her. Like too many people nowadays, he played fast and loose with sacred rules — skipping prayers, mixing taboo foods, not making offerings or helping neighbors build new houses. She'd snapped at him over his disgraceful behavior too many times to count. But she never imagined he'd defy the law on breaking millstones.

Sadly, that was only the beginning. That same week, Alev began forming mud bricks and carrying them up the hillock. He said he was building a new settlement, where people had more freedom. He also began claiming certain goats around town as his — as if someone could own an animal! Worse still, he marked off a bed of wild emmer near the hillock and carefully weeded it. The grains thrived. But when others helped themselves, he drove them off, yelling that he hadn't put in all that work for poachers to enjoy the spoils. His selfishness shocked Darga.

Exactly a year ago, she walked over to beg Alev to stop, but he refused. In fact, he claimed that other people in town were approaching him at night, encouraging his rebellion.

—Do you think I got this millstone here by myself? I had help.
—Who?
—You'll see. They've promised to join me here.

Darga returned weeping, and spent a long evening alone. What happened next was inevitable. Around midnight, she heard distant screams. She hobbled up the ladder and across the roofs and splashed through the river over to the hillock.

Even before the sight fully registered, she crumbled to her knees in the dirt, squeezing her eyes shut so hard it hurt. But she could not squeeze the image from her mind—Alev sprawled across the millstone in the moonlight, both of them equally broken. The only mercy, however small, was that his face was so swollen she barely recognized him.

When she could breathe again, she opened her eyes and dragged herself forward. The assailants had left, but she could see the angry purple dents of hammer blows on his chest. Some perverse instinct made her want to touch one, and as soon as her fingertips found the hollow, she lost it. She fumbled for his hand—it felt like pulp—and let all her frustration and love for him come pouring out in angry sobs.

After collecting herself, she began dragging Alev's body back to town for burial—the obligation fell to her. She didn't make it even fifty steps before Çekiç and the twins arrived to block her path. He raised his hammer like a staff and decreed that Alev's sins were unprecedented, so his punishment must be, too—forced delay of burial.

Darga protested. The law required immediate burial.

Çekiç shrugged. He acknowledged that the decree broke a moral law, but claimed that the violation was necessary to fulfill a higher, more important principle: he'd heard the same whispers Alev had of simmering rebellion, and here was the way to stamp it out.

To Darga that was no morality at all; you can't pick and choose laws—that's why they're laws—especially with the dead. Moreover, she pointed out that her own soul would be imperiled and punished if she didn't fulfill her obligations. But Çekiç would not budge, and he and the twins dragged the limp body back to the millstone.

The first two days were the hardest for Darga, when the vultures came. She barely left her home, barely ate anything. Every time she imagined the beaks tearing his flesh, she felt the pain of it herself, piercing her legs, her stomach, her breasts. After the birds picked him clean—she could no longer hear their cries—her suffering ebbed into a dull ache. She felt his absence most keenly at night—her bed platform felt weak, ready to cave in, without him down there. Otherwise, she refused to think about Alev at all until a week ago, when she started preparing the burial shroud.

The thought of the ruined shroud now leaves her dejected. It's part of her obligation—a body buried without its shroud, its second caul, cannot be reborn in the afterworld. She wonders if she can finish a new one today. Maybe Binya can help.

She suddenly hears shouting above. A moment later, there's a small stampede across her roof. It sounds like Sersem and Sirkin returning from somewhere. Darga would normally ignore them, except she hears one of them say Alev's name.

She ascends the ladder to find the twins on Çekiç's roof next door, holding a block of dusty black obsidian. She's shocked to see it—it disappeared last year from a public storage pit outside of town. The twins swear they found it on the west hillock, buried near the foundation of Alev's ruined home. Çekiç turns to Darga.

—So your brother isn't just a rebel, he's a thief!

Darga knows Alev would never take any obsidian. Would he? She wonders fleetingly if Çekiç planted it. But before she can protest, Çekiç makes an announcement.

—Pending an investigation, I'm suspending Alev's burial indefinitely. And remove *her* from my sight.

His minions step forward and grab Darga. But she surprises them—she scratches one's neck and kicks the other's shin. In the scuffle that follows, Sersem nearly wrenches her tunic off and Sirkin wrestles her down to the dirt roof and pins her.

Only Binya's pleading with Çekiç earns her release. Darga rises to find herself half-naked, her tunic torn. Çekiç sees this and slaps Sersem—ostensibly for enabling lewdness, but really, Darga suspects, to show him up in public.

Binya hustles Darga away to the latter's home, and by the time they've descended the ladder there, Darga has made up her mind. Binya finds a needle and some sinew thread and tries to mend the tunic.

Between the tear and the burns, the garment looks disgraceful, but it will have to suffice. Binya asks Darga what else she can do.

—Help me make a burial shroud. Today.

Binya looks stricken.

—You heard Çekiç.

—I heard our public enforcer of morals breaking moral laws. But I have obligations, too. Alev deserves burial.

Binya drops her eyes and whispers that she can't. Darga is so disgusted that she yells at her to leave.

Hurt, Binya starts to ascend the ladder. She stops halfway up.

—Promise me you'll leave Alev's bones alone.

Darga tells her to leave again, but Binya begs.

—If you'll leave now, then yes, I promise.

When Binya goes, Darga wipes her feet with the leather rag and steps up to the clean side of her home. She eases herself onto her knees on the raised earth platform of her bed, and says a quick incantation.

She rises painfully, then reaches over to the corner to retrieve a cow scapula hafted to a stick. She begins pounding the earth bed, breaking it into pieces and shoveling the dirt aside.

A foot down, she finds what she's seeking—the skeleton of her father. There are more bodies beneath his—her mother, two grandparents, great-grandparents for several generations.

She'll hide Alev among them. She promised Binya she wouldn't go after her brother's bones. It's a promise she has no intention of keeping.

The enduring fascination of Çatalhöyük lies in the glimpse it offers of a completely different way of organizing urban life. No streets, no public spaces, no doors or windows, every building identical inside. A typical home had an oven on the south wall beneath the skylight-door—the dirty area, which was sometimes marked off with a red line. Next to this, up a low step, sat a clean area with bed platforms, wall murals, and mounted cattle skulls covered in plaster, so-called bucrania. A small, built-in bin served as a granary for storing food. Ultra-traditionalists like Darga could not even conceive of another way of arranging a home.

However fascinating, though, the architecture of Çatalhöyük raises some obvious questions in the modern mind. For instance, given the lack of windows, weren't the homes awfully gloomy inside? I learn the answer—no—on a trip to the Çatalhöyük site in modern Turkey. Archaeologists there have built four replica houses, each a dozen feet tall. Entering them, I expect to find a depressingly dim interior, but am pleasantly surprised at how bright and warm they are. The skylight-doors catch sunshine all day, making them superior to windows, and the white plaster walls reflect the light brilliantly. (The light shifting along the walls may have acted as a sundial, too.) The plaster comes from marl, a soft white mineral that smells earthy and forms a thin putty when blended with water. The people of Çatalhöyük were, as one archaeologist put it, "plaster freaks," slathering the stuff on every surface to keep things bright and counteract the accumulation of smoke; some houses show a hundred separate applications. They were rebuilding freaks as well, regularly toppling their walls to erect new homes on higher foundations.

Visiting Çatalhöyük also allows me to take in the view from the 70-foot mound where the city lies buried. Nowadays, the region is dry and dust-choked, with a summer sun as merciless as any desert's. But thousands of years ago, the land was marshy and lush, with a muddy river meandering by. Archaeologists picking through old trash heaps on-site marvel at the abundance of food. People here ate pistachios, waterfowl eggs, hackberries, plums, chickpeas, acorns, and more. They hunted boars and ducks and aurochs—the gigantic ancestors of modern cows—and supplemented them with domesticated sheep and goats. In fields beyond the marshes, wild grains grew, helped along with occasional hoeing or weeding, horticulture on its way to agriculture. The nearest timber was likely dragged in from the foothills of nearby mountains. Because the soil near Çatalhöyük lacks hard rock, people would have trekked there for obsidian and millstones as well.

A modern reproduction of the ancient city of Çatalhöyük, which lacked streets and public spaces. Residents got from place to place by crossing their neighbors' roofs. (Copyright: ZDF/Terra X/Film Produktion Stein e.K./Alexander Hogh/Martin Papirowski/Timm Westen, Roxana Ardelean/Golem Studio/Alexander Leuck/Frauenhofer Institute for Graphical Data Processing/Christofori and Partner.)

Still, however commanding the vista, I find the Çatalhöyük site a bit dull. It's mostly traditional, dig-in-the-dirt archaeology. Moreover, it's static; it lacks the dynamism of Darga's time, when people were constantly remodeling and rebuilding their homes. So to get a sense of Çatalhöyük as a living place, I visit another Neolithic site six miles away called Boncuklu.

Boncuklu ("bon-JUHK-loo") means "the place with beads"; the ground there used to sparkle with prehistoric jewelry every time it rained. The collection of huts dates to 9200 BC, which makes the homes there some of the oldest structures on Earth. And while these homes differed from those at Çatalhöyük in certain ways—they had doors, and were not packed as tightly together—there are more similarities than not. People at both sites used the same building materials—mud bricks, reeds, logs, a shitload of plaster—and decorated the interiors with murals and bucrania. In fact, DNA evidence from buried bodies suggests that the people at Boncuklu were the ancestors of the first Çatalhöyük settlers; they perhaps even abandoned Boncuklu to found it.*

Unlike at Çatalhöyük, the Boncuklu dig site boasts an active experimental archaeology program. The first thing I see upon arriving is an artificial swamp of reeds and pale green water, built to mimic the ancient landscape; it's a haven for buzzing dragonflies and frogs. (Which were apparently a nuisance once. A middle-aged guard at Boncuklu tells me that his mother could barely sleep at night as a girl because the frogs were so loud. Since then, intensive farming has lowered the water table in central Turkey and parched the land. The frogs vanished.)

The Boncuklu team also builds replica homes with mudbrick walls. They form the bricks by mixing clay from a local riverbed with cut straw as temper, then pat the sludge into rectangles to bake in the sun. The bricks are flatter and wider than modern bricks, but are surprisingly robust for being just mud; they ring solidly when I rap my knuckles on them. Still, for protection from the elements, brick walls in ancient times were covered with an inch of regular mud. The roofs consisted of wooden beams covered by reed mats, with several more inches of mud slathered atop.

Under the intense sun, the layers of protective mud on the Boncuklu homes are crumbling when I visit, full of cracks and infested with spiders and sprouting plants. Happily, though, a team of young archaeologists is refurbishing them, and they let me help. Given that I'm essentially doing their work for them, I'm sure they feel like Tom Sawyer duping his friends into whitewashing Aunt Polly's fence, but I have a blast. The replacement mud comes from a kiddie pool–sized pit they've filled with the thickest, stickiest goop I've ever had the pleasure of

---

* Unlike Çatalhöyük, Boncuklu featured a communal work area, where people gathered to do daily tasks: weave baskets, pack mud bricks, knap stone tools, tan hides, and, well, take dumps together. Excavators know this because they've found thousands of coprolites—dried human turds—piled up in spots. Every so often they'd cover these steaming pits with reed mats, then start over. "Crapping was quite a social activity," suggests Douglas Baird, co-director of the dig. "You'd squat down next to someone, ask how their day went." Baird's team has even found bodies buried in the loo.

mucking around in. It's dark gray with the consistency of frosting and the density of cement. If you stuck your foot in there, you'd definitely lose a shoe. For an hour, I scoop up fat handfuls and *splat* them onto the wall. There's little sense in smoothing it out: Aesthetics were a minor consideration in Neolithic cities, especially at Çatalhöyük, where no one could see your outer walls. People invested in the interiors instead.

Mudbrick homes have a reputation for being cozy in the winter and mild in the summer. Sure enough, it's 90°F outside at Boncuklu but a dozen degrees cooler in the replica houses. Still, that reputation needs qualifying: Light a fire inside the homes, and they turn into mud ovens. Despite their chimney holes, they also get horrendously smoky, as I learn the hard way one afternoon. As part of an experiment, archaeologist Gökhan Mustafaoğlu kneels down in the dirty half of one replica home and lights a bundle of sticks and reeds, common local fuel.* Although I'm standing three yards away, I soon feel waves of blistering heat. Then the smoke starts. This does have one benefit: there are scads of mosquitos lurking in the rafters, and at the first hint of smoke they hightail it for the exit. But I don't get much time to enjoy the bug-free life, because within a minute I can barely breathe. The setup appears to defy physics: the white smoke rises straight up toward the ceiling hole, only to bounce off and curl back inside. Before long, the back wall disappears in a haze, the smoke is that dense. Mustafaoğlu drags me outside for my own good, but I'm determined to experience what Darga did and duck back in after a few minutes. I'm rewarded with a glimpse of beauty: the shaft of sunlight streaming down from the skylight looks

---

* In another experiment, Mustafaoğlu and a colleague roast a cow bone over some coals, so we can sample the marrow. The project goes comically awry. First, the tibia they procure is approximately the size of a bridge pylon and proportionately sturdy. To compound the problem, the soil around Boncuklu and Çatalhöyük lacks much hard rock, so we're forced to use crumbly sandstones as the hammer and anvil. After a few whacks, both have shattered, while the tibia has nary a scratch. We do eventually crack the sucker open, but it takes vastly more work than expected. (Darga's goat bones were slenderer and easier to break.) Incidentally, the marrow was not the blended puree you get at bistros. It was mostly liquid, with stringy membrane bits mixed in. Nutritious, but hardly gourmet.

angelic, refulgent with white smoke. But an instant later my eyes water over and I feel so light-headed I have to grope my way back outside.

So how did people at Boncuklu and Çatalhöyük endure all this smoke? No one knows. During fine weather they likely cooked on their roofs, or at least started fires up there and transported the coals below. Perhaps they also devised crude flues from reed mats to channel smoke out, although there's zero archaeological evidence for this. The most probable answer is that they simply inhaled a lot of frickin' smoke — buckets and buckets of it, the equivalent of smoking several packs of cigarettes daily. Some skeletons from Çatalhöyük have black marks all over the ribs. Darga and her ilk likely hacked their way through life with diseased lungs.

But the most macabre feature of life at Boncuklu and Çatalhöyük was where they buried their dead: beneath their beds. In modern times we associate basement burials with creepy true-crime headlines, but in Neolithic Turkey, it was a sign of affection. When someone died, their relatives would break up the home's earthen bed platform with bone shovels and expose their previously buried ancestors. When the hole was big enough, they'd add the new body, bound in the fetal position and wrapped in a leather or linen shroud. The upshot is that people snuggled up for bed each night with uncles and cousins and great-grandmothers lying just inches below their heads. One building excavated at Çatalhöyük had sixty-two bodies inside. As one archaeologist said, "Çatalhöyük is as much a cemetery as it is a settlement."

As for *why* people buried relatives in their bedrooms, we'll never know for certain, but the ritualistic nature of the practice suggests some sort of supernatural belief, perhaps involving the veneration of ancestors. (As Darga's kindred spirit, Antigone, would later say, "We have to please the dead far longer than the living.") The practice also probably helped establish property rights. Remember, these were the first cities in history, the first time people could say they were *from* somewhere. But the idea of owning land, as opposed to merely wandering across it, had to be invented and justified. Burying bodies may have been crucial to establish such rights — skull and bone as deed and title. Indeed, skeletons moved from house to house sometimes, following the owners. (*Rest in peace* was not a precept people subscribed to there.)

Around 6000 BC, for reasons unknown, a new, splinter settlement was founded on a hillock across the river from Çatalhöyük. Over time, the homes there grew bigger and more elaborate, with second stories and walled courtyards. People also began using fancy decorated pottery and stopped burying the dead beneath their beds. Most striking of all, homes on the hillock had far larger bins to store food, a sign of people accumulating wealth by stockpiling grain. For a place as ruthlessly egalitarian as Çatalhöyük—where all homes were roughly the same size, with the same goods inside—these seem like ominous signs, the abandonment of a community ethos. But of course people didn't simply wake up one day in 6000 BC with completely new ideas. There must have been generations, perhaps centuries, of smoldering tension before this. Maybe a few brave souls like Alev even made early forays to the western mound. Regardless, starting in about 6500 BC, archaeologists see signs in Çatalhöyük that people seemingly wanted more space and individual freedom, which they ultimately got—along with all the strife and conflict that accompany such desires…

※ ※ ※

Darga sees the surprise on Çekiç's face when she arrives with her pouch of offerings. He's helping some workers dismantle the roof beams for his new home, breaking up the mud mortar with his hammer. He yells, *If you're trying to bribe me about Alev, it won't work*. Darga says nothing, merely descends the ladder to meet Binya.

The new home is being constructed next to their current one, using an abandoned shell that's been sitting empty ever since its roof collapsed. Çekiç says their current home has a poor foundation, necessitating the switch. But Darga can't help but notice that Çekiç's new home, when rebuilt, will rise higher than all his neighbors'. Higher than any home in town, in fact.

The first step, undertaken last week, involved knocking the

abandoned structure's ten-foot walls down to three feet, and scattering the rubble to create the foundation for the new home; clean soil was packed into the gaps. Then a phalanx of workers—eager to curry favor with Çekiç—rebuilt the walls higher, stacking mudbricks and smoothing the plaster on the inside with bone scrapers. Several bucrania skulls were mounted as well. Today, the workers are transferring the roof from the old house, tearing up the reed mats and timbers and mortaring everything together with fresh mud. Çekiç also needs to transfer his ancestors' bones. To this end, a dozen women are kneeling in the dirt with scapula spades, sorting ribs and vertebrae and teeth into piles.

Darga finds Binya and hands her the first two offerings from the pouch, clay figurines of voluptuous matrons; each has ample hips and is cupping her pendulous breasts in her hands. Darga even carved tiny eyes and ears and belly buttons with obsidian flakes. Binya gasps and clutches them, which makes Darga smile. Then Binya's face falls.

—I can't break one. They're too beautiful.

Darga laughs until she starts coughing.

—It took less time than cooking an egg.

Binya raises an eyebrow, but Darga nudges her over to the grain bin, where Binya solemnly places one figurine inside. The matrons represent abundance and fertility, and the clay they're made of comes from the earth the same way grain does. Placing the figurine inside the bin will therefore fill it with grain. Then Binya crosses to her new hearth. She grips the other figurine, hesitates, and snaps its head off. She'll bury it in the hearth later for good luck. She hugs Darga in gratitude.

It's a short-lived moment. Çekiç has sneaked down to spy on them, and grabs the broken figurine from Binya. He drops it next to the hearth, flips the hammer off his shoulder, and proceeds to crunch it into dust. Binya howls at this, but Çekiç just smiles.

—It's supposed to be destroyed, right? I'm helping it along.

Darga is annoyed enough to consider withholding the last offering. But she reminds herself why she's here. She reaches into the pouch and holds up the clay leopard.

They're the apex predators of the nearby plains—equal parts

majestic and deadly. Darga finished it yesterday, but spent a half hour this morning detailing it, scratching spots and teeth and claws into the soft, unfired surface. (She's left the area around the eyes undone, however—it's taboo to depict those. Her people believe that if you see a leopard's eyes, you die instantly.)

It's the most gorgeous figurine Darga has ever sculpted, and Çekiç grabs it greedily from her fingers. He gestures to his wife for a needle. Binya locates one, then holds the leopard so Çekiç can stab the soft body—once, twice, three times, plus the head. A ritual sacrifice, to give Çekiç the leopard's luck and power during hunts. They'll bury it in the floor later.

Despite his hostility before, Çekiç claps his hands when he finishes the sacrifice, insisting that all the roofers look at Darga's work. They gaze down respectfully. Darga despises Çekiç, but can't deny a swell of pride at his crowing over her statue. It's a complicated moment.

And one Darga hopes to take advantage of. She speaks loud enough for everyone to hear.

—I'm glad you like it. In exchange, I'd like one of your soaked hides to tan.

She sees Binya cringe, staring at the ground and shaking her head. Çekiç is instantly on guard, and asks her why.

—A new tunic. Your minions tore mine. If it tears any further, it will be an affront to public morals.

—Those hides are for our home.

—I just need one. And I'll start a new one soaking for you.

Çekiç seems not to believe her. But he can feel everyone's eyes on him. *One hide*, he announces. He claps his hand again.

—Everyone else, back to work.

Darga wastes no time in leaving. Based on the sun, it's already midmorning. She hands Binya her pouch, and asks her to keep her fire stoked for later.

She crosses several dozen rooftops to the edge of town. They rise or fall every twenty feet—jutting up here, sinking there, with sharp reed edges fringing each. Their dirt covering feels warm on her toes. People sit cross-legged on each one—knapping blades, weaving linen. Several eye her suspiciously. Rumors must have spread about the stolen obsidian. Darga holds her head high anyway. No matter how many times she tells people she disagreed with her brother, they don't listen. What more can she say?

At the edge of town, she descends a ladder, then makes her way down the grass to the bottom of the mound. There, in a pile, hunters leave deer hides for anyone to claim. Darga fights through the horde of flies and selects one, then finds the giant pots she helped Binya fill with water and ash the other day—a caustic solution that loosens fur. By now, the liquid inside the pots is cloudy, with sludge at the bottom. Darga reaches into one, pulls out a sopping hide, and replaces it with the fresh one.

She drapes the wet hide over a wooden post and wrings it as best she can, twisting the folds around a stick. Then she selects a sharpened auroch rib from a pile of tools and scrapes off the hair.

When the hide is denuded, she begins the laborious process of scraping off the excess layers of skin. It takes two full hours, by which point her own skin feels like leather in the hot sun.

This task complete, she makes her way over to another pot and opens the lid. Inside, in cool water, sit several jiggly deer brains. She plops one into a pit lined with rawhide, and adds a gallon of liquid from the pot. Ignoring her hip, she mashes the brain into slurry with her feet. When it's frothing on top, she drops the hide in and massages it with her hands, working the brain grease into the skin.

When it's saturated, she drapes the hide over a rope stretched above the pit, anchored between two wooden stakes. She wraps a section around a stick again and wrenches it until her shoulders strain, letting the brain slurry stream back down into the pit. She wrings every section of the hide like this, then pulls it down and starts massaging more brain juice in.

After five cycles of soaking and wringing, her arms burn. She curses Binya for not being here to help. But she can't rest, because the last steps are the longest and most tedious of all. She drapes the hide over the wooden post yet again and begins rubbing it back and forth as vigorously as she can manage, to stretch and dry it. Every inch needs working. Dozens of people pass by as she sweats in the fading sun. No one offers assistance.

It's nighttime before she finishes. The hide is ghostly white now, softer than feathers; she can smoke it tonight while she eats. (Although her hands are so sore—twisted into claws—she fears she might not be able to grip a spoon.) As she trudges up the mound, her hip throbs. But in a way she's glad for the pain; it's a reminder that she's fulfilling her duty.

Back home, she adds fresh reeds to the fire Binya kept going, and sets up the wooden smoking rack. Dinner is duck eggs whipped with emmer. Then, with one eye on the shroud, she settles in to dig up more of her ancestors' bones. Four bodies emerge before she sees one whose skull reminds her of Alev. It will make a perfect substitute.

"Yeah, deer guts are not super appealing."

He can say that again. I'm standing next to Caelan Dunwoody, gazing down at the hide of a 90-pound whitetail. It ain't pretty. The side opposite the fur is slick with blood and guts, and smells faintly of crab that's turned. Yet Dunwoody—a grinning, broad-shouldered, sunburned twentysomething who's an expert on the prehistoric art of processing animal carcasses—promises me that adding even more blood and guts, in the form of brains, will alchemize this hide into something so clean and supple you could swaddle a baby in it. It's one of ancient humankind's most amazing magic tricks: tanning leather.

My brain-tanning lesson takes place at Moose Ridge, a wilderness school in Maine run by Dunwoody's mother that teaches survival skills and runs archaeological experiments. It's aptly named. The campus sits on a grassy knoll surrounded by trees, and a moose called Bertha really does wander through now and then. The school attracts the kind of people who flaunt body hair and disdain shoes, even on gravel. The morning I arrive, two women have scooped up a roadkill fawn to butcher; another one speed-dials the game warden to secure permission. Besides Dunwoody, I meet several other dedicated tanners. One tells me, "If I had a house fire, the first thing I'd grab is my hides. Social Security card? Who needs that. Laptop? Replaceable. Those hides are everything." They all despise commercial leather as sterile and drab, lacking authenticity. Indeed, they embrace even the imperfections of their hides. Every flaw adds "character"—a high compliment.

As my arms would learn, tanning is perhaps the most intensive labor that ancient people undertook. The process varied from culture to culture across the globe, depending on the climate and available animals, but the general outline was the same as what Darga did: soak and scrape the hide to get the gunk off, then infuse the skin with fats (or tannins) to make it more waterproof and durable.

Dunwoody and I are tackling a deer, among the easiest animals to tan. (The difficulty increases markedly with the density and thickness of the skin. Bears are challenging, pigs near impossible.) To remove the fat and muscle clinging to it, I drape the skin over an angled log and scrape it with a two-foot shard of wood, using the same basic motion as a rolling pin. It's slow going, as the hide keeps slipping off the frame. And unlike with Dunwoody's long, smooth strokes—each of which scours away an enviable amount of skin goo—my slow, clumsy scrapes remove mere wisps of flesh. I also fret about tearing the skin—although that turns out to be impossible. However delicate-looking, raw skin is cartoonishly stretchy. You can yank it and wrench it and twist it and gouge it, and it simply will not tear. I've never looked at my skin and thought *wow*, but Mother Nature is a brilliant material scientist.

After it's scraped clean, ancient people soaked the hide in water mixed with a caustic substance like wood ash. (Some people tested the solution's concentration with a raw egg; if the solution was strong enough, the egg floated just under the water's surface.) Normally this soaking step takes days, but Dunwoody has presoaked another hide to save us time. Soaking a hide removes mucus from the skin and loosens hair. It also changes the skin's color, from a raw pink-white to a tawny brown. I pull this hide out of its bucket. It feels slippery, like bleach does on your fingertips, and it's arm-bendingly heavy from all the water it's absorbed, like a huge beach towel dropped in a pool.

We start the tanning process by removing the hair, which turns out to be my favorite step. I simply drape the hide over the log, run the wood shard over the surface, and presto—huge handfuls tumble down onto the grass. Compared to scraping guts, it's shockingly quick and easy, and it gives me the same satisfying feeling that cleaning the lint out of a dryer trap does. I could depilate deer all day.

The fun ends quickly after that. With the guts scraped off and the hair gone, we could stop working here and let the hide dry into a perfectly serviceable material—rough, tough rawhide, similar to a dog chew. But we want soft leather, which requires removing unwanted layers of skin. Skin consists of the epidermis on the outside, the dermis beneath (including the "grain"), and the hypodermis on the inside. The dermis layer, minus the grain, will become the leather. We therefore need to remove the other layers; if left intact, they'll block the fats that need to penetrate the hide later. So it's back to scraping. The hypodermis comes free quickly, but the grain and epidermis are stubborn; they peel off only in tiny chunks that look like grody blister skin. Worse, the task is endless: no matter how much I scour one spot, more comes off. I despise Sisyphean chores like this (scrubbing the tub, scraping paint), and already miss the quick magic of removing hair.

At last, all the clingy extra skin is gone. The remaining dermis is a translucent white, similar to a thick caul. The scraping also squeegeed the hide, forcing water out and making it far lighter. At this point, several hours in, we break for lunch. (Beyond all the neat skills taught at Moose Ridge—tracking, tanning, foraging, making mukluks—the school is renowned for its delectable fare. During my stay I sample fried wild mushrooms, field-green spanakopita, pickled milkweed blossoms, homemade pesto, and more.) While we eat, other tanners trade recipes for raccoon and swap tips on rendering moose oil. Afterward, our bellies full, we begin the least appetizing aspect of the process: mashing in the brains.

The overall goal of tanning is to manipulate certain fibers in the dermis layer—namely, the collagen fibers, which give skin its suppleness. In short, leather is soft and supple when those fibers are open and lofty, as opposed to collapsed. Tanning is essentially the art of using chemicals to keep them open and lofty over the long term. Nowadays, industrial tanners achieve this state with chemicals containing chromium. But for much of recorded history, people used oak bark or other vegetable matter rich in tannins (hence the term *tanning*). I've even heard of people tanning skins with aged urine, which I tried at home and which proved to be the most spectacular, and foul, failure of this whole book.\*

The most ancient tanning substance is fat, especially brain fat. Compared to vegetable-tanning—which involves little more than soaking a skin in bark water for a month or two—fat-tanning is fast but labor-intensive. It also produces lighter leather that's ideal for clothing or tents that need dragging from place to place. The first step in fat-tanning involves emulsifying the fat (blending it with water), and brains are

---

\* I tried urine-tanning a salmon skin, a process used in prehistoric Scandinavia, Canada, and Japan to make salmon-leather boots and clothing. In short, I pee into some Tupperware, plop a salmon skin inside, lock it in my spare bathroom to age, and try hard to pretend it doesn't reek like a New York City alley in August. After a week, the scales work loose and float to the top, a positive sign. But over the next few weeks the urine turns cloudy, then moldy, and the smell somehow keeps getting worse—an eye-watering punch of ammonia. This persists even though I change the urine weekly to keep things fresh ("fresh"). To my dismay, the skin also shrinks in size, deteriorating from a postcard to a postage stamp.

The recipe I'm using insists that I touch the skin every few days to gauge its progress. It's perhaps the slimiest thing I've ever felt, a cross between pudding skin and phlegm. Still, one magical day, I pick up the postage stamp and—sweet Jesus, it feels like leather. It's pebbly, and so flexible I can stretch it like rubber. I'm thrilled. But my triumph is short-lived. I change the urine, and when I crack the lid a few days later, the skin has disintegrated into microscopic ribbons, as if run through a shredder. Crushed, I flush the whole smelly mess away.

I later learn that, technically, urine-tanning isn't a true tannage. It merely strips oil and grease off the skin and exposes the dermal layer that sucks up the tanning agents. So while urine-soaking can be a stage in tanning, you always need to do more. All that work—all that reek—for nothing.

convenient because they already contain water and are therefore naturally emulsified. Mayonnaise is another emulsified substance, which means you could theoretically mayo-tan a hide as well.

There's an old saw among tanners that every animal has exactly enough brains to tan its own hide. Which seems macabre, but tidy—a perfectly balanced entry in the cosmic ledger. Sadly, I don't get to test this maxim at Moose Ridge, since they don't have any deer brains handy; we use goat brains instead. (Dunwoody shies away from using deer brains anyway, given the prevalence of chronic-wasting disease, the cervid equivalent of mad cow. This neurological disorder has never shown up in humans, but as Dunwoody says, "We don't want to be patient zero.")

Fresh brain is softer than most people realize, the consistency of custard. (If you dissected any animals in high school and remember the brain as rubbery, it was fixed in preservatives.) The goat brains are gelatinous pink masses that look troublingly like fish bait, and I dump them into a bucket along with eggs and oil for more fatty goodness. I then add water and whisk the mix to further emulsify it. It turns a creamy yellow, like hollandaise sauce. Dunwoody instructs me to keep beating until we have a good foam on top, "like a meringue." Despite their softness, parts of the brains stubbornly refuse to break down, clinging to the whisk like strings of snot.

At peak meringue, I plop the hide into the bucket and start massaging it. It greedily sucks up the fatty hollandaise until it's heavy again. I pull out the soggy, dripping skin and slop it over a makeshift wringing rack—a skinny wooden pole running at chest height between two pines. I take the hanging ends and twist them around a stick, like a medieval torture device, then plant my feet and pull until my shoulders strain. Brain juice gushes out. I aim as much as possible into a bucket for later rounds of soaking. (People way back when likely caught the juice in ceramic pots, pits lined with rawhide, watertight baskets, or old canoes.) Indeed, Dunwoody informs me that I have to repeat the whole soaking-and-wringing cycle seven times. He's strict on this. Then he leaves me alone to toil.

The wringing process proves half gorgeous, half disgusting. On the plus side, the stream of brain juice catches the afternoon sunlight and sparkles in the air like liquid diamonds. It's lovely. That said, the direction of the stream always catches me off guard—especially when it heads my direction. Like a Vaudeville seltzer gag, I get squirted in the eye, in the mouth, even a million-to-one shot into my ear. As someone who despises the feeling of grease on my skin—I refuse to use even lotion—I sure didn't need brain slop dribbling down my face.

I try to do seven rounds, sincerely. But it's been a long day, and after a few more soak-squirt cycles, I'm spent. My fingers are wrinkly with brain juice, my limbs feel rubbery. The juice looks worn-out, too. It's more the gray-yellow of pus than sunny hollandaise now, and there are dead flies floating in it, kamikaze casualties. I just can't face another round. So Caelan, if you're reading this, I'm sorry. I told you I did seven rounds but quit after five. If your hide falls apart someday, I take full responsibility.

It's late afternoon by now, and given that I'm not under Darga's time crunch, I hang the hide up to finish the next morning.

When I return, I'm surprised to find that the color has changed overnight, from bright white to a marbled white-and-khaki. Today's task involves stretching the hide, a crucial step. We've already infused the skin's collagen with liquid fat. But as they dry over time, collagen fibers close ranks and squeeze fat out; if this keeps up, the hide will devolve into something stiff and scratchy, like fuzzy rawhide. Stretching the fibers prevents this by disordering them and keeping them from bunching together. This locks the fats in place and keeps the skin supple.

We stretch the hide two ways, lengthwise then side to side. The first method involves hanging the skin over a cable or rope and working it back and forth, the way you dry your back off with a towel. Given the weight of the hide, it's hard to do this solo, but Dunwoody and I, each holding one end, slip into a nice rhythm of shimmying. Tanning was a communal activity in ancient times, and it's at this stage that Darga would have missed Binya's help the most.

After we work the hide for a spell, Dunwoody shows me the stretch marks on it, which look identical to those on human skin. Removing these marks requires working the hide the other way, side to side. We do so by draping it over a post and yanking it back and forth in a shoeshine motion. Dunwoody repeatedly urges me to pull harder—harder—and I'm nervous about tearing the skin again; it's bulging grotesquely. But skin trumps muscle. The stuff is indestructible, made by Acme.

We alternate several times between cable and post. As heat and friction dry the skin, it changes color again: the marbled khaki reverts to wedding-dress white. We can also see scars emerging from the deer's life—tick bites, gouges from antlers, scrapes from brambles. "Character" aplenty.

After two grueling days of scraping and wringing and stretching, I'm gratified to hear that the last step, smoking, is passive. You simply twist or sew the hide into a loose bag, hang it over a fire, and let the insides billow with smoke. Smoke keeps bugs from devouring the leather, and also induces chemical reactions that preserve the suppleness when it gets wet; when unsmoked leather meets water, it once again deteriorates into a crusty, rawhide-like substance. (There are a thousand paths to rawhide or its equal, but everything has to go right to reach leather.)

We hang the hide on a rack and get a nice, smoky fire going. Because I have other activities scheduled at Moose Ridge, Dunwoody sportingly volunteers to watch the hide to make sure it doesn't catch fire. Alas, one of my scheduled activities involves trepanning a skull—a type of ancient neurosurgery. (See chapter 8.) And the marvels of trepanation prove too tempting to Dunwoody; he begins wandering the twenty yards back and forth from the firepit to the picnic table where I'm working. Each trip lasts a little longer. An hour in, one of the other tanners screams. Five heads whip around to see flames crawling up the side of the hide.

A *Three Stooges* scene ensues, as a barefoot Dunwoody sprints to the firepit, yanks the hide down, and beats the flames out. Fats catch fire readily, and several inches of hide are scorched and ruined. But Dunwoody laughs it off and gamely declares this to be more character. It helps his mood that, despite my clumsiness (and skipping steps), the hide turns out beautifully. The smoke has shifted its color one last time, from wedding-white to rich manila. It's pillowy soft, too, and smells wonderful; we bury our noses and *mmmm* over the smoked wood.

Running my fingers along the surface, it's almost impossible to comprehend the transformation that's taken place. We'd started with a gut-smeared pelt smelling of crabs; now we're holding a garment softer than baby skin. My limbs are weary, and my own skin still reeks of brain juice. But holding leather, real leather made the ancient way, is a damn fine compensation. There's a reason Darga's people used it for things like burial shrouds, despite the great sacrifice of time, labor, and sometimes even life…

Basket in hand, Darga steals across the rooftops, trying hard to suppress her cough. Thankfully, the chorus of frogs should mask her footfalls. She'll make her way back by another route, too, to be extra cautious.

On the other side of the river, her stomach clenches. She hasn't visited the west hillock since last year. The sight of Alev's skull doesn't distress her; like everyone at Çatalhöyük, she's seen her share of human bones. But she cringes to see the damage to it: weathering, gouges from vulture beaks, several loose teeth.

Working quickly, she removes a clay figurine from the basket—a vulture—and smashes it with a piece of the shattered millstone. She's retaking power from the birds that consumed her brother. The rite requires only one strike, but she keeps going anyway, each additional blow fortifying her. After a year of feeling unsteady, as if she might lose her balance with every step, she feels like she's touching solid earth again. She even picks up one of the larger fragments and crushes it with her own hands.

But however gratifying, there's little time to linger. She starts swapping bones from inside the basket, whispering apologies as she does so. It shames her to disinter her ancestors, but her immediate duty lies with Alev. Hopefully, she can rebury these bones later.

She's almost finished when a growl cuts through the frogs—quite near. A tingle shoots up her spine. She looks up, and ten yards away sees the moonlit glint of two eyes.

Darga stops breathing. The leopard's tail snakes back and forth, and its ears are pinned to its head—not a friendly sign. She considers hurling a piece of broken millstone to drive it back. But given that she's already seen its eyes, she's as good as dead. She remains frozen, trembling, holding Alev's skull.

Before she can think what to do, the leopard darts forward and pounces.

She gasps and ducks—only to feel the leopard swish past her. When she finally turns, she sees the glint of another set of eyes twenty yards away. The two leopards nuzzle a moment, then trot off together. Darga exhales hard, then breaks down coughing. She wonders if it's a cub reuniting with its mother—leopards are mostly solitary, but cubs return home sometimes if they're struggling to find food. Regardless, she's not waiting around to find out. She drops Alev's skull in the basket and hurries toward the city, stopping only to grab some clay from the riverbank and fresh marl from a nearby pit.

The moon is setting now, and as she tiptoes across the roofs, her mind is still churning over the encounter with the leopards. What did it mean? Preoccupied as she is, she doesn't see that Binya's rooftop isn't quite finished. Bare beams are exposed, and she steps awkwardly and wrenches her foot.

She manages to suppress a cry, at least mostly. But she goes down hard on her hip and Alev's skull spills out of the basket. She tries to rise quickly.

—Darga?

The sound is every bit as chilling as the leopard growl. Darga looks up—and sees Binya's head peeking out of the skylight.

Darga hisses at her to go back down—Çekiç will hear. But Binya is already hurrying over. And however grudgingly, Darga is grateful her cousin is here. Her hip stings and her knees are scuffed, and she's wobbly after her long day in the sun. Binya replaces the skull in the basket and guides Darga home. She doesn't even chastise her for so obviously breaking her promise.

She gets Darga down the ladder and helps her pile all of Alev's bones into one corner. Then she gets Darga into bed—or at least to the clean area, given that Darga has excavated a mound of dirt from the bed platform to get at the skeletons beneath. She wraps Darga in her buckskin blanket and says she'll wait to make sure she falls asleep.

Darga has just about nodded off when she hears her name again.

—Darga?

The question is not gentle this time. And it isn't Binya asking.

—What the hell are you up to?

Darga cowers and scoots back into a corner, trying to hide. But there's no escaping the face that appears in the skylight. The bearded glare of Çekiç.

---

The practice of burying loved ones in the bedroom, and not very deeply, raises some obvious questions. For one, didn't the houses smell?

To answer this question, intrepid archaeologists at both Çatalhöyük and Boncuklu have run experiments that involved burying animal carcasses beneath plaster floors. The Boncuklu team used sheep, buying them from local farmers whenever one died naturally. Digging the holes for the bodies produced far more dirt than expected—a good pitcher's mound's worth, which then had to be scooped back in. But Boncuklu archaeologist Douglas Baird said the dirt suppressed the smell of rot quite well, at least at first. After a few days, the patch of earth settled and sank, leaving cracks for gases to escape. Baird says he didn't mind the mild odor seeping out, although others did. He even speculates that the odor played a role in funeral rites, with the fading smell symbolizing the fading presence of the person on earth.

Most skeletons at Çatalhöyük are plain old dusty bones, but

occasionally decorated skulls turn up, with sculpted clay noses and cheeks rouged with ochre or cinnabar; the exposed teeth form eerie rictus grins. In perhaps the most spectacular discovery ever at Çatalhöyük, archaeologists unearthed an adult woman hugging a red skull to her chest. No one knows why she was clutching it so tightly.

Beyond the skeletons, other exciting discoveries at Çatalhöyük involved art and arson. One figurine depicts a matron sitting regally on a throne, each arm resting on a leopard. Another shows a pregnant woman in front and an exposed skeleton in back. Memento mori. Murals were popular, too, mostly geometric figures and scenes of hunters attacking wild game. (The animals often had erections for some reason, and the hunters seem frankly cruel, baiting the prey by yanking on their tails or tongues.) These are the first pictures in history on human-made surfaces, as opposed to rock walls—the first canvases. The oldest known depiction of a drum appears in a Çatalhöyük mural as well. We don't know what exactly people used for brushes,* but we do know there's a correlation between the number of burials in a home and the number of layers of paint. This suggests that people repainted the walls during funeral rites.

That notion fits quite nicely with most archaeologists' view of art at Çatalhöyük as a sort of ritual magic—a spell. Recall Binya shattering the figurine or Çekiç stabbing the leopard for hunting luck. Art could also perhaps protect groves from harsh weather, save sickly children, or redeem a disgraced brother. Art didn't simply stir up memories or evoke a sense of beauty. Art at Çatalhöyük made things happen.

Then there are the fires. Burnt timbers and charred plaster are common throughout the dig site. These aren't just smatterings of ash, either, little cooking hearths; some homes are thoroughly scorched. (Observers have joked that archaeologists there could probably get hired as arson investigators by the time they finish their research.) Given

---

* Stone Age people used pretty much anything you can think of to make pigments: ochre, clay, bird droppings, soot, and a whole slew of colored ores. Brushes included moss, fur, chewed-on twigs, and feathers.

the city's tight spaces and flammable roofs, some of the fires were probably accidents. Others, however, might have been very much intended...

The mess of Darga's home ends up saving her. When Çekiç hurries down the ladder and sees the open bed-grave, his eyes go wide with glee—he thinks he's caught her. But she claims she unearthed the bodies days ago, in anticipation of getting Alev back. Çekiç scowls, clearly not believing her. He begins marching around the room, grabbing bones and studying them in the firelight for any telltale sign that they're Alev's. But there are so many bones lying about that he can't prove anything. He finally slams a femur against the wall—her father's, she thinks—and grabs Binya's arm. She yelps as he storms off with her.

Alone, Darga creeps up the ladder with an obsidian blade. For safety's sake, ladders in town are bound to roof beams with sinews; she saws through them now, so she can yank the ladder down if needed. At this point, she badly wants to sleep. But it seems unwise. Çekiç could return tomorrow in the daylight, and the vulture pecks on Alev's bones would be a dead giveaway. He'd surely seize them, perhaps seize all her bones. She has to bury them tonight.

But the funeral rites must come first. She crunches some marl into a pot and adds water to make paste, then dabs it onto the walls with a cow scapula and smooths it with stone. She's sad to cover her and her brother's handprints from their father's funeral, but doesn't have time to dwell; they disappear beneath a streak of white.

She paints while it's still wet. For the past year, she's been envisioning a hunt, one of Alev's favorite activities. But after tonight, she abandons that idea; there's only one choice now. She crunches some ochre into a turtle shell, then urinates into it and adds the white of a duck egg, both of which help bind the mineral to the plaster. She moistens her chewed twig brush with her lips and begins.

Residents of Çatalhöyük decorated their homes with murals and buried family members beneath their bed platforms. (Credit: Murat Özsoy, Wikimedia Commons.)

She has the leopards face each other, almost nuzzling nose to nose. They have claws and spots, and tails scything overhead. She hesitates with the faces. But despite the taboo, she adds the eyes, trying to capture every detail she can remember. At their feet, she paints Alev's body. These solitary creatures coming together over her brother speaks powerfully to her. After the tumult of the last few hours — or really, the whole last year — a sense of calm wraps itself around her like a mantle. She hadn't realized all the tension she was carrying in her back and shoulders. It releases now like a cramp; even the ache in her hip dulls. She looks the portrait up and down, and after a little thought, adds a slim moon in the upper corner. She's pleased with the results; she feels a sense of perfect balance.

At last, she can finish the rite. Alev's body never got a chance to decompose in the ground beneath their home. Indeed, the whole point of burying bodies at home is for the flesh to melt away and become part of the soil, further cementing their claim on it. So she has to give Alev new flesh. She coats each bone with thick clay from the river, then lays them one by one on the supple leather shroud, arranging his body in a fetal position.

It's dawn by the time she finishes; the frogs have ceased croaking. Darga adds some chopped reeds and dung to her fire for warmth, then sits down to start the skull. She's dismayed to see several teeth missing, but can't help that now. She takes great pains sculpting the likeness—his pudgy nose, the dimple in his right cheek, his thick mustache. Seeing him in the flickering firelight raises a lump in her throat. After all this time, after all the anger he caused, she's restoring him to life—only to let him decompose again soon.

She's so absorbed in recollection that she almost misses the soft crunch of footsteps overhead.

—Darga!

It's Binya, crying out. She's quickly stifled.

Darga sets the skull down and hobbles for the ladder. She yanks it aside not a moment too soon. Seconds later Çekiç and his minions appear at the skylight.

—Give us his bones, Darga.

She backs away again into the corner. When she doesn't answer, Çekiç yells down.

—Do you deny having them?

—I have nothing unlawful.

—Then what did I find on my roof?

Çekiç tosses something down; they scatter on the floor like pebbles. Darga creeps forward and looks.

Her heart drops. It's Alev's missing teeth.

—Darga, don't be stupid and jeopardize yourself. He was a criminal. He hated what you love.

*That doesn't negate my duty,* she thinks.

When she doesn't answer, Çekiç barks at the twins to unlash the ladder from his home and bring it over. Darga steels herself. If they lower the ladder, she'll kick it away.

But then what? She's trapped down here, too exhausted to fight. She wastes several seconds thinking—or fretting, really, *telling* herself to think. Until her eye falls on the answer.

She kicks open the earth oven, exposing the fire inside to the air. It swells red, and she feeds a long bundle of reeds inside, extending a trail toward the middle of the room.

By now the twins have returned. They start lowering the ladder. When it's nearly to the floor, Darga screeches like a demon to stop. The twins do, at which point Darga jerks the ladder from their hands. It clatters down onto the fire. Çekiç cuffs the two men for their stupidity, and yells at them to find another one.

Darga hurries. She darts to the corner, grabs her blankets, and dangles them in the flames. The fats inside flare up spectacularly.

She lays them on the ground near the ladder to burn, then rolls a reed mat into a brand to light the ceiling. Çekiç is screaming above, with Binya's hysterics cutting through. But Darga barely hears them; she's working with purpose now.

Within a minute, the fire is too far gone to contain. She finds herself hacking so hard she can barely walk. But between spasms, she smiles; she always thought her lungs would be her demise. Calmly now, she scoops up Alev's skull, clutching it to her chest. Then she lies down beneath the watchful eyes of the leopards, wraps them both in the shroud, and waits for the fire to consume them.

# EGYPT — 2000s BC

In Çatalhöyük, people practiced low-key, casual horticulture, but the size and density of the societies that arose over the next several millennia demanded something more: intensive agriculture. This shift to sedentary farming allowed for population booms, but it also exacerbated social disparities and gave rise to entirely new problems. Hunter-gatherers were generally egalitarian, with little formal distinction between people. And whatever tensions were smoldering in Çatalhöyük, most people at least aspired to an ideal of equality. By the time of ancient Egypt, such pretenses had been abandoned. Vast differences of wealth and power were on display.

Ancient Egypt was not the first civilization in history, but it was the grandest and most enduring, and it continues to fascinate us today. We marvel over the glittering treasures and enigmatic mummies. There's also the pyramids, the most stunning achievement of the ancient world. Not until four thousand years later, with the construction of cathedrals in medieval Europe, did humankind build anything taller.

Dozens of professions were necessary to build a pyramid. You needed crews to move blocks, masons to work stone, blacksmiths and carpenters to forge tools, craftsmen to paint walls and sculpt sacred objects, bakers and brewers to feed everyone. Chief among these were the scribes, who documented everyone's work and whose ability to read and write made them immensely powerful. Building a pyramid, then, was a society-wide effort, and as we'll see in this chapter, contrary to popular belief, those who toiled away on the pyramids were not persecuted slaves but proud laborers working for the greater glory of Egypt.

Of course, not everyone skulking around the pyramids had such pure or selfless motives. The treasures of Egypt that astound us today were no less stunning back then, and no less coveted. Which meant people were no less tempted to steal them, even at the risk of their lives...

Amon, the royal brewer-baker for His Mightiness the Pharaoh—the Strong Bull and Golden Falcon, He Who Is Chosen of Ra and True of Voice—awakes at dawn with a parched mouth and the same knot in his stomach he went to sleep with. He gropes around near his cot for a cup of beer and takes a long drink, then lies back down and stares at the cracked mud ceiling. It's the morning of the heist, and he still doesn't know if he can go through with it.

After donning his kilt, he ties his only remaining bit of jewelry around his neck, a cheap jasper pendant, then sets off. Everything in his neighborhood is a different shade of beige, from the packed-earth street to the mudbrick huts to the reed window shades that mercifully block the sun on the harshest afternoons. Given the cloudless sky above, he fears today will be one of them.

Several minutes later, he reaches a two-story estate—not beige, but gleaming white. It's surrounded by a wall whose square arch entryway has two recumbent lions atop. Several lush palm trees frame the second story, and a leather canopy on the roof—at the perfect height to catch the breeze coming off the Nile—provides a shady refuge. Amon glances around for guards, then sneaks around back and hops the wall. He skirts a small blue pool and heads straight for the nearest window, where he lifts the reed mat and peeks inside. Edrice is lying on a gilded bed with a mattress and lavish decorated footboard. He hisses her name, but she merely mutters and rolls over.

The mountain of her belly never fails to tear his heart in half. He spent two years buying her gifts he couldn't afford—colored glass beads, gold amulets, small jars of perfume. He ruined himself, and for what? A few chaste kisses. He has no idea who the father is, and he's humiliated to admit that he still loves her.

—Amon, what are you doing?

The voice is coming from outside. He flinches to see Khnurn round the corner of the home, already glowering. He's short, bearded, and chesty, with a growing girdle of fat on his belly, the sign of a man who's rising in the world. But what Amon notices now, always notices, is how his dark brown eyes never blink—they simply lock on and bore in. As usual, he's holding a sheet of papyrus. As chief scribe and assistant overseer of His Mightiness the new pharaoh, Khnurn's job is to track all of the goods that go into making the pyramid for His Mightiness the late pharaoh, and his papyrus checklists and schedules keep every facet running smoothly. He's also Edrice's father, and has accused Amon several times of getting her pregnant, to Amon's keen embarrassment.

—I have a gift for Edrice, sir. I know she's been ill.

He removes a turquoise scarab from the linen purse around his waist, tilting it to glint in the sun. Khnurn snatches it up.

—I'll make sure she gets it.

He drops it into his own purse, leaving Amon fuming. Khnurn has intercepted and stolen at least half his gifts for Edrice.

Before Amon can protest, Khnurn shoves a checklist into his face, which annoys Amon since he can't read.

—I still want to go over the discrepancies I found last week. You're sure someone isn't stealing grain from you?

—I assure you, sir. No one is stealing from me.

—You can say that with complete confidence?

Amon can, because he's the one stealing grain, sneaking sacks out at night to barter for gifts for Edrice. Years ago, Amon's father used to point out the wretched gamblers in town, the desperate ones who knew they were ruining themselves but indulged their vice anyway. However contemptible they once seemed, Amon is doing the same thing to himself for Edrice.

Amon repeats his denials, until Khnurn snorts and says he has to leave.

—I have a tight schedule before we seal the pyramid tonight. But we're getting to the bottom of this. Tomorrow. Otherwise, I tell the Overseer.

Amon cringes. His Mightiness the late pharaoh of course took a keen interest in the construction of his pyramid-tomb over the past two decades, until his death a year ago. But given all the demands on a pharaoh's time, the actual day-to-day construction of the pyramid fell to the Overseer. He's a giant spider of a man, six-and-a-half-feet tall with drooping eyes and a lisp no one dares make fun of. He's a bastard son of the late pharaoh, by a concubine the queen despised. His Late Mightiness elevated the Overseer to master builder over her objections, and the royal family retaliated after his death by confiscating the Overseer's inheritance. The Overseer finished the beautiful pyramid anyway, ahead of schedule and under budget. But he's as bitter as Khnurn is greedy, and as unforgiving as the desert sun. Had Amon simply lost the grain, he'd still face a hundred lashes; if the Overseer finds out he's been stealing, he won't live to see another full moon. Amon's only hope right now is the heist—there's no other way to pay back what he owes quickly enough.

The merchant who's been buying his pilfered grain has a warehouse on the edge of town, and Amon has heard he's eager to sell the property: as a native of the southern part of the kingdom, many days' travel away, he reportedly longs to return home now that the pyramid is complete. But grain isn't easily portable; it's too bulky. So if Amon can swipe some smaller, luxury items from the pyramid before it's sealed up, he's sure the merchant will cut bait and accept a trade.

He knows it's a ludicrous plan in some ways: thieving to make up for earlier thefts, like trying to bet your way out of a gambling debt. But if he can just be bold for once, and pocket a few choice items, he can still wriggle out of this—and maybe even resurrect the life he's always hoped for.

Before Khnurn leaves, Amon drops his eyes.

—If it pleases you, sir, may I see your daughter?

The disgust on Khnurn's face is visceral.

—Ten minutes. Then meet me at the brewery. We have a tight schedule.

Inside her room, he rustles Edrice awake. She groans to see it's him.

Heavyhearted, he makes her drink some beer—she needs the nutrients—then brushes her black hair with an ivory comb decorated with ibises and elephants; it cost him two sacks of pilfered flour. She turns to him.

—Will the Overseer be taking you into the tomb later?

Amon shudders and changes the subject.

—I think your father will be. You know, you look lovely today.

—I smell like dung and my feet are fat.

Amon protests that they don't look that swollen today. She counters that he must think her feet do look fat most days. Amon sighs. She's always wrong-footing him like this. Why does he put up with it? His brother wouldn't. It's one of the few ways he wishes he was like Abukar.

After thirty minutes, he tears himself away, feeling oddly relieved to go. Outside, he hurries toward the worker village, a grid of low-slung mudbrick shops with thatched roofs. Its diminishment makes him gloomier every day. At the peak of pyramid construction, he could barely walk down the street for all the crowds. Even at night, the cacophony never stopped—millers grinding flour, masons chiseling stone, smiths pounding metal, cooks butchering goats, construction workers shouting and rolling sheep-knuckle dice and laughing with their mouths full. No more. After two decades of work, His Mightiness's burial chamber in the pyramid will be closed at dusk. A few last workers will seal the entrance with polished limestone next week, then scatter along with everyone else. As such, the streets and mudbrick workshops are all but silent now. There's also a new shade of beige everywhere—the sand piling up against the sides of the abandoned buildings.

Amon's bakery looks even sadder. He once managed hundreds of men and women who baked thousands of loaves every day, plopping the dough into clay molds and cooking them in firepits. He employed a dozen potters just to replace cracked molds, and a half-dozen grubby teamsters to haul in donkey dung for fuel. He even employed two hobbled old men just to pick up scraps of crust to keep birds away. Now, there's just a rump crew of crones, tending a few small fires for the last remaining workers' meals.

He ducks inside their hut, and his feet sink into the soft layer of ash underfoot. Upon seeing him, one woman asks Amon where his girlfriend is. Then they all laugh; they're convinced he's the father. More brusquely than he means to, he orders them to fetch the special funeral offerings he ordered yesterday. But his irritation softens upon seeing the loaves; they turned out brilliantly. They're baked in the shapes of animals: geese, cattle, gazelles, flamingos, hippos. They're different from normal, plain funeral loaves, but he hopes they'll find favor with the gods—and more importantly, Khnurn.

He walks on to the brewery a few hundred yards away. There are no more guards watching the grounds these days—why bother. The brewery's divided into dozens of plots with scores of ceramic vats, each one a yard wide and a yard tall. Today, most of the vats sit empty, with crows perched on the crumbling rims.

Still, a few last trusted workers are making beer in one hut. The air inside is thick with steam and pungent with cinnamon and rosemary. In the corner, two women squash cooked grains with their feet. Amon asks the head cook, a woman as stout and neckless as a beetle, for some of the special date beer intended for His Mightiness's tomb. While she fetches a cup, he glances at his reflection in a vat of warm water nearby. His lips are slight, his eyebrows thick, and his lashes long. Edrice complains that his skin is creamier than hers; he fears she finds him womanly. He certainly doesn't see the face of a man hard enough to rob the pharaoh's tomb.

—Amon! Where's the beer? We're behind schedule.

Khnurn again. He's standing at the door, his silhouette hazy through the steam. Amon accepts the cup from his cook and heads outside.

The first sip tastes sweeter than normal to Amon but not unpleasant, with a nice fizz. A second sip convinces him. This will be a delicious new treat for His Mightiness in the afterlife.

When Khnurn takes the cup, he complains about the flies on top—as if you could brew beer without flies getting in! Indeed, he demands a reed straw before drinking any, which makes Amon roll his eyes. Finally, the scribe sips. Amon holds his breath as he swishes the liquid around.

The suspense is short-lived. Khnurn crinkles his face and spits.

—What is this?

—It's a new recipe and...

—No, no, no. I need two jars of *traditional* beer. You're not doing anything fruity with the funeral loaves are you?

Amon sighs and shakes his head no. He'll find some plain loaves.

Khnurn checks his papyrus, and gives Amon a half hour to gather the bread and beer and meet him at the temple. Then he hands Amon the cup, slopping the beer on his chest and kilt. As the scribe stalks off, Amon can't help but wonder—it's habit, addiction—whether Edrice would find the animal loaves amusing. That might win him a few minutes of happiness.

Given the secrecy of the sacred rites, Amon cannot approach the mortuary temple where the high priests are blessing the funeral loaves and beer. Instead he paces around outside another temple a quarter mile away. He's wearing a robe for his own upcoming ceremony, as well as a wig made of animal hair, with date palm fibers woven in to provide volume. The extra layers make the sun seem even more sweltering than usual. It doesn't help that he's pacing outside the temple where His Mightiness's mummified body was prepared. It smells faintly of death, and a phalanx of unsmiling guards keeps watch over him from the temple steps. He hasn't even done anything yet, yet he already feels guilty.

At last Khnurn descends the steps of the far temple with the sanctified loaves and jars of beer. Amon hurries along a stone causeway to meet him. By tradition, the royal brewer-baker must deposit the offerings at His Mightiness's sarcophagus, and however obviously Khnurn wishes that Amon were not that man, he will not break tradition.

They walk single file toward the pyramid. The sight of it blinds Amon: aside from the entrance, all four faces are plated with polished white limestone that flashes in the sunlight. He's left dazed, his legs wobbling as he ascends the steep steps. At the entrance, he slips on the slick paving stones and nearly drops the jars, earning a sharp rebuke from Khnurn. He mutters something about the heat, but truth be told, it's not just the sun that's getting to Amon. The heist is now upon him.

The corridor leading to the burial chamber inside is dark and cramped, just four feet high. It's also uphill, and before long Amon's back is straining from bending double.

At last, the corridor levels off and they can walk upright. At the end of the passage loom three 6,500-pound granite portcullises that will be dropped at dusk, after the priests deposit the last offerings, the sacred mummies. Amon holds his breath as he ducks warily beneath.

Only to have his breath stolen away a moment later. He's of course witnessed the burial treasures being transported through the village, but only piecemeal. Seeing them all together, piled to the ceiling in some spots, dazzles him. Gilded couches with lion armrests. Daggers made from meteorites. Alabaster vases. Bowls full of gold rings. Ivory pomegranates, painted ostrich eggs, elaborate decorated sandals and harps. As well as—Amon's throat tightens—the most precious treasure of all, blue-glazed jars of scented oils and perfume.

—A worthy tribute, no?

Even Khnurn seems humbled. They share a quiet moment of wonder before the scribe clears his throat.

—I did well here. Get to work.

Amon places the bread at the foot of the thick sarcophagus and recites a prayer to the god of grain. The beer requires more ritual. One jar will remain full, but the other must be decanted into a dozen cups around the room. Amon fills the first, kisses the foam on top, and brings it to the designated spot. There, he kneels, recites a longer prayer, and sets it down.

He repeats this with a second cup, and a third. He senses Khnurn watching him, ready to pounce if he deviates from protocol. But the routine never changes. Pour, kiss, kneel, pray. Pour, kiss, kneel, pray. By the sixth cup, the restless scribe is studying his checklist.

*It's time.* As Amon kneels with the next cup, he eases open the flap of the purse beneath his robe. His fingers move as slow as the sun. He's inches away from a blue-glazed jar of perfume.

A cough from Khnurn nearly explodes his heart. Amon's hand darts back, and he risks a glance behind him. But the chief scribe hasn't noticed; he's still perusing the list. Amon nevertheless retreats to regroup.

And so it goes. Time and again, Amon kneels and sets a cup down, willing himself to act. Then he quails and scurries away. When preparing the next cup, he curses his cowardice and redoubles his resolve—only to wilt the same way. For the second time today, he wishes he were more like his brother, the grifter. Abukar would not shrink.

Amon's soon down to the last cup. But before he has to test his mettle again, a voice echoes from the corridor, a voice with a lisp.

—Khnurn, are you there? The priests want to change the schedule with the mummies.

Amon sees Khnurn's shoulders tense. He turns on his heels and hurries out.

For once, Amon doesn't hesitate. He snatches up the nearest perfume jar—then a second and third. He can trade two to the merchant to square himself, and keep the third to spoil Edrice.

But his bravery is no more sturdy than the foam on the beer. What if Khnurn notices the missing jars? Amon takes several steps backward to see if their absence looks conspicuous—and bumps right into the second jar of beer. With his hands full, he can't do anything but gape. It wobbles and crashes, and the sacred liquor gushes out.

Amon scrambles to replace the perfume jars, nearly dropping those as well. He darts back and looks around, his hands still shaking. Besides the spill, nothing else looks amiss, praise the gods.

Still, something's not right. It's the beer. It's running toward the corner of the room, but it isn't pooling there; it's trickling through a crack. And from the sound, it seems to be falling somewhere below, splashing down like a waterfall.

He doesn't have time to investigate. A moment later, the scribe and Overseer enter. The goliath looks furious when he sees what happened.

—That was elixir for His Mightiness! I should remove the same weight in blood from you!

Amon tries to apologize, but ends up sputtering. His entire back is slick with sweat. He might be a dead man now regardless.

The Overseer finally tears his glare away from Amon and turns to Khnurn. He asks whether they can prepare another batch in time.

Khnurn doesn't answer. His eyes are locked onto the corner where the beer is trickling away. And even through the haze of his own fear, Amon can see that the scribe's face looks oddly pale.

The Overseer notices, too, and follows Khnurn's gaze. His frown shifts from anger to confusion.

—Where's the beer running to?

Now it's Khnurn's turn to sputter.

—There's a tunnel there—down there—beneath the chamber. For bringing in bigger treasures.

—I ordered that tunnel filled with rubble.

Khnurn squirms. Amon has never seen him so agitated. The brown eyes that normally lock on and never let go have dropped to his feet.

—We did, but there's—what I mean to say, there must be some space left over.

The Overseer grunts. He looks skeptical. But whatever his misgivings, he simply orders Amon and Khnurn to consecrate a new jar. They nod and scamper off.

But as they're crouch-walking down the corridor again, Amon's mind keeps churning over what he just saw. Khnurn seemed...frightened. And the tunnel he mentioned beneath the chamber can't be filled with rubble. From the sound of it, the beer was falling quite far. That means the tunnel must still be passable—a fact that Khnurn lied about. Why?

In the harsh daylight outside, Amon glances over at him, and receives a glare no less scorching than the sun. It sends his mind whirring. There could be any number of reasons the tunnel hasn't been filled: lazy workers, a lack of help, simple oversight. But one explanation, however unlikely, begins picking at Amon's brain. Perhaps it's his own paranoia—his guilty conscience making him see things that aren't there. Yet he can't help but wonder.

Is it possible that Khnurn, the chief scribe for the great pyramid of His Mightiness the late pharaoh, has plans to break into the burial chamber himself?

From the most wretched servant to the most exalted prince, people in ancient Egypt ate bread and drank beer with every meal. These staples were so vital to their diet that, in Egyptian hieroglyphs, the combined symbols for *bread* and *beer* actually meant *meal* or *sustenance*.

During pyramid construction, bakeries the size of football fields stood near the worker villages. Every morning, battalions of men and women would grind up bushels of emmer—the main grain eaten in Egypt—into flour on stone hand mills called *querns*. Still more bakers mixed and kneaded the dough, probably with their feet. To bake the bread, rather than waste time making thousands of mud ovens, crews used conical clay molds. They'd dig holes in the ground, fill them with glowing embers, drop the molds in upside-down, plop some dough in, then cap each mold with a second one and heap hot ash over the top. The endless rows of these devices made the lot behind the baking huts look like giant egg cartons.

A model bakery and brewery from an ancient Egyptian tomb. The workers are grinding grain, preparing mash, and filling beer jars. (Copyright: Metropolitan Museum of Art.)

Timing was critical. The glowing embers had to be ready the same time the dough was, and the bread had to finish just as the construction crews and other laborers were lining up for meals. Given the immense scale, there was a factory feel to the operation, and someone like Amon was as much a foreman as a baker, equally concerned with workflow and worker training as he was flour quality or seasoning.

To learn the ins and outs of Egyptian bread, I make a trip to Los Angeles to visit Seamus Blackley, a computer programmer and self-described "gastro-Egyptologist." He's been obsessed with Egypt ever since seeing his first mummy on *Scooby-Doo* as a child. Today, he's a graying redhead with bright blue eyes and a ponytail. Among the many fascinating lines on his résumé, he invented the Xbox gaming system in a previous life. His company now does "computational holography," he says. "And if I tell you more, you'll have to sign a non-disclosure agreement." I don't press.*

Blackley's first foray into heirloom food involved medieval European bread. "My challenge to myself was to be as good at making bread as the average twelve-year-old in the twelfth century," he says. Egyptian bread was a logical next step, although at first he largely improvised.

"I made this bread with supposedly ancient yeast, and a lot of people asked questions" on Twitter about its authenticity, he recalls. "They basically said, *You're full of shit*—including my wife. I wanted to be not so full-of-shit." So he reached out to an Egyptologist and a microbiologist for guidance, and recruited some colleagues. ("I employ a lot of people who are chemists and such, so I have a nerd army I can call on.") Eventually, he hopped a plane to Egypt to collect yeast from ancient pottery, using sterile swabs and other microbiology equipment to gather samples. Friends also made him replicas of the conical bread molds, using representative clay. Finally, Blackley built a pharaonic-era firepit in his

---

* Instead I ask about the giant metal dinosaur skull sitting near the front door. Blackley tells me it's Cathy the *Acrocanthosaurus*, a sort of mini *T. rex*. Not only is it authentically life-sized—a yard long, with porcelain teeth the size of ice-cream cones—it's biomechanically accurate as well, with a linear actuator to provide thousands of pounds of bite-force. Blackley's company built it for a television show to re-create a puncture wound found in an ancient fossil. Less historically accurately, a second show commissioned Cathy to chomp on a dummy human skull filled with red glop, just to see what would have happened if a dinosaur had ever snacked on one of us. Imagine putting a stick of dynamite in a jar of raspberry jam, and you get the idea.

backyard to bake with; he even sourced acacia wood from Arizona for kindling, similar to what Egyptians used. It took a year of practice, he says, to "make bread that doesn't suck."

Archaeologists have documented roughly forty different types of bread from Egypt's five-thousand-year history, everything from biscuits to baguettes to pitas. Some bakers even made loaves in fancy shapes: flowers, phalluses, fruit, obelisks, birds, cattle, crocodiles, gazelles, fish. ("Those breads would kill on *The Great British Bake Off*," Blackley says.) For everyday meals, though, most people ate the conical bread. In his office, Blackley shows me his replica mold. It's scorched black and much heftier than I expect — fifteen inches across and probably twenty pounds.

As a treat, Blackley has also baked a loaf for me to sample. It's a foot wide and sand-colored with a springy crust. It consists of just a handful of ingredients — salt, yeast, coriander, emmer flour — and its blunted shape reminds me of NASA space capsules from the 1960s.

A loaf of ancient Egyptian–style bread baked inside a replica mold. (Copyright: Sam Kean.)

Blackley apologizes for not making toppings for the bread, like leeks in beef tallow, which he says was an Egyptian favorite. He needn't have worried: it's hard to overemphasize how delicious this bread is. It's spongy and chewy and has a scrumptious sourdough tang, with the coriander sneaking in late to tickle the tongue. It would draw raves in any New York or Paris bistro. And mind you, I'm eating a two-day-old loaf warmed up in the company microwave; fresh out of the mold, it would have been orgasmic. Blackley notes that bakers in ancient Egypt had strong incentives to cook well. "Have you been to Giza and seen the pyramids? The blocks are a hell of a lot bigger than you'd think in the pictures. Bread was the primary currency used to pay those guys, and also their food. You wouldn't want to give the guys who could move those blocks a bad meal."

Beyond bread, Egyptian laborers were paid in beer as well — 1⅓ gallons daily, roughly ten pints, which they happily sucked down given that temperatures on the hot sands could reach 130°F. (One scholar estimated that it took 231 million gallons of beer to build the largest pyramid.) Even children drank beer, largely because the alcohol killed microbes and rendered it more sanitary than the water in rivers (a.k.a. their sewers). Egyptian doctors also prescribed beer as medicine to remedy coughs, constipation, swollen eyes, and upset stomachs.

Based on residues in ancient pots, amber or mahogany beers were probably the most common in Egypt, although records do list other varieties, such as celery, dark, date, iron, and sweet. As with bread, the Egyptians mostly brewed with emmer grains (along with some barley), which would have made their beers wheatier and creamier than modern varieties. The Egyptians did filter their beer, but it still would have had husks of depleted grains floating in it, so some people likely drank through reed straws to avoid a mouthful of chaff with their morning quaff.* Egyptian beer also lacked hops,† so it wouldn't have had the

---

* Sometimes the straws got quite elaborate. In the Caucasus, yard-long gold and silver straws have turned up; they were so ornate that archaeologists initially mistook them for scepters.

† Hops never appeared in beer before the Middle Ages. After that, they became ubiquitous, even mandatory at times: under penalty of law, you couldn't sell beer in some districts if it lacked hops. When Anchor Brewing in San Francisco made a batch of Egyptian-style beer in 1989, the brewmaster invited some guests in to observe the process. But when it came time to add hops, he made everyone turn around,

bitter undertones of modern brews. Instead, it probably leaned more sour, in part due to the fruit flies that swarmed their open vats and introduced bacteria that convert alcohol into vinegar. (Other microbes that produce sour tastes might have wafted in as well.) Some scholars have described Egyptian beer as alcoholic porridge, or a "sour barley milkshake."

Eager to sample this beer, I dug up a recipe online began brewing a gallon myself. All beer starts with malted grains, grains that have been germinated and dried. Brewers today dump these directly into hot water. But some scholars believe that ancient Egyptians performed another step first: forming the grains into crude "offering loaves" and gently baking them at a low temperature; only then did they crumble the loaves into water, usually while reciting prayers. This created a symbolic connection between bread and beer, with one transforming into the other.

Admittedly, my offering loaves don't look very appetizing: less like bread than birdseed patty-caked together with wet sand. But they smell delicious coming out of the oven. I say my prayers, crumble them into a pot of water, and simmer them. This gives the enzymes in the grains a chance to break down the long-chain starch molecules inside into simple sugars. After two hours, I'm left with a pale white liquid, which turns brown when I add cardamom, cumin, rosemary, coriander, and cinnamon. (Scholars don't know what exactly the Egyptians used to flavor beer, but all these spices were available to them.) After a hard boil, I try a sip, and coo with delight. It tastes like sugary oatmeal runoff. I decant this into a milk jug, then measure the ABV with a device called a refractometer—around 2.3 percent, probably typical for beer then.[*] To start the fermentation, I add a packet of yeast. Ancient brewers either

---

feigned adding a handful, then pocketed them. He didn't want to risk anyone else getting fined or arrested for abetting the "crime" he was committing.

[*] Depending on how watery they were, Egyptian brews could have reached 4 percent ABV, and adding fruits or other sugars could have boosted that to 8 or 9 percent. But in general, the Egyptians did not drink beer to get drunk. If they wanted to get smashed, they drank wine.

added the yeasty foam from a previous batch as a starter, or else relied on wind-blown yeast or yeast from the skins of fruits they added for flavor.

After a week fermenting in my spare bathroom, the beer looks like muddy apple cider, with a quarter-inch of sludge on the bottom. I sniff it, and the rush of bubbles instantly clears my sinuses. Then I sip. All I can taste at first is a smack of sour, a *pow* right in the kisser. Eventually, a little citrus emerges, but sadly, none of the other spices come through—no rosemary, no cardamon, nothing. Still, those spices might be adding flavor, even if I don't taste it. David Falk, an Egyptologist and amateur beer-maker whose recipe I'm following, compares beer spices to the vanilla in ice cream. Falk once made a batch of ice cream without vanilla, which we normally don't think of as a dynamic taste—quite the opposite. But Falk says he immediately noticed its absence; the frozen sugar milk just tasted blah without it. The same goes for beer, he tells me: "You need something in there to offset the blandness." Hidden flavors still contribute to taste.

In all, the description of Egyptian beer as a "sour barley milkshake" didn't ring true to me. Mine tasted closer to Kombucha—a perfect thirst-quencher after a hot day piling up pyramid stones in the sun.

Historically, people have made alcoholic beverages from pretty much anything they could get their hands on: bananas, tree sap, corn, cactuses, rice, pumpkins, even horse milk. But grain-based beer played an outsized role in the human saga, to the point that some archaeologists believe it fueled the rise of the first complex civilizations in history.

This argument starts with the fact that both beer and bread require grain. Traditionally, archaeologists assumed that people started gathering wild grains to make bread, and that beer was a happy by-product. But to other archaeologists, that idea doesn't make sense—mostly because making bread was a gigantic pain in the rear. Imagine you're a hungry lad or lass 10,000 years ago. You could easily gather some nuts or roots for dinner, or hunt some game and fill your belly that way. What you probably wouldn't do is spend a few hours hunched over in the hot sun picking tiny grains off stalks while bugs eat you alive—especially because you then need to spend several more hours grinding that grain into flour—followed by still more time building a mud oven to actually bake the flour into bread. I mean, I love bread, dearly, but does all that effort seem worth a few dinner rolls?

Beer, on the other hand, would have been worth the squeeze. Beer gets you buzzed. It's fun! It also enhances social bonding and was frequently linked to religious festivals in prehistory. Beer-making is far less tedious, too, since you don't need to grind flour. Just warm some half-cracked grains in a pot of water, dump in some spices, and presto.

As for linking beer to the rise of civilization, people probably started making beer initially from wild grains for seasonal religious ceremonies. Over time, in order to worship more regularly, they began cultivating favorable varieties of grain in certain areas. As this work developed into dedicated farming, they began to settle down near their plots, and given the heavy investment of labor, they began eating the grains as well. Over time, as farming techniques became more sophisticated, food surpluses became common, allowing for population growth and the development of cities—plus the division of labor that cities require, including specialized roles like artisan, priest, scribe, warrior, and merchant. (For the first time, people had professions.) From there, it was only a modest step to complex civilizations like Egypt. All that glory and grandeur from a little thirst.

It's worth noting that not every archaeologist accepts this proposed sequence of events. Indeed, some of the more genteel types find it horrifying—all the art and poetry and stirring architecture of the world's great cultures springing from plebes swilling pints. Regardless, the spread of farming in general was inarguably linked with the rise of civilizations, and Egypt's growth in particular would have been unthinkable without vast grain farms for bread and beer. Again, people consumed both at every meal and often got their wages paid in them. Tombs were also crammed with beer pots and bread loaves for the hereafter. Ancient Egyptians couldn't imagine life or death without them.

Of course, Egyptian tombs included many other goods as well—and the more elite the person, the more fabulous the treasure. Despite their importance to daily life, it wasn't the beer and bread in tombs that people risked their lives to steal...

After consecrating and depositing another jar of beer, Amon hurries over to an embalming workshop on the outskirts of town that belongs to his brother, a junior priest. For a business built on death, the grounds are surprisingly lively. There are baboons tied to stakes, hooting like mad, as well as bleating rams, cages of squawking birds, even a few snapping crocodiles. All future mummies.

Beyond the bestiary lie heaps of crushed white powder, some anthill-sized, others the size of small sand dunes. They're piles of natron—a drying agent—with animal corpses beneath. Abukar's assistants are shooing rats away from one. Other assistants are digging through a mound with their fingers, picking out maggots and beetles. When Amon passes, he catches the distinct odor of rot.

At the entrance to his brother's underground workshop, a woman emerges with red eyes and shaved eyebrows—a sign of mourning for a pet cat. Amon offers condolences, then climbs down the ladder himself.

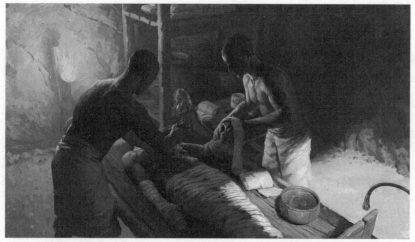

Egyptian priests mummified people in underground workshops, a process accompanied by cleansing rites and prayers. (Copyright: Nikola Nevenov.)

The room reeks of incense. As his eyes adjust to the dimness, he sees different workstations in each corner: wooden tables at waist-height for ibises and baboons, low earthen platforms with drainage channels for sheep. Some assistants are doing initial preparation on bodies—slicing open flesh, removing organs. These fresh creatures will be buried outside under natron for a month or two. Other assistants are putting the finishing touches on mummies—massaging the stiff flesh with fragrant oils, binding the limbs in linen bandages.

—Amon, my second-favorite brother!

It's the same joke every time. Amon looks his brother up and down. Despite being elbow-deep in corpses each day, Abukar never has a speck of blood or filth on his robes. Even his fingernails are clean. Like a dandy, he carries a fat jar of scented almond oil with him wherever he goes, to keep his curls in place. He's grown plump lately, which Amon has to admit suits him.

Only with great reluctance did Amon secure his ne'er-do-well brother a job in the worker village, trapping animals to help make mummies for nearby temples. Against all expectations—Abukar had never worked an honest day in his life, preferring grift and petty scams—he thrived at the task. He then won appointment as a junior priest somehow (Amon suspects a bribe), and hustled his way into a commission to mummify the pets in His Mightiness's menagerie, for burial in the pyramid itself. Gallingly, unlike Amon's dwindling prospects, business at Abukar's shop is booming, as pilgrims from all over the kingdom have started streaming in to gape at the latest pyramid and buy mummies as votive offerings.

 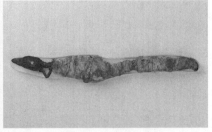

The Egyptians mummified vast numbers of animals, including cats (*left*) and crocodiles (*right*), often as offerings for temples. (Credit: Rama, Wikimedia Commons.)

Abukar kisses Amon on both cheeks.

—I've a surprise for you. Or for Edrice, actually.

From a nearby cedar chest, he retrieves a tiny bundle of linen and unfolds it.

—It's a beetle mummy. I know she's fond of scarabs.

The gesture touches Amon. Abukar alone has never mocked or scorned him over Edrice. Still, he knows his brother.

—A real beetle? Or is there a stone inside?

Abukar chuckles, and assures Amon it's real—only the best for family. Amon then asks if they can speak privately. Abukar raises a curious eyebrow, and steers them toward a locked back storeroom.

Inside, Amon winces to see all the evidence of fake mummies. Clay sculpted into "skulls," reed "skeletons" to wrap, filed stones for faux teeth.

—Looks like business is thriving.

Amon means this sarcastically, but Abukar nods with satisfaction.

—It is. I owe you, brother.

—Are you putting fakes in the pharaoh's tomb? Tourists are one thing, but...

—Everything going into His Mightiness's tomb is real, I assure you.

Amon doesn't know whether to believe this; a lifetime of misgivings dies hard. Then again, has Amon been acting any more honorably lately?

Abukar directs them to two shabby stools, then waits as Amon fumbles about, unsure where to begin. The story finally spills out. His skimming grain. The failed heist. The beer glugging through the crack, and Khnurn's lie about the tunnel beneath. Amon then adds his growing suspicion that Khnurn plans to break into the burial chamber himself and loot it.

Worst of all is Amon's new fear: What if Khnurn *realizes* that Amon has detected the lie? If the chief scribe thinks Amon is onto him, he could be in serious danger.

Amon isn't sure why he's unburdening himself to his brother—a lack of other options, perhaps. But thefts and intrigues are also Abukar's element. Amon hopes that, somehow, he'll know what to do.

What he doesn't expect is for Abukar to start laughing. A big, belly-and-curls-shaking chortle that nearly topples him from his stool.

—Brother, I never knew you had it in you. What a stew!

Amon can barely hide his fury. He's on his feet a moment later, heading for the door. He rattles the locked handle and turns back.

—Open it.

His brother begins nonchalantly fixing his curls with oil, which makes Amon even angrier.

—Open it.

—I'm happy to. But then what? You stomp out, and Khnurn will still arrest you, for stealing the grain if nothing else.

Amon is silent. His hand falls from the handle. Abukar continues.

—Did you even think to check over your shoulder on the way here? What if Khnurn sent guards to seize you the moment you step outside?

—On what pretext?

—Does he need one?

Amon swallows hard.

—So what do I do?

—Get the jump on him. Rob the tomb first.

Abukar explains. He proposes concealing Amon inside the pyramid before the senior priests seal it up at dusk. Afterward, he can grab some perfume, sneak out the tunnel beneath the burial chamber, and swap it with the merchant first thing in the morning.

Amon, obviously, has some objections. First, he has no idea where the tunnel goes. Abukar counters that it must lead to an exit somewhere; Khnurn wouldn't be planning a robbery otherwise. Amon then points out that if Khnurn sneaks in later and finds anything missing, he'll know Amon did it.

Abukar shrugs.

—If you unload the goods right away, there'll be no evidence. Besides, if Khnurn denounces you, the Overseer will wonder how *he* got inside the tomb, and why he lied about filling the tunnel. He can't expose you without exposing himself.

Every objection Amon raises, Abukar answers. By the end of their conversation, Amon doesn't exactly feel confident, but he's impressed with his brother's shrewdness. And despite his qualms, his mind keeps circling back to one irreducible fact: that if he does nothing, Khnurn will nail him anyway for stealing grain. However slim, this plan gives him a chance.

But there is one point he isn't clear on.

—How do I conceal myself inside the pyramid? Once the priests deposit the mummies tonight, no one else is allowed in.

Abukar puts down his jar of almond oil and picks up a nearby bandage.

—Then I suppose we'll have to make *you* a mummy.

Cultures throughout history have mummified their dead, and a handful still do today,* but Egyptian mummies remain the most iconic. Unfortunately, the Egyptians wrote down virtually nothing about their embalming process. This leaves experimental archaeology as one of the few avenues available for understanding mummification, and several practitioners have indeed re-created mummies in modern times. In most cases, they work with animals, but a few intrepid souls have mummified human beings, most famously when Bob Brier and Ronn Wade did so in 1994.

Wade grew up wanting to be a mortician like his father. After a stint as a medic in the Vietnam War, he became an anatomist and eventually the head of Maryland's state anatomy board. Brier also has training in anatomy, but is an Egyptologist by training and passion. He's accumulated so many books on Egypt over his life that he rents a second apartment just to accommodate them.

Brier and Wade selected their mummy from the pool of people in

---

* For more on cultures still making mummies today, visit https://samkean.com/books/dinner-with-king-tut/extras/notes/. I also have loads of bonus pictures and other bonus material there.

Baltimore who donated their bodies to science. Ultimately, they settled on a seventy-six-year-old Caucasian man who died of a heart attack. His identity remains secret, but a bit crassly, Wade nicknamed him E. M. Balm.

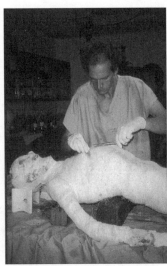

Egyptologist Bob Brier helped create a human mummy in 1994 by dehydrating the flesh with a mineral called natron (*left*). After the flesh dried, Brier wrapped the mummy in bandages (*right*). (Photo courtesy of Pat Remler.)

For authenticity's sake, Brier and Wade used replicas of pharaonic-era tools and materials, including linen wraps, an oddly wide wooden embalming table, and copper and obsidian blades — although they quickly abandoned the copper ones, which couldn't cut flesh well. Before starting on their mummy, they practiced one important step on other cadavers: extracting the brain. Instead of using full cadavers for this, they obtained some decapitated heads leftover from a medical school's plastic surgery class. ("They were looking a little weird," Brier recalls. "They'd had facelifts and such.") From some scant references, Brier knew that Egyptian embalmers removed the brain by inserting a hooked rod through the nostrils, but the details were vague. Brier and Wade first tried scooping the brain out with such a rod, but the tissue proved too soft and wouldn't come out. They finally took to squirting water up the cadaver's nose, then used the rod to whisk the brain into a slurry. After that, it poured right out. "Like a milkshake," Brier says. "A strawberry milkshake to be exact."

Skills honed, the duo began making their mummy in May 1994. The first step involved removing his organs.

Different organs met different fates in Egypt. Unclear about the purpose of the brain, embalmers typically threw it away. The heart, in contrast, was left in situ; it was considered the seat of all thinking, emotion, and intelligence. Abdominal organs were carefully extracted and preserved. Following this protocol, Brier and Wade made a 3½-inch incision in their cadaver's abdomen and removed the spleen, liver, gallbladder, lungs, and twenty-two feet of intestines. Given their size, extracting the liver and lungs required some creative geometry and determined squeezing. The most difficult part involved detaching the lungs from the heart while working blind inside such a tiny hole.

With the organs removed, the pair cleaned the abdomen with palm wine and myrrh, then stuffed frankincense into the skull. This was an important ritual step to prepare the body for the afterlife, and also helped kill microbes and mask bad smells. Ancient embalmers used other sacred substances as well, often imported from Europe and Asia at great cost—pistachio resin, beeswax, castor oil. Ramses the Great's mummy had peppercorns from India shoved up his nose.

Next, Brier and Wade dehydrated the body using natron, a mineral of equal parts salt and baking soda that forms naturally in Egyptian wadis, or dry gullies. Like a sponge, natron sucks the moisture out of flesh, leaving it too dry to support bacteria, maggots, beetles, and other putrefying agents; the leftover tissue is essentially jerky. (Fully committed to authenticity, Brier dug the natron himself in Egypt, and recalls that sneaking hundreds of pounds of unidentified white powder through customs at JFK Airport was one of the more ticklish aspects of the project. Luckily, he was traveling with a film crew, and could hide the powder in suitcases amid their equipment.) In their lab, Brier and Wade placed the mummy's spleen, lungs, liver, and intestines into bowls and covered them with natron. They also packed 29 linen bags of the powder into the body's empty torso, laid the body on top of 211 more pounds, and dumped 583 additional pounds over it. They kept the body in Wade's old office, with the heat cranked up to 104°F and dehumidifiers running night and day to simulate Egyptian air.

Over the next five weeks, the natron on top turned crusty and brown from absorbing bodily juices, forcing Brier and Wade to crack through it with an iron rod. (Today Brier remembers the odor as acrid but not unpleasant, although news reports at the time say he and Wade donned surgical masks against the smell.) Regardless, the sight of the body beneath thrilled Brier. As it dries, the skin of mummies tightens and shrivels, especially on the face and scalp. The lips retract to reveal the teeth, and skin with less melanin turns brown-yellow. Brier always wondered whether those changes resulted from the immediate mummification process, or from several thousand years of weathering in Egypt's arid climate. One glance at his mummy and Brier knew the answer: even after five weeks, "he looked just like Ramses the Great," he recalls, with leathery skin, a beaky nose, and wispy hair sticking up. The embalming process, not time, made the iconic mummies we know today.

Beyond changing the body's appearance, the dehydration process left the limbs as stiff as tree branches, and dropped its weight from 188 pounds to just 79. (Thirty-one pounds of that represented the removal of organs.) The organs drying in the bowls withered as well, which helped explain another mystery of Egyptian mummification: as other archaeologists have noted, embalmers typically placed the organs in so-called canopic jars, funerary vessels with slim necks—so slim that it seemed impossible to fit the larger organs inside. But the natron shrunk them down enough to slip right in.

After removing him from the natron, Brier and Wade gave Mr. Balm a full body massage with lotus, cedar, and palm oils, another step that, while important ritualistically, also had pragmatic benefits—restoring flexibility to the joints, making the mummy easier to handle. This accomplished, they wrapped the body in linen bandages. (Embalmers in ancient times started with the hands and feet, wrapping each digit separately, then proceeded to the arms, legs, and torso. The penis was individually wrapped as well—or, if embarrassingly shriveled, a codpiece of stiff linen was bound in place.) At this point, they let the mummy dry for three more months in the arid office, which dropped its weight to 51 pounds. Afterward, they added several more layers of wrappings. In between the layers, they slipped magic amulets and scraps of papyrus with spells on them, a common practice in ancient times.

For the past three decades, the mummy has been lying in a metal casket in Maryland, stored at room temperature. Brier and Wade have partially unwrapped it twice to check for rot, but found nothing amiss. "He's dead and well," says Brier.

Among archaeologists, Brier and Wade's experiment proved controversial. One critic fumed, "It's macabre and…tasteless, and I don't think there's a great deal of scientific value." Less emotionally, another pointed out that donating your body to science "should not be a blank check for any kind of experimentation" that researchers dream up. It's a fair point. Brier and Wade, however, deny any implication that they mistreated the body: "We're treating this man like a king," Wade said at the time. Brier adds, "In fifty years, he's going to be in a lot better shape than I will be." Plus, it's not true that the experiment lacked scientific value. It revealed much about Egyptian mummification that we simply didn't know before—what blades embalmers used, how to remove the brain, how much natron was needed, even why embalming tables were so wide back then, four full feet—to accommodate the huge pile of natron necessary to dry the flesh.

Given all the controversy, few modern archaeologists have dared embalm human beings, but several have mummified animals, a practice common in ancient Egypt. For some hands-on experience in this realm, I decide to seek guidance from the world's leading expert on animal mummies, Salima Ikram, an Egyptologist at The American University in Cairo. She suggests I mummify a fish, so I pick up a half-pound red snapper from the grocery store with the head, tail, and eyes intact.

I start by making some bootleg natron, blending six cups of baking soda and salt with a few cups of water. I dry the resulting paste in my oven until it resembles chalk, then crumble it into powder. Next, I wash the fish inside and out with white wine, which foams up as I scrub every cranny, including under the tongue. After patting everything dry, I pack the interior with tiny linen gift bags of natron, set it on a bed of more natron in a casserole dish, and heap another few pounds atop.

Alarmingly, I can smell fishy odors through the natron the next day. But it doesn't smell rancid—more like parboiled fish. Six days later, when I unearth the snapper to check on it, I get another scare—little black flecks in the powder. Are they bugs? Putrid tissue? I don't know, but the flesh shows no sign of rot; it looks like smoked trout, tan and stiff. Considering that it's been sitting unrefrigerated for nearly a week during a sultry Washington, DC, summer, it's in darn good shape.

After eighteen days, the snapper is half its original weight. Following Ikram's protocol, I massage the flesh with castor, lettuce, and myrrh oils to restore pliability. The oils turn the fish an attractive gold color, but do little on the flexibility front: when I try bending the body, I cringe to hear the skin and muscles crack; several tears appear. (Perhaps I dried the fish too long, or perhaps mammal flesh absorbs oil more readily.) Nevertheless, the process impresses me overall. After wrapping my mummy fish in linen, I place it on my shelf, and it's still there today. Last I checked, the eyes and skin were intact and there were no signs of decay or untoward odors. It was shockingly easy, then, to turn even fish—an animal proverbial for smelling horrible—into an inoffensive relic. No wonder the Egyptians considered mummification a sacred, even magical process.

Beyond fish, the Egyptians also mummified baboons, hippos, pigeons, rams, shrews, lions, beetles, lizards, owls, bats, hyenas, raptors, and more. Some of the cruder animal mummies looked like sock puppets or piñatas, but the best embalmers set the beasts in lifelike poses: monkeys on their haunches, cats with their tails curled, snakes coiled to strike. They often painted faces on the outside wrappings, too, for extra verisimilitude.

Animals were mummified for several reasons. Some were cherished pets, including cats, dogs, gazelles, and horses. Some were sacred beasts that religious cults worshipped, including bulls. Some were cuts of meat that were placed in tombs for the deceased to snack on in the afterlife—ribs, steaks, roast chickens, even gizzards and other organs, perhaps for mummy gravy.

The most common mummies were so-called votive mummies. People bought them for devotional reasons, like votive candles today, and buried them at temples as offerings to gods. The Egyptians produced votive animals in staggering quantities. One excavated graveyard contained four million ibis mummies; another held seven million dogs. Mummification was a high-volume, high-growth industry.

Perhaps inevitably, the high demand for votive mummies also led to a thriving industry of fraud. Sometimes embalmers simply cut corners. They might split the bones of one creature into three or four supposedly whole mummies, so people got at least a little magic. Other embalmers cheated people completely, either by passing the bones of common animals off as rare ones or by withholding bones altogether in the case of hard-to-capture creatures like raptors and crocodiles. One modern study found that one-quarter of all animal mummies contained zero organic remains. For faux eels, embalmers might coil up some thick bandages instead. Other fake mummies were fashioned from reeds or potsherds, sometimes with stones added for heft. It helped if you posed the beast convincingly and decorated the outside with some flair. Indeed, the most visually lifelike mummies often had the least biological material inside.

While most cases of mummy fraud were probably just someone out to make a buck, other instances are real head-scratchers. Take the "dog" mummy found with nothing but a human long bone inside, or the "baby crocodile" made up of a human finger bone. Perhaps the embalmers simply made mistakes in these cases, or maybe these were more scams—but if so, why use human bones? What was the angle? It makes you wonder whether, as with Amon and his brother's caper, something more sinister was afoot…

They make Amon a crocodile. A few hours before dusk, he lies down on a cedar table, legs extended like a tail, while Abukar arranges broken potsherds around him to simulate a dried croc's tough skin. Then he begins wrapping. Amon's arms remain bunched on his chest, clutching a blade to cut himself free later. He'll breathe through a reed straw.

As the wrappings reach his neck, Amon begins feeling claustrophobic. He tries taking deep breaths, but his chest feels too constricted. Wild thoughts flash through his mind.

—What if I have to pee? Or sneeze?

—I wouldn't advise it.

Before Abukar wraps up his mouth, they go over the plan again. Abukar's men will carry Amon and the other votive mummies to the pyramid. There, the high priests—clad in elaborate robes, their heads freshly shaved—will deposit them all in the burial chamber. After saying a final prayer over them, the priests will withdraw, and the granite slabs of the portcullises will slam down, sealing the chamber off. Amon will then cut himself free. The tomb will be dark, but he saw enough of it earlier today to grope around and find a few jars of perfume. With one of the ceremonial daggers, he can pry loose the slab in the corner and drop into the tunnel to make his way out. Generously, Abukar promises to meet him outside and make sure the coast is clear. Amon appreciates this, but also suspects that he plans on reentering himself for more loot.

With everything settled, Abukar finishes wrapping Amon. It's easier to breathe through the reed than he expected, little sips of air. The odor is another matter. Abukar treats his bandages with juniper and cypress oils, and however pleasant they smell normally, they sear Amon's nostrils up close. They also feel greasy on his face. He hopes the priests move fast.

A half hour before dusk, Abukar's crew trundles into the back room and starts hauling mummies to the pyramid. Two fellows grab Amon by the ends as if they're moving furniture, and drop him on a wooden sledge outside, banging his head. Several more mummies are tossed on top, each one landing with a thud.

It's a bumpy ride to the pyramid. Although it's muffled, Amon can hear his brother greet the senior priests. They don't seem inclined to chat, and before he's quite ready, Amon is being dragged up the steep steps to the entrance.

High priests are generally not the most robust fellows, having never done manual labor in their lives. The two carrying Amon grunt and wheeze and complain about his weight the whole way up the cramped corridor inside. They even drop him at one point, though they catch him before he bangs his head again.

Amon is part of the first wave of mummies deposited, and has to lie there mutely in the dark while the priests leave to grab more. He's starting to sweat inside the wrappings now, and the smell of juniper and cypress is getting nauseating. It takes well over an hour before every last creature in His Mightiness's retinue—baboons, snakes, gazelles, foxes, his beloved birds—is piled inside. And with that, the priests finally begin their prayers.

In normal circumstances, a plebeian like Amon would be put to death for spying on such rites, and despite the danger he's in, he can't help but be curious. Unfortunately, much of the ceremony remains muffled; he can only tell when the priests chant or sing. Even worse, it drags on and on. An hour passes, then another. Although he bites his lip, fighting to remain alert, he eventually nods off to sleep.

Some unknown time later, a peal of voices shocks him awake, a shouted prayer for the pharaoh's well-being.

—You are young again! You live again! You are young again! You live again—forever!

Several long minutes of silence follow. Amon feels flutters of panic. Has something gone wrong? Have they noticed him sweating through the bandages? Eventually, though, he hears the scraping of sandals as the priests withdraw one by one. A minute later, the sound fades to nothing. Then the first portcullis slams down, with a *boom* loud enough to rattle his spine. Two more portcullises follow, leaving his ears ringing.

Then, irreversibly, he's alone.

He makes himself count to one hundred, just in case. But even by ninety, he's slicing the bandages with the blade in his hands. He works so frantically he cuts himself, but barely pauses. By the time he frees his face, he's dizzy. He spits out the reed and takes several gulping breaths, as if he's been held underwater. Then he sneezes, twice, which strikes him as funny somehow. Indeed, he starts laughing deliriously. He's made it. The plan worked.

At least so far. He wriggles out of his cocoon and stands up to stretch, feeling as stiff as a real mummy. Then he takes a few tentative steps. Navigating in the dark is harder than he expected. He trips on something, and knocks into what feels like a throne. He finally drops to his hands and knees and resorts to crawling. Soon, his fingers find some bottles of perfume, which he shoves into his purse. He could likely grab more, but he wants to quit this place as soon as possible.

He gropes around for a dagger. When his fingers close around a handle, he crawls to the corner and works the blade into the crack around the slab there. It's sticky from the beer earlier, but lifts with a little muscle. He peers down and studies the tunnel floor in the dim light below—a drop of five feet. Easy enough.

He's so eager to flee that it takes him a moment to wonder why there's any light at all. Is it morning already, and he's that close to the exit? No. As he stares down, he sees the light flickering. Like an oil lamp.

Holding his breath, Amon waits until he hears it—footsteps, then voices. It's two men, arguing. One sounds like the chief scribe.

The other, he realizes, has a lisp.

We live far closer in time to Julius Caesar and Jesus Christ than either of them did to the construction of the first Egyptian pyramids, in the 2600s BC. Those first pyramids predate even Moses by a millennium. Heck, there were still dwarf woolly mammoths lumbering around on islands off Siberia.

The largest pyramid covers an area the size of ten football fields and once stood taller than a 48-story building (it's lost a bit off the top since antiquity). Its 2.3 million blocks weigh 5,000 pounds on average, with the heaviest weighing 36,000 pounds.* Given its construction time of twenty years, crews would have needed to slot a new block into place every five minutes, day and night, for two straight decades—and with such precision that, in many cases, you can't slip a knife blade between adjacent blocks. Builders decorated the facades, too. Today we see only the skeleton of the pyramids, but in ancient times they were covered with slabs of polished white limestone that glistened in the sun for miles.

Archaeologists estimate that 20,000 to 30,000 people worked on the pyramids. The majority provided logistical support—making and sharpening tools, sewing and mending clothes, digging latrines, cooking bread, brewing beer. In all, perhaps just 5,000 workers actually cut stone and dragged blocks into place. What's more, the popular image of those workers—slaves driven by whips—is almost certainly wrong. Archaeologists have uncovered graffiti from work crews chiseled into pyramid stones, messages declaring themselves the "Vigorous Gang," "Enduring Gang," and "Khufu's Gang." In other messages, crews brag about returning home "in good spirits, sated with bread, drunk with beer, as if it were the beautiful festival of a god." It's hard to imagine beaten-down slaves having such a high esprit de corps. In truth, most of the workers were probably farmers helping out during fallow seasons.

All that said, the workers were probably not wholly free. Pharaohs sponsored major state projects throughout Egypt's history—pyramids, canals, mines—and judicial records speak of people (or their families) being punished for running away from such jobs. And while not a pyramid, the construction of a royal tomb in 1156 BC occasioned the first recorded labor strike in history, when workers threw down their tools and marched on government buildings to demand prompter payments of bread and beer. Clearly, everything wasn't hunky-dory. Overall,

---

* A temple built in ancient Lebanon contains a trio of even larger stones, roughly 750 tons—1.5 million pounds. Again, archaeologists have zero clue how they were transported.

maybe the best analogy here is to compulsory military service in certain countries today: many young people drafted into the ranks love the work and take pride in it, while others feel oppressed and long to flee. Similarly, pyramid crews were no doubt made of individuals of varying temperaments, who thrived or chafed in turn.

Still, even with an enthusiastic workforce, you might be wondering how on earth they built such vast structures—especially given their lack of wheels, pulleys, and iron tools.* Archaeologists wonder, too. The Egyptians wrote down nothing about pyramid construction, perhaps to keep their techniques a state secret. So, as embarrassing as it sounds, archaeologists have no clue how they were built.

That hasn't stopped them from speculating. Historically, most archaeologists assumed that workers pushed the blocks up ramps, perhaps on log rollers. However nice rollers look on paper, though, modern experiments have shown that they stink for moving heavy objects. Big stone blocks can crush the wood, and unless the logs are perfectly smooth and uniform, they slip and rotate unevenly, causing all sorts of headaches. There's also the hassle (and danger) of running the logs at the back of the procession around to the front while the rest of the crew strains to hold a 5,000-pound block in place—much less the 36,000-pounder. And even if you eliminate rollers and imagine the crews dragging the blocks up ramps on sledges, building the ramps themselves would introduce all sorts of problems, some of them potentially insurmountable.

I learned about these problems on a trip to Columbus, Mississippi, to visit an Egypt enthusiast named Roger Larsen. Larsen works as a carpenter in his day job, although for twenty years he published an alternative weekly newspaper in town. He's in his mid-seventies, and bears a resemblance to Larry David—glasses, balding on top, puffy gray curls. He has the sunken chest and ropy legs of a hardcore cyclist, and he in fact regularly competes in 100-mile bike races across the South.

---

* Oddly enough, kingdoms adjacent to ancient Egypt did have the wheel, and presumably the Egyptians were exposed to the technology. But for unknown reasons, they did not employ wheels during the construction of the pyramids.

We meet on a sticky Mississippi morning in his cavernous shop—16,000 square feet crammed with lumber, dusty lathes, old windows and doors, *Titanic*-era chandeliers, piles of books (*Testing in the Atomic Age*, a Frederick Douglass biography), two broken pickup trucks, and what looks like an old water tower. I also see cycling medals hanging here and there, as well as socks, shirts, and other laundry, and I realize with a start that Larsen lives here. At one point he takes me into the plywood shed where he sleeps alongside his "roommate," an auburn-colored hen named Buffy. She has a crippled leg, and I watch Larsen chew up grapes and chunks of Parmesan cheese to feed her. "Hens make the sweetest pets," he says between mouthfuls. He keeps five stray dogs in his shop, too, as well as another hen and rooster, all of whom are barking and clucking and cock-a-doodle-dooing my entire visit. Add the train rumbling by within five yards of the back wall, and it's the noisiest interview I've ever conducted.

Although he's never visited the place, Larsen got bit with the Egypt bug several years ago, and after studying the pyramids closely, he's convinced that the builders did not use ramps to move the stone blocks, at least not to the top. That's because any ramp with a gentle enough slope to push blocks up would need to be both incredibly long and—here's a point worth emphasizing, since most scholars neglect this dimension—incredibly wide. So long and wide, that the ramp's volume would be equal to building *a whole second pyramid*, or possibly more, making an already daunting job nigh impossible.

Larsen demonstrates this to me with an experiment. In his shop, he points to a mound of sand, two-thousand pounds' worth. Nearby, he's built a 24-inch plywood pyramid, painted white. He's penciled two lines onto one of the pyramid's faces. The first line is 5½ inches up from the ground. That represents the height at which a little over half the blocks (54 percent) would be in place in a full-scale pyramid.[*] A second line, at

---

[*] It might sound surprising that more than half the blocks would be in place at less than a quarter of the height. But it's simply a function of how wide pyramid bases are, and the fact that most of the volume resides there.

9½ inches up, represents roughly three-quarters (78 percent) of the blocks in place. He tells me we're going to use the sand to build ramps to both of those target heights. The ramps will have a slope of 1:12, meaning every inch of rise requires twelve inches of run: at the 5½-inch mark, the ramp will extend 5½ *feet*, and so on. Then we'll compare the volume of sand needed to reach each height to the volume of the pyramid itself.

Many scholars doubt that the Egyptians used ramps to build the pyramids, given that the volume of sand needed for a ramp would have exceeded the volume of the pyramid itself, as shown in this demonstration in Roger Larsen's workshop in Mississippi. (Copyright: Sam Kean.)

We start by weighing out sand, using a white plastic bucket and a scale that I'd expect to find in an old boxing gym, with lever arms and iron cubes you clatter back and forth. Each bucketful weighs fifty pounds. Based on the density of sand, thirteen buckets equal one pyramid volume.

As he scoops the first bucketful of sand, Larsen harps on a favorite gripe of his, how close-minded he thinks academics are. He's tried for years to get a hearing for some of his theories, on both the impossibility of ramps and on alternative block-raising schemes, but says that credentialed professors snub him. "They've got a mental block. They're so invested in their ramp theories that they can't see beyond them." He often says—in what's clearly a joke, but one that can't help his credibility among people already wary of cranks*—that if he had to choose between the pyramids being built with ramps or built by aliens, "I'd pick aliens."

Each time Larsen fills the bucket, he hauls it over toward the plywood pyramid and dumps it out. Meanwhile, I build up the ramp, sculpting the sand and letting it find its own slope. It ends up taking five buckets of sand to build a ramp to the 5½-inch line, a bit less than half a pyramid's volume. Given that 5½ inches also represents roughly half the blocks in place—1.25 million for the biggest pyramid—this doesn't seem terrible to me.

But my optimism wanes as we resume hauling sand. Overall, it takes eleven buckets to build a ramp to the second pencil line. In moving from 5½ inches to 9½ inches, then, we've more than doubled the amount of sand required, but we haven't doubled the number of *blocks* that would be in place—that's risen just 24 percent. Put another way, we've gone from 1.25 million blocks in place before to just 1.79 million now—a gain of only 540,000 blocks for more than twice the amount of sand. The returns are diminishing.

And they only diminish faster as we build higher. Pitying the work crews of old, Larsen mutters, "This is where the work begins for these poor guys." He has some old railroad ties lying around, and he jams one behind the plywood pyramid to anchor it. Then he starts scooping and

---

* Given the problems with ramps, some "thinkers" have indeed proposed that ancient aliens built the pyramids. Marginally more seriously, other scholars have suggested that the Egyptians used massive kites to fly blocks up, or long-lost "acoustic levitation" technology that lifted blocks using intense sound waves. Some Egyptologists refer to the adherents of such theories as "pyramidiots." I mention this to explain why many archaeologists are wary of, or even hostile toward, people like Larsen: they deal with a lot of screwballs, and pushy, stubborn screwballs at that. It's often just easier to dismiss anyone who lacks professional credentials.

dumping, scooping and dumping, over and over. Our goal now is to reach the pyramid's top. We hit fifteen buckets, then twenty. I offer to take over scooping, but Larsen won't hear of it. I make this offer partly out of fairness, but mostly because I realize this experiment isn't going to end anytime soon: despite all the additional sand we're adding, the ramp is barely growing higher. A close examination reveals why.

As anyone who's ever built a sandcastle knows, there's a limit to how steeply you can pile dry sand. Make a pile too steep, past about 35 degrees, and the sand starts tumbling down. That's what's happening in Larsen's shop. Each bucketful we dump onto the ramp mostly skids down its sides. The only remedy is to make the ramp significantly wider, so the sand below can support the growing height above. And this is Larsen's point. A 1:12 ramp to the top of the Great Pyramid would already be 5,700 feet long, more than a mile. But based on the physics of sand (or dirt, or any similar material), the ramp would need to be immensely *broad* as well, far broader than the pyramid itself. You simply cannot build a tall, skinny sand ramp. It would crumble.

We soon hit twenty-five buckets, then thirty. Larsen is sweating through his gray T-shirt. Somehow, amid all the chandeliers and watertowers, he owns just one shovel, so I can't help him fill buckets. I ask if I can take over, then plead. Larsen refuses—he's determined to bury the ramp theory himself. Thirty-five buckets, forty. The dogs are howling like mad.

It eventually takes us (him) 48 buckets. In other words, it took 37 extra buckets just to emplace the last quarter or so of the blocks. Overall, then, building a ramp long enough and wide enough to reach the tippity-top of the biggest pyramids would require moving and piling enough material to build not just a second pyramid, but nearly *four* additional ones. And that's actually an underestimate; a tuckered Larsen finally stopped scooping and dumping sand a good inch short of the top.[*]

---

[*] As Larsen admits, people could challenge this experiment on different grounds. Perhaps you could get away with a 1:10 slope, or use rubble made of particles larger than sand, particles that "lock together" to form steeper slopes. But no matter what, you'd still need to pile up multiple pyramids' worth of material. Moreover, if the Egyptians did build ramps out of anything but sand, there should be archaeological

Things look even bleaker when you consider that the walking surface leading up our ramp is little more than a precarious spine, a single roadway. Given the frantic pace of pyramid-building—one block every five minutes—ideally you'd want multiple lanes with room for multiple crews to maneuver. But widening the road would of course require even more material. Crews couldn't drag the blocks over bare sand, either; they'd sink and get stuck. So every one of these mile-plus-long roadways would need to be paved with stones or timbers. And you'd have to keep tearing up the roads and repaving them at higher elevations every time the ramp grew.

Afterward, as we stand and sweat, Larsen shows me some sketches of proposed ramps from different archaeology texts. Having just assisted on this experiment, I start laughing when I see them. Every last one is comically skinny, as if made of metal. Larsen shakes his head, wondering aloud why no academic has ever built a scale model to test how much volume a ramp actually requires. "Anyone can do this, it's just no one has. I don't think they've thought things through."

Larsen doesn't just want to tear down ramp theories; he has his own pet notions of how the pyramids got built. So after some Deep South BBQ for lunch, he drives me to a quarry fifteen miles outside of town to show me.

The quarry pits are filled with tropical-looking aqua water, and a flock of geese are idling on the surface. After parking, we tramp through a thicket of trees with sticky mud underfoot and giant spiderwebs looming in our path. We emerge to find what he calls his pyramid device, a ten-foot-tall machine sitting at the top of a steep slope. Sadly, it's fallen into disrepair, but Larsen explains how it works.

---

traces of it nearby—mountains of rubble. Archaeologists have found some potential ramp rubble, but nowhere near four pyramids' worth.

Many scholars also propose building spiral ramps that, instead of running in a straight line, wrap around the faces of the pyramid. These ramps would not need to be as wide as straight ramps because the pyramid itself would provide support on one side. Even if spiral ramps cut the amount of material in half, however, building two extra pyramids is a heck of a lot of work. Spirals also introduce other problems, like the pain of turning multiton blocks around corners—not an easy task.

Roger Larsen shows off his "pyramid device," a hypothetical machine to raise stone blocks. Workers marched down steps (*top*) while heaving a wooden beam. Through a series of levers and ropes (not shown), this action would drag a large block (*bottom*) up the pyramid's face. (Copyright: Sam Kean.)

In short, the device uses levers and ropes to drag blocks up the slope. To operate it, a crew of workers stands atop a staircase on either side of a tall wooden frame. Each crew is holding a large beam, eight inches thick and twenty feet long. The end of the beam is attached to an intricate system of ropes inside the frame; those ropes are in turn attached to a sledge holding a block. As the workers descend the stairs, they heave the beam down and forward. And as the beam arm moves down, it acts as a lever, tugging on the ropes and hoisting the sledge up a wooden track, a few feet each stroke. When the workers reach the bottom of the stairs, a ratchet-system holds the sled and block in place while they hike back up to repeat the process.

The key to Larsen's machine is its ability to haul blocks straight up the face of the pyramid. As a result, there's no need for an external ramp.

Larsen has orchestrated several test runs of the device. His crew consisted of friends and local halfway-house residents that he sweet-talked into helping. Over the years, they've raised drums of water, mining carts, even Larsen's two-ton Dodge Dakota pickup. He opted for the truck, he admits, solely for clicks. "I figured the pickup would be visually dramatic," he says, and would generate buzz if he posted videos online. Unfortunately, as the truck neared the top of the slope, the rope holding it snapped—under a full load, the rope's as taut as a banjo string—and his truck went careening off and barely missed skidding into a pool of water below. There was nearly a human casualty, too: "My friend Corky, he went and almost did a Superman" off the slope when the rope snapped, Larsen admits. "He got skinned up a little bit."

After that, Larsen decided to "take the redneck aspect out of things" and swap the pickup for a 4,500-pound block of concrete and marble. To prepare, he and his team greased the sledge track with trowels of lard. The Mississippi heat made for a grueling afternoon, but after roughly two hours of marching up and down the stairs, Larsen's ragtag crew could cheer: they'd successfully dragged the block thirty feet up the slope.

Now, pyramid faces are 612 feet long, not thirty, and two hours is far longer than the five-minutes-per-block pace needed for pyramid construction. But Larsen points out that, once his team eliminated the slack in their ropes and got some practice using the machine, they could raise the block two feet every stroke—not bad. Moreover, Egyptian engineers could have placed dozens of identical machines around the perimeter of the growing pyramid, allowing them to lift fifty or more blocks at once. Because the machines were stationary, foremen could have erected canopies as well, to shade workers from the brutal sun.

Again, however, to Larsen's bemuse- and disappointment,

professional archaeologists have shown zero interest in his contraption. Once, as a workaround, he tried writing an article for a popular archaeology magazine. When the pitch got rejected, he offered to buy an ad instead. But even in an era of plummeting revenue, the magazine told him to keep his money—a decision that, as a former publisher, baffles Larsen. He says he understands the need for skepticism in science, but worries that it's hardened into something pettier. "They've already written their books and articles, and the last thing they want is for some amateur out in Mississippi to come up with something they haven't thought about."*

Regardless of how the pyramids got built, they served as tombs and monuments for pharaohs. The burial chambers inside were also stocked with food, daily items, and luxury goods to nourish and entertain pharaohs in the Great Beyond.

Sadly, due to extensive looting over the centuries, virtually all of those goods have vanished. But we can surmise what the burial chambers might have held, at least to some degree, based on a few underground tombs in other areas, especially the Valley of the Kings three hundred miles south of the Giza pyramids. King Tut's tomb, for example, contained diadems, a silver trumpet, gilded beds, ivory rattles, gold cobras and vultures, a dagger made of meteoric iron,† six full-sized chariots, and perhaps most valuable of all, exotic perfumes and scented oils.

Tomb thievery was a serious problem in Egypt as far back as the 2600s BC. As a result, pyramid engineers incorporated security measures into their designs. These included doors with "execration

---

* To be clear, I'm not endorsing Larsen's device as *the* solution to the enigma of the pyramids. Just because he raised some blocks in modern times, that doesn't mean the Egyptians used such a device in ancient times. My guess is that the Egyptians used ramps, either straight or spiral, to place maybe half or two-thirds of the blocks, then switched to lifting devices to finish off the top.

† Meteor weapons appear in Chinese and Turkish history as well, and the people of Greenland began producing iron harpoons and blades in the AD 700s from a meteor that struck five thousand years ago.

texts"—curses—promising to unleash terror on anyone who dared pass. ("They that break the seal of this tomb shall meet death by a disease which no doctor can diagnose.") More practically, engineers filled interior tunnels with heavy rubble and erected sham walls to further conceal passages in and out. They also installed massive stone portcullises that slammed down and sealed off everything behind them.

Given such obstacles, most tomb robbers were probably professional thieves who chiseled their way in through hundreds of feet of stone, chunk by tedious chunk. But archaeologists have also found evidence of inside jobs. Inside one mastaba (a sort of early, truncated pyramid), excavators in the 1920s found an undisturbed stone sarcophagus with a massive lid. The sight thrilled them; they could only imagine the treasures inside. But when they heaved the lid off, they found little beyond some stray bones—and a hole, which led to a tunnel. Given the unlikelihood of someone burrowing blindly through solid bedrock and popping up in the exact right spot to rob the tomb, the tunnel was likely in place during construction, and left unfilled with rubble for easy access.

According to some contrarian scholars, such thievery provided a boost to the Egyptian economy, because instead of expensive goods sitting uselessly in tombs, they returned to circulation and stimulated commerce, however illicit. But officials in ancient Egypt were not as forgiving as economists today. Punishments for robbery included being beaten, impaled, roasted alive, beheaded, behanded, and/or benosed. But to some people, especially the desperate, the risk of robbing a royal tomb was worth the prize...

As the voices below grow louder, Amon eases the floor slab back into place, and for the second time that day, finds himself fumbling to replace jars of perfume. Then he gropes his way over to the crocodile-mummy bandages and drapes them over himself. A minute later, he hears the slab being lifted, and Khnurn and the Overseer climb into the chamber.

No longer bound inside the wrappings, Amon can hear everything with alarming clarity. It sounds like the thieves—his suspicions about Khnurn were correct—are filling sacks as they prowl the chamber. He can scarcely breathe as they pass by him, and fears his trembling will betray his presence.

But all is not well with the thieves. The Overseer begins chastising Khnurn.

—You've taken enough. I saw you slip those perfumes into your waist-purse.

—You've taken twice what I have.

—I am taking strictly what I'm owed as the son of a pharaoh. This monument wouldn't exist without me.

—Well, *I'm* taking some on commission, for letting the son of His Mightiness in on this.

—Do not mock me, Khnurn.

—You also owe me for Edrice, for the mess you've put us in with her pregnancy.

The bite of an asp could not have stung Amon more sharply. That explains why Edrice was always asking questions about the Overseer. Khnurn, meanwhile, has been accusing him of seducing his daughter while knowing who the real father was the whole time. Amon feels humiliated, an utter fool. Every subsequent word from the men's mouths—his love's father, his love's lover—grates his wounds. And he can't do anything but lie there and listen.

At least, not until Khnurn says something more.

—Besides, I should get a premium for pinning this on that idiot brewer. You don't want him prowling around Edrice anymore, do you?

Amon jolts beneath the bandages. Until now, he planned to just lie still and sneak out after Khnurn and the Overseer depart, but he suddenly realizes how much danger he's in. He and his brother had assumed that they had Khnurn cornered: if the scribe denounced Amon for stealing anything, he'd be exposing himself for lying about filling the tunnel. But Khnurn and the Overseer are now aligned—aligned against him. They could easily plant evidence at his home, or bribe some merchant to swear that Amon sold him illicit goods. If the authorities even bothered holding a trial, it would be Khnurn and the Overseer's exalted word against Amon's. After his inevitable punishment, there wouldn't be enough of his body left for even the crows to pick at.

Amon isn't sure what his brother would do in this situation, but he knows Abukar wouldn't just lie there. Perhaps he could attack the thieves—grab a ceremonial dagger and hack them down. If he's quick, it just might work. He lowers the bandages and risks peeking out.

Only to realize it's impossible. The pair are practically standing on top of the pile of daggers; there's no way to snatch one without betraying his presence.

But in surveying the scene, he does spy another opportunity. The giant and the scribe have their backs to him, still squabbling. An oil lamp burns on the floor nearby. And bless the gods, there's a full cup of beer within reach.

He'll get only one shot. He memorizes his escape route. Then, stealthy as a cobra, he reaches out, grips the cup, and flings the sacred beer across the flame.

One hiss, and the tomb is plunged into darkness. Amon throws off the bandages and pads toward the corner while Khnurn and the Overseer bicker in confusion.

—I told you not to let the lamp go out!

—I heard a splash. Something happened.

Amon is already swinging his legs over the exit. Unfortunately, he didn't count on the ladder. Khnurn and the Overseer must have climbed up it, and Amon crashes into it as he jumps down, landing hard and wrenching his knee.

The tomb above explodes in curses. Amon tries to hurry off—and runs smack into a wall. It's a dead end. The tunnel must run the other direction.

He hears the thieves stumbling into furniture above. But before he makes a run for it, some instinct tells Amon to stop. He'll never outrun them with a throbbing knee. But what if he stays put? Khnurn and the Overseer will presumably run the other direction in pursuit of him, toward the exit. If Amon waits, maybe he can slip out unscathed afterward, then find his brother and figure out what to do.

Sure enough, Khnurn and the Overseer drop down—one thud, then another—and scurry off the other way. Amon once again counts to one hundred, plus a little extra this time, then goes limping after.

To his relief, the tunnel has no branches or forks. It ends in a small hole near the floor, barely bigger than a beer jar. He wriggles through, and stands to see a false wall behind him, with one small block pulled aside. He's now in the cramped entrance corridor, and can see light ahead.

Again, instinct tells him to be cautious. What if Khnurn or the Overseer is lurking there? He creeps forward.

—Amon?

Panic flings him against the wall. His throat vises shut, and he squeezes his eyes closed like a child.

—Brother, is that you?

Amon looks, and sees a plump, curly-haired silhouette against the mouth of the tunnel. He limps forward to find Abukar holding out a hand in caution.

—Mind the oil. It's slippery.

In the first streaks of dawn, Amon can see a slick sheen on the smooth paving stones. It smells of almonds. Despite the warning, he nearly slips anyway.

—What the hell's going on?

—I was prowling around, waiting for you, when I saw Khnurn and the Overseer approaching. I hid and let them pass, then prepared a surprise.

Abukar points. Holding onto the wall for support, Amon gazes down the steep stone steps. Crumpled at the bottom, bloody and battered, lie the bantamweight scribe and the giant overseer, their necks twisted at impossible angles.

Amon looks at his brother, who laughs and throws an arm around him.

—Praise the gods I had my hair oil with me!

—What if I'd run out first? That would be me down there.

—A mercifully short death compared to the alternative. Come, let's see what loot they got.

Abukar helps him down the steps. Alas, the thieves aren't holding their sacks anymore; they must have dropped them in their haste. And the jars of perfume in Khnurn's waist-purse have shattered in the fall. Abukar looks at Amon.

—Dawn is coming. Can we get back inside the tomb before anyone sees us?

Amon shakes his head no. His leg is throbbing, and his doughy brother would never fit through the tiny hole in the false wall. He desperately wants to leave anyway.

Abukar nods sadly, but manages a smile.

—Do indulge me for a moment.

He approaches the scribe's body and plucks up a broken jar, dabbing the last traces of liquid on his wrists and behind his ears. He inhales deeply.

—It's exquisite. Try some.

Amon merely looks around in dejection. He's alive, but what does he have left to live for? He gained no treasure, and has no livelihood anymore. He half-hopes he never sees Edrice again, either. Abukar studies him.

—You know, with all the pilgrims coming these days, I'm looking to expand. Come into business with me. It's good, honest work.

—Honest work?

—It makes people happy, which is a lot more than I was doing before.

—I know nothing about embalming.

—I'll handle the mummies. You can run a café for refreshments. We'll call it Abukar and Amon's Oasis. *Buy a mummy and your first beer's free.* I can see it now. Lines for a mile.

He spreads an arm toward the sun on the horizon.

Amon smiles in spite of himself. He can almost, almost, see it too.

# POLYNESIA — 1000s BC

An important theme of the past few chapters has been concentration—the concentration of wealth, of people, of power, into tighter and tighter spaces. But on the opposite end of the planet from Turkey and Egypt, human history developed along different lines. In Oceania, the guiding principle was dispersal.

The sheer scale of Oceania boggles the mind. It stretches over a quarter of the Earth's circumference, and its easternmost point, Easter Island, actually sits closer to Greenland than it does to its westernmost point of Oceania, near mainland Asia. Anthropologists traditionally divide Oceania into three regions: Melanesia to the west, Micronesia to the northwest, and Polynesia to the east. In truth, Melanesia and Micronesia are not coherent entities; the languages and cultures vary too much among the different islands. The cultures of Polynesia, in contrast, do share many important traits. Above all, its first inhabitants were the best sailors the world has ever known.

Polynesia forms a rough triangle stretching from Hawaii in the north to New Zealand (Aotearoa) in the southwest and Easter Island (Rapa Nui) in the southeast. The triangle contains more than a thousand islands spread over 16 million square miles—bigger than Russia, Canada, and the United States combined. The first settlers there trace their origin to the ancestral (non-Chinese) people of Taiwan, who began exploring the archipelagos east of them circa 1500 BC and then pushed on into Polynesia three centuries later. The details of that expansion, including the exact dates, are murky, but the trips began as short jaunts and gradually grew into epic voyages of two thousand miles or more without touching land.

The vessels for the longest voyages were called canoes—a term that belies their size and might. These were not little barks for splashing around a pond on Sunday afternoon. They were massive ships, some over 100 feet long. (When Captain James Cook sailed through Polynesia in the late 1700s, his crew was chagrined to find that a few of the native canoes stretched longer than their *Endeavour*. Cook also estimated that the ships were one-third faster than his.) So when you see the word *canoe* in the context of Polynesia, think *big*.

A model of an ancient Polynesian canoe (*left*), a Fijian *drua*. A life-sized re-creation of a drua (*right*). Notice the hut. (Left copyright: Francis Pimmel. Right photo courtesy Fiji Museum.)

Polynesian canoes were built without a single nail or metal tool, and they generally followed a common design. Similar to catamarans, they consisted of two parallel wooden hulls up to three inches thick. Besides keeping the canoe afloat, the hulls provided storage space for food, water, ropes, tools, cages of pigs, chickens, and dogs, and crops to plant on the next island. The hulls were connected with wooden crossbeams lashed together with coconut-fiber ropes. Above the beams sat a wooden deck where people spent most of the trip; the decks also commonly had a hut for shelter. Finally, rising above the deck, a mast or two supported the sails. The ships could take years to construct, because they were as highly engineered as any airplane—every component crafted and optimized to boost speed, power, and performance.

Using such ships, ancient sailors reached Fiji, Tonga, and Samoa around 1000 BC. In those three island groups, especially Tonga and Samoa, a distinctly Polynesian culture and language emerged. And in settling each new island, these pioneers found something exceedingly rare in human history—truly virgin territory, unspoiled and untrod by human foot. Eden. As for why the ancient Polynesians pushed so far, they likely did so for the same reasons that people have always struck out for new lands: they craved adventure and glory, and sensed there was more to life than what their homes could offer. Unfortunately, for all their noble aspirations, they often transported their ugly, vicious problems with them into paradise...

Loa awakes to a flying fish smacking her in the face. It leaps right up into the ship and slaps her, shattering her predawn dream of drinking from a fresh coconut. Grumbling, she tosses the flopping fish up onto the deck where the coral hearth once sat, then rolls over on her mat to resume sleeping.

It's no use. She just lies there, feeling the gentle rock of the ocean. Automatically, she notes the waves are coming from the left, striking the hull perpendicularly—a swaying she knows more intimately than even her first memories. Under other circumstances, it would be a comfort. But now, each lap only nags her with worry. *Will our drinking water last? Can we fix the sail?* Even the scent of the leaf mat, the smell of home, sends a pang of anxiety through her. She swallows painfully, her throat raw. Then the itching starts, from the crusty stripes of salt on her skin.

It's day three adrift, and still no sight of land.

When she sits up, Little Kako flutters down to her forearm. The parrot's tethered to a thin rope, and she strokes his soft green chest feathers. Feeling a sudden loneliness, she rises and climbs up from the hollow hull where she sleeps most nights and sneaks past the dozen women dozing on deck. Inside the hut there, she finds a hanging basket, peels back the tapa bark cloth covering it, and picks through the nest of damp moss for seeds. She shouldn't be doing this: they're crop seeds, for their new island. But who knows if they'll ever reach it now. Kako flares his wings in excitement, showing the blue feathers at the tips.

—Speak first, she tells him.
—Good girl.
—See any other birds today?
—Be patient.

She smiles. It's an uncanny imitation of Big Kako, the grandfather who raised her after her parents drowned. She closes her eyes and keeps the parrot talking. For a moment, she can all but see her grandfather, with his jiggling potbelly and bloodshot eyes. Since his death three days ago, she's talked to this bird more than any of the other women onboard. Only one of them, Eponi, even thought to console her.

At last, Loa opens her hand. Little Kako gobbles the seeds with his orange beak—he's probably tired of fish, too. When they leave the hut, he darts upward, straining against the tether, testing it, before settling on the mast.

Loa studies him as she pulls her black hair back. She stands five feet tall and has thick triangles of shark teeth tattooed on her arms and legs. She has darker skin than any other woman onboard, from being outdoors so much. Unlike them, she doesn't consider herself pretty. She's not fat enough.

While the others sleep, she leans out and studies the dark sea. Waves hiss, and another flying fish smacks into the red hull before tumbling down. Just beyond where it splashes, she notices something an arm's length beneath the water's surface—flickers of blue-white luminescence, like aquatic shooting stars. It seems like a mocking reflection of the sky above—a sky whose actual stars have been hidden behind clouds for days now, making navigation all but impossible. The wind has been no help, either, barely fluttering the bark cloth pennants atop the mast. There aren't even birds around—which might mean they're far from land, or might simply mean another storm is coming and the birds have hunkered down. Her grandfather, the master navigator, was just starting to teach her to read such signs, but he didn't teach her nearly enough.

There's something about these streaks of light, though. Each one darts in the same direction, straight at the ship. A memory tickles her mind. She turns to the bird.

—Didn't Big Kako mention this once?

—That's my girl.

The more she thinks, the more she's sure of it. He told her that you can find land somehow by observing these faint blue flickers. She closes her eyes and tries to summon the memory.

But before she can, a noise intrudes—someone gulping water. She dismisses it as a drowsy remnant of this morning's dreams: after three days of strict rationing, thoughts of water dominate her mind.

But the noise doesn't fade. Curious, she tiptoes around the battered hut. She fears a fellow castaway is going mad and drinking saltwater.

It's much worse than that. Loa sees a figure holding a bamboo tube to her lips, gulping down their limited fresh water. The face is shadowed, but Loa can see the woman's regal lacework tattoos and swollen womb. There's also her elegant, and undamaged, bark cloth gown. The recent storm ruined all the other women's dresses, smearing the colors. Only Mala-Wife-Of had a second.

Her grandfather always said Loa lacked patience, and she has yet to prove him wrong. She runs up and cracks Mala in the face, sending a spray of water across the deck.

Little Kako jumps up and starts squawking, his wings snapping like sails. Mala-Wife-Of crumples and screams—laughably, given that Loa stands six inches shorter and weighs a hundred pounds less. The scream nevertheless rouses the other women onboard—Mala's attendants—who spring up and rush to her side.

Loa ignores them, and removes the plugs from the other bamboo tubes hanging near the hut. She's horrified to find them all empty. She wheels on Mala.

—We're out of water. She's been stealing it for days.

Predictably, Mala wails in response and begins the litany that inspired Loa's private, mocking nickname—that she's Mala, Wife of Chief Sioni. How dare someone accuse her of theft. Loa cannot hold her tongue anymore.

—Sioni's the reason we're in this crisis. We never should have taken this sacred canoe.

Mala instantly stops crying, her face twisted in hate.

—My husband died trying to fix the idiocy of your grandfather. Kako's lucky he drowned. I'd have executed him for sleeping through a storm otherwise.

Probably fortunately, the attendants get in front of Loa before she can tear Mala-Wife-Of's hair out. They move her aside, and Eponi, a stout woman with a flat nose who's been kind to Loa, cautions her to leave things alone. Loa sputters in anger.

—We're not going to punish her for drinking our water?

—She's pregnant. With our future chief.

—She's going to get us killed, her son included!

When no one answers, Loa screams in frustration and stomps off.

But not far. Back home, their island was a hundred miles around, with dozens of pockets of trees and mountain clefts to disappear into. Here, in the middle of the sea, she can retreat just forty feet. At least there's a lower level. She hops down from the deck into the large hollow hull. Little Kako flutters down to join her.

—Mala's a witch. The rest are useless, she spits.

—Canoe-fillers.

Loa laughs ruefully. It was her grandfather's sharpest insult—canoe-fillers, the inept.

Despite the laugh, she doesn't feel better. And to think, she'd been excited for this voyage—her grandfather was finally going to teach her some real navigation. But one mistake, and now he never will.

Frustrated, she goes to check the coconut-fiber fishing lines nearby, and sees one is taut and straining. She hauls the catch in, a snub-nosed mahi-mahi. It lands in the hull with a smack and begins flopping around, its glittering green scales shifting colors in the dawn light—flashes of copper and blue and silver. A rainbow in fish form.

As the mahi-mahi exhausts itself, Loa studies the weather. The haze looks darker today; perhaps there's another storm pending. That's bad enough for a battered ship, but it also wipes out all chance of spotting land in the distance. They can determine east and west from the glare of sunshine amid the haze, but orienting yourself does no good if you don't know where land is. They might have drifted a hundred miles past their new island, or even be drifting back toward their homeland.

Not that it's really home anymore. Last month, Chief Sioni's long, hopeless war to conquer the rest of their island finally ended in disgrace. Her village, instead of reigning supreme, was left barely able to feed itself. Worse, they became pariahs; fear of retaliation hung over them like storm clouds.

So Big Kako proposed an escape. Based on clues he'd seen on his voyages, he felt certain that undiscovered land lay over the horizon, a few hundred miles at most. Chief Sioni's eyes lit fire at the news: they all knew the sagas of epic sea voyages from the past, and he declared that it was time to found a new race on a new island. He commanded his warriors to steal two large canoes from their neighbors, and insisted on taking his sacred red chief's canoe as well. When Loa's grandfather pointed out the obvious—that the chief's canoe was dangerously small for an open-ocean journey—Sioni smiled and said he wasn't worried, because Big Kako himself would be helming it. That's how Loa ended up aboard, along with Sioni, Mala-Wife-Of, her attendants, and a small crew.

If the storm had swept in during the day, they might have had a chance. As it was, Loa woke up that night to winds like screeching birds and waves like angry whales. (She still cannot understand how her grandfather missed its coming; she fears that Mala was right—he must have been asleep, an unpardonable sin for a navigator.) On the two main canoes in the distance, the alert crews sprang into action—lowering sails, jettisoning the heavy coral cooking hearth, dumping the cages of chickens, pigs, and dogs overboard. Things went less smoothly on the sacred canoe. Lines got snarled, and the lashing wind tore the sail. Worse, the gusts began shoving them sideways, until the other canoes were lost to sight.

All six men aboard the sacred canoe died. Two got crushed against the giant steering paddle by a wave; two others got tangled in lines and swept overboard. Sioni died in even more spectacular fashion. Despite the tear in the mainsail, the winds were strong enough to fill it up and lift the whole ship, slamming the prow down over and over. The canoe would have been bashed to splinters if Sioni hadn't shinnied up the mast and used his whale-tooth necklace to sever the lines. When the sail dropped, he raised a fist and screamed in triumph. Then a wave bucked him overboard. He landed hard, smacking face-first, and never emerged—vain, foolhardy, and brave to the very end.

Loa's grandfather died last. Several bamboo tubes of water washed overboard, as did their backup water supply, a net with a hundred coconuts. There's a saying among Loa's people: *Once overboard, always overboard.* Big Kako dove in anyway. He even reached the coconuts before a wave crashed over him. A few seconds later, the coconuts bobbed up; he did not. It was a valiant, stupid thing to do, and left them stranded without a navigator. Mala's accusations sting Loa in part because they're true.

When the storm lifted, they found themselves floating under a dark haze. The one surefire navigation trick Loa learned from her grandfather was to watch for birds, since certain species never stray far from land. But they haven't seen a single one, and they've been adrift ever since.

The mahi-mahi has finally finished flopping in the hull. Loa's tempted to eat him. They have firm meaty flesh, much more savory than the oily flying fish they've been subsisting on. But in looking over its shimmering scales, she suddenly gets another idea.

As she thinks it over, Kako leans down and playfully nips her forearm with his beak. Her people believe strongly in birds as emissaries of the gods; after all, they can fly to the heavens. She takes his attention now as a sign—he's nipping her shark-tooth tattoos.

Before this morning, she assumed they had enough water for several more days—time for the haze to clear and reveal land. But Mala-Wife-Of's selfishness has plunged them into a crisis. Loa sees just one option.

She nuzzles Kako, thanking him, and turns to climb up to the deck. If they're going to survive this, they need to catch a shark.

Given their warm climate, ancient Polynesians eschewed leather and instead made clothing, blankets, wedding gowns, burial shrouds, and more from a bark cloth called tapa. On a trip to Hawaii I get the chance to make tapa—or *kapa,* as Hawaiians call it. My teacher is one of the great living tapa/kapa artists, Dalani Tanahy, whose pieces have appeared in the British Museum and Smithsonian Institution. She's tall, with broad shoulders and dark, shaggy hair. She lives on a farm in Oahu a hundred yards from a stunning beach. Her workspace sits in a lush grove of banana and paper mulberry trees, the latter being the most common source of tapa.

The trunks of young paper mulberries are the thickness of broom handles, and look more like bamboo stalks than proper trees. Tanahy snips me a footlong piece with gardening shears. I kneel on a woven mat and use a clamshell to scrape off both the outer bark, and a thin, parrot-green layer beneath. I'm left with a stick the color of apple flesh. Tanahy then unzips a bag of tools and fishes out a knife with a blade made of a shark tooth. With it, I make a shallow incision along the stick's length. I'm trying to isolate a layer of soft fibers called the bast. The bast looks the same apple color as everything else, but after I make the incision and pick with my fingernail, it comes free as easily as a banana peel. I now have a twelve-inch by three-inch sheet whose texture resembles wet cardboard crossed with palm leaf.

A Polynesian woman (*left*) prepares to pound tree bark into cloth. Bark cloth artwork (*right*). (Credits: Sémhur, Wikimedia Commons, and Verodemortillet, Wikimedia Commons.)

Next, we pound the tapa to thin it. Tanahy hands me a dense hardwood club called a *hohoa*. I grab it at the end like a truncheon, ready to rain down blows. Tanahy stops me and scoots my fingers up to the middle. Hohoas are for tapping, not beating. I drape the bark sheet over a rock anvil, and begin.

With each tap, the bast widens a little, as the mulberry fibers smoosh outward. It's a delightfully musical process, like playing a woodblock. *Tunk, tunk, tunk, tunk.* Once I've pounded the bast top to bottom, I hold the widened sheet up. Its mottled cream-and-white surface reminds me of brain-tanning deerskin. It's like tree leather.

I fold the sheet over itself and repeat the process several more times. *Tunk, tunk, tunk, tunk.* After a half hour the sheet has browned like an apple, but has widened into a thin handkerchief of kapa. While it air-dries, Tanahy opens up about her background. She grew up in San Diego, a dedicated surfer, but has deep ancestral roots in Hawaii. At one point she declares that, when her time comes, she wants to be buried in a kapa shroud with all her beaters.

Once the handkerchief dries, I decorate it with wooden stamps, using inks Tanahy made from lilies, turmeric, berries, and coconut milk. I also use a paintbrush to add rainbows and curlicues; the pigments run on the soft kapa like watercolors. Tanahy smirks at my designs, which depart from traditional motifs. Much like Polynesian tattooing, kapa art relies heavily on abstract and geometrical shapes—which Tanahy calls her "two nemeses," since she hates math and prefers figurative art. ("I'm like, 'Where's the puppy?'")

As my lesson wraps up, Tanahy tells me that Polynesians could make tapa sheets of any size by pounding smaller sheets together at the edges. (Some cultures also used a starchy glue.) I rub my fingers along my bark handkerchief and imagine it as a shirt. The texture, like thick tissue paper, might feel odd on modern skin. Then again, I'd rather wear tapa than scratchy old cotton tunics.

Undecorated tapa also found use protecting shoots and seedlings on voyages between islands. Indeed, beyond navigational ability, a green thumb was the single most important skill for ancient Polynesian settlers. Collectively, they carried two dozen vital crops—now called canoe plants—from Asia to the islands of the Pacific. Examples include the paper mulberry, bananas, breadfruit, wild ginger, sugar cane, and coconuts.

Another important crop was taro, a starchy root that Polynesians steamed and pounded into a purple paste called poi. After my tapa class, I stop by a little shack to sample some. All I can say about poi is—well, it's there. I enjoy the vibrant color, and the goopy texture recalls the clay I ate with potatoes in Peru. But even the clay had more taste. I keep ladling poi onto my tongue, trying to capture any hint of flavor. It doesn't help. There's just no there there.*

I have far more fun eating another staple of the Polynesian diet, coconuts. But I don't learn to process coconuts in Hawaii. Instead, I travel to Palau, an island chain 4,500 miles west of Hawaii that's technically not in Polynesia, but Micronesia (more on why I chose Palau below). My lesson takes place on a sandy beach on one of Palau's majestic Rock Islands. They're huge gumdrop mounds of shaggy green foliage, amid shimmering lagoon water so bright aqua it looks fake.

My coconut sensei is Jefferson Nestor, a barefoot, potbellied motorboat captain with a shaved head and wide grin that's missing a few teeth. He grew up eating coconuts on Palau, and first teaches me how to dehusk them. He wriggles a sharp wooden stake into the sand, pointy end up. Then he grabs a coconut off the beach and jabs it down— *thunk*—a few inches off-center. Gripping it in both hands, he wrenches it hard, like he's breaking someone's neck in a kung-fu movie. The stake tip tears through the husk, and a third of it drops off. In a flash, he hoists the coconut off the stake, rotates it, jabs it down again, and re-wrenches. Another third snaps off. One more jab-and-wrench, and he's exposed the hard wooden nut inside. It's taken him thirty seconds.

Nestor dehusked his first coconut at age nine using a piece of rebar. His mother screamed at him for doing so unsupervised; I soon learn why. When it's my turn to dehusk, I impale my coconut on the stake in a mere two tries, then attempt the kung-fu move. But while I can feel the fibers tearing, nothing snaps free. To get more leverage, I hunch over the coconut and prepare to throw my body weight into the twist. That's when Nestor grabs me. "Think about what will happen next." I look

---

* In fairness to poi, I find potatoes and rice bland, too. Poi also lacks the cultural resonance for me that it would have for someone who grew up eating it.

down at the coconut beneath my stomach, imagine the sharp stake tearing through, and cringe. "The stake will pierce my guts?" Nestor nods. That's why he got in trouble for cleaning coconuts without supervision. If you don't know what you're doing, you can disembowel yourself.

Not all Nestor's lessons are so dark. Ten minutes later, when I finish dehusking my coconut, he shows me how to crack the wooden nut in the center. First, he finds the three black dots on one end, like bowling-ball eyelets: ⦂•. He then grabs a jagged prism of coral from the beach and, holding the nut up, smashes the dots. The shell shatters in half. Fluid gushes out, but he quickly rights the shell and removes the top. *Voilà*, a cup of fresh coconut water on the beach.

Nestor knows other tricks, too. For a quick snack, we shred some of the dense coconut meat inside with a scalloped clamshell. We also find a coconut that's sprouting a stalk, and crack it open to find not water, but a soft white ball called the pearl. It's the germinating coconut embryo, and Nestor has me pop it out and bite it like an apple. The texture is like soggy Styrofoam, but there is some there there this time, hints of sweet yummy coconut milk. Not bad.

Post-coconuts, Nestor and another Palauan guide, Mac Sasao, take me out spearfishing and snorkeling. Incredibly, we swim with a baby sperm whale at one point, watching it tumble underwater like a giant drumstick. Later, Sasao dives beneath the surface with an empty water bottle, which he squeezes to make a crinkling noise. I'm confused about why, until he points. I stiffen to see several sharks approaching, attracted by the sound. Turns out they're as curious as kittens. It's a neat stunt, however unnerving. And it's fully in line with ancient Micronesian and Polynesian practices. They had their own tricks for attracting sharks, and good reason to do so...

Loa finds the other women on deck, huddled around Mala-Wife-Of. One fans her with palm leaves, another dabs her pregnant belly with a wet sponge. Meanwhile, the stout Eponi is plucking out another woman's eyelashes to divine their fortune: if the lash comes away easily, people believe, they're near land. The yelps of pain are not encouraging.

Loa interrupts to announce her plan. With their water supply gone, they need to catch a shark. Who will help her?

Mala-Wife-Of declares the idea madness. Loa insists it's safe, explaining that her grandfather taught her how. But the mention of her grandfather dooms her case. Even the kind Eponi ducks her eyes.

Loa snorts and hops back down from the deck into the hollow hull. There, she draws in a slack fishing line and uses the turtle-shell hook to pierce the shimmering mahi-mahi and slice around its neck. Then she straddles the fish and wrenches the head like a coconut husk. It snaps off the second time.

Holding the fish upright, she leans out over the hull; the water lies a few feet below. With a deep breath, she tilts the fish so the blood and bodily fluids spill like oil on the surface. Little Kako flutters down to watch.

Within minutes, several blue shark fins are slicing through the chop; they look like canoe sails. Kako edges away, while Loa tries to calm her own heart. The truth is, her grandfather never taught her how to catch sharks; she simply overhead some snatches of conversation. All she knows for sure is that she needs to grab its tail.

When the next shark approaches, she dips the mahi-mahi into the water. The first bite shocks her. It's so clean she's not even sure the shark got anything—until she raises the fish and sees a wedge missing.

When the shark makes another pass, she dangles the fish just out of reach, until its long nose breaches the air. She tries not to look inside the mouth with its dozens of gnashing knives. She lets the brute get a nibble, then raises the food inch by inch, coaxing it out of the water as it propels itself higher. Loa senses the other women peeking over the deck from above, watching her. But she doesn't dare take her eyes from the shark. She's focused on the tail fin, timing its swings.

When the shark pivots and turns to dive back below, she drops the fish and snatches the shark's tail with both hands. Then she twists around and heaves, using her shoulder for leverage. The shark's heavier than she feared, a few hundred pounds. It scrapes the red paint off the outside hull, and she nearly drops it. Then she hears a thump beside her. It's Eponi. She grabs on, and with a final lurch, all three of them tumble onto the deck.

However fearsome below the waves, sharks are helpless if they can't thrust with their tails. Loa is in fact stunned to see how little fight this one puts up — less than the mahi-mahi. It doesn't even writhe around, just lies there panting.

Proudly, Loa looks up at the other women. Then, reveling in their gasps, she runs a hand along the shark's skin; it's as rough as dried sand. She even bops its nose.

The bite surprises her. It's comparatively feeble, a last wheeze, and she snatches her hand back quickly. She might have thought it minor, even, if not for the screams above her.

For a moment, everything is chaos. As if awakened, the shark begins churning its tail, bashing the hull. Kako swoops in bravely, pecking at its eyes. Eponi drags Loa back and eases her to the floor, then tears a shred from her dress to bind Loa's wounded hand.

—Can you move your fingers?

Loa can, although it hurts. The gashes reach down to the bone. Eponi leans in, hissing.

—What were you thinking, showing off like that? You're our only hope of surviving this.

Loa scowls. But there's nothing to say.

Several minutes pass until the shark stops moving. However wobbly, Loa pulls herself up with her good hand and grabs a wooden stake nearby, for opening coconuts. She clubs the shark, twice, to make sure it's dead, then pokes its eye. No living creature can remain still after an eye-poke.

Satisfied, she begins probing the teeth. Sharks recycle teeth so frequently that some are always loose, and Loa wriggles a serrated one free from the upper jaw. Sitting down again, she vises the stake between her knees and starts sawing a slot into the end. Eponi offers to help, but Loa refuses; the other women stare down silently.

She didn't catch the shark for meat; shark flesh requires bleeding and soaking or it tastes of ammonia. She caught the shark because it stores drinkable fluids in its brain. But to reach them, she'll need to dissect it.

When the slot is ready, she tosses the dulled tooth into the ocean. She then works another one loose and jams it into the slot. A little rope hafting makes a crude but workable knife.

Sharkskin is difficult to pierce, so Loa works the blade into a gill slit and begins sawing upward; it splits with a quiet zipping sound. When she's worked enough skin loose, she pulls back a flap over the pink flesh. She then cuts deeper, exposing the skull. It's made of cartilage, so it takes only a minute to saw open a lid. Shark brains are modestly sized, but the skulls have large spaces inside, ventricles full of fluid. Reaching into the cavity, she squishes the lumpy brain with her fist, watching the liquid pool at the top. She motions for Eponi to grab a half coconut shell hanging nearby, then fills the cup partway and hands it to her.

Eponi looks dubious. But after a slow, reluctant swallow, she nods up at the other women.

They file down one at a time. There's enough brain fluid for a few mouthfuls each, albeit only after wringing the brain like a greasy sponge. Loa also digs the knife blade between the vertebrae, and extracts a gelatinous goo with more liquid. Only Mala-Wife-Of refuses to partake—which is only fair, Loa figures, after stealing their water. Still, Loa knows she needs Mala on her side to survive this ordeal. So she gouges out the most delicate prize of all, the shark's rubbery eyeballs, and holds them up as a peace offering. Mala accepts without comment.

After everyone has drunk, Kako flutters down and squawks, *That's my girl*. Loa feels she's earned it—she's no canoe-filler. But with the adrenaline of her injury wearing off, she can't ignore her aching hand anymore.

—Any signs of birds?, she asks the others.

They shake their heads no. Eponi then offers to help Loa up to the deck, but Loa declines. She wants to lie in the hull alone and rest. She cradles Little Kako and curls up far from the shark, letting the rocking of the waves lull her.

However exhausted she is, the pain won't let her sleep. She drifts instead into a sort of lucid dream of her grandfather, fueled by Little Kako's voice.

Part of her, she can admit now, is angry at him—for diving overboard, for leaving them adrift, for not teaching her more. But she's too spent to stay angry long. Her mind drifts back to some of her earliest memories with him, those long lazy fishing days at sea. Sometimes he would lower her into the water and turn her left and right, making her feel the waves against her body. Her first navigation lesson. After the fifth or sixth session, she grew impatient. She didn't want to just splash around. She wanted him to *tell* her things, reveal secrets. He laughed and said it didn't work that way. Some knowledge you can only absorb...

Loa's body goes rigid in the hull. Just like this morning, with the underwater flickers, something's tickling her mind—vague but insistent.

Another memory burbles up, of coming across Big Kako on the beach one morning. He was instructing another navigator. They'd laid out shells like islands, and were drawing currents in the sand with sticks. Loa sneaked close enough to eavesdrop, and overheard them discussing how the currents hit the islands and curve around them, the waves warping like a bent stick. That idea of bending currents always stuck with her. And lying in the hull now, she can feel the memory in a way she couldn't before—bodily. She shushes Kako and blocks out the ache of her hand, remaining as still as possible.

Something takes shape. Early this morning, while trying to get back to sleep, she'd felt the waves rocking the canoe from the left, striking the hull perpendicularly. Based on the sun, the ship's oriented the same direction now. But the waves are striking at a different angle, a bit askew, lifting the front of the canoe a fraction before the back. It's subtle, but it's there.

She thinks back to the map of clamshells and curved lines in the sand. She imagines herself at different points on it. How would the angle of the waves change from place to place? Could she trace that difference back to an island?

A moment later, she struggles to her feet. Her throat is so parched she can barely speak. But a smile softens her face. She points a shaky finger.

—There's land that way, Kako. I'm sure of it.

Not all ships in Oceania were as enormous as the Polynesian canoe. The smallest, around twenty feet long, were fishing barks for daily use; they usually consisted of a canoe hull lashed to an outrigger for stability. Medium ships consisted of large hulls lashed to either outriggers or secondary hulls. These ferried people and troops around on island chains like Hawaii, or served as sacred vessels for chiefs.

Loa's vessel had asymmetric hulls—one large, one small—a design that dates back more than three thousand years. After their construction on land, these ships were often transferred to the water by rolling them over the corpses of sacrificial victims, leaving a trail of crushed bodies. (In absorbing the men's blood, the ships allegedly absorbed their power, too.) On the water, they sailed incredibly fast, slicing right through the waves. Observing one in modern times, a British colonial official said, "It barely skims the surface, and... becomes a veritable flying machine." But however glorious in calm weather, such ships rode low and often got swamped in storms. As a result, only rarely did they venture into open ocean for long-distance travel.

For those sorts of journeys, Polynesians built giant canoes with two equal-sized hulls that could stretch 110 feet long—the dreadnoughts of the Pacific. These ships could carry a hundred people and cover 150 miles a day in fair wind.*

---

* In a contrary wind (blowing into your face), a voyage of 100 straight-line miles might require 350 miles of actual sailing, because you have to zigzag back and forth by tacking or shunting. With their ability to sail in all types of wind, fair or contrary, ancient Polynesians and Micronesians far outpaced Europeans, who lagged millennia behind. In the 900s, when a Viking chief tacked into the wind to flee King Olaf I of Norway, Olaf grumbled that the man must have used sorcery. Olaf later caught the fellow and executed him by jamming a viper down his throat.

I saw one such dreadnought in drydock in Hawaii, a modern re-creation that local sailors sometimes take out on the water. While approaching it, I got the same feeling I did upon seeing sequoias and whales up close for the first time. Its sheer Brontosaurian bulk—hulls like tree trunks, ropes as thick as arms, a steering paddle with a blade as big as a kitchen table—is overwhelming. How can something so big still move?

Given that the ship is not an archaeological specimen, I ask a docent at the center if I can climb up on deck to look around. She shoots me down. Still, hoping for some vicarious insight, I ask her whether she's ever climbed up. Her eyes flash wide, and she shakes her head no. "I'm afraid of heights," she says. I have to admit I laugh—afraid of heights on a boat? But glancing up again, it doesn't seem so far-fetched. The top platforms sit a dozen feet aboveground, and the sails and rigging would have stretched far higher. You could easily get vertigo and tumble.

In ancient times, building such canoes was part ritual, part science. When Hawaiian sailors were selecting trees for hulls, they watched to see which ones the 'elepaio bird (a small brown flycatcher) would alight on and peck at. In Hawaiian lore, the female goddess of the canoe, Lea, often takes the form of this bird, and pecking a tree showed disapproval. More prosaically, the builders probably noticed that flycatchers peck trees to get worms, and vermin-ridden trees make poor hulls. After selecting a tree to chop, the carpenters might lay their tools "to bed" in a sacred spot, then "awaken" them in the morning by dipping them in the sea—pure sacrament. They'd then fell the tree with lumberjack skill, making sure it landed on the largest branches to cushion the fall and prevent cracks—practical physics at its finest.

The hulls of small and medium canoes were hollowed out from single trees using adzes (thick chisels attached to axe handles), then sanded with coral and breadfruit leaves. Candlenut or another oil sealed the hulls and made them waterproof. The hulls of giant, deep-ocean vessels—which carried livestock, food, and water on voyages—were fitted together from planks with sticky breadfruit sap as caulking. Coconut-fiber ropes lashed the hulls together, and sailors sometimes attached fluttering coconut leaves at the waterline to determine the direction of currents.

Above the hulls sat a planked deck and hut, as well as one or more masts for sails. Polynesian sails generally looked like inverted shark fins—triangular, with the tips pointing down. They were woven from pandanus leaves, long botanical slivers that resemble iris or daffodil fronds; up close, the sails had the texture of baskets, but with the suppleness of leather. To judge the wind, sailors affixed streamers of feathers to the sails, or pennants of bark cloth.

Beyond their general form, however, the details of Polynesian canoes varied from culture to culture—especially in terms of frills. Some islands adorned their vessels with intricate carvings or totems. Some hung up tufts of grass or inlaid the surfaces with snail shells. Others added accents of yellow, green, black, red, or white paint. These ships had flair.

Aside from the elegant ships, what really made Polynesian sailing special was the skill of the navigators. Rather than view the high seas the way I do—a vast scary expanse of blank death—they saw something like an interstate highway system, with navigable "water bridges" connecting every island to its neighbors. They could chart open-water routes up to 2,500 miles—from Washington, DC, to San Francisco—using clues so subtle that most of us wouldn't even notice them.

The Polynesians didn't have compasses, so they relied on stars to establish and maintain direction. But their skills went far beyond simply knowing that Polaris points north. Stars rise and set throughout the night, and different stars are visible in different seasons, so navigators needed a mental catalog of hundreds. Most stars don't rise straight up toward the zenith, either; they follow an arc across the sky. As a result, a star that rises due northeast or whatever will remain due northeast for only an hour or so. Navigators therefore had to "hand off" the direction from one star to another as the night progressed. They actually spoke of "five-star" or "seven-star" journeys, and multi-night journeys were common. The series of stars the canoe followed was sometimes called the star path.

Stars off the path played a role, too. Clouds can obscure the sky for days on end and frustrate navigation. In those conditions, if a tiny patch of sky suddenly cleared up, navigators had to identify the stars there instantly to get their bearings. Given the vast distances involved, a mistake of a single degree could cause you to miss an island completely and go adrift. One Māori navigator I spoke with described each journey as a stark pass/fail test: "You passed if you got to the island and you were still alive. You failed if no one ever saw you again."

Still, however useful, stars can lead ships only to known islands. To discover *new* islands, navigators used other clues.

Imagine you're sailing on the open ocean. How would you know there's an island beyond the horizon? The most obvious clue comes from clouds. Clouds tend to drift fast over water and drag over land, so a traffic jam of clouds on the horizon could well mean land ho. Clouds can also dip downward toward land in certain conditions, as if draining into a funnel. Their colors can change subtly as well—gaining a green tint above lagoons or a pink hue above reefs.

Noddies (*left*) and terns (*right*) have been guiding Polynesian sailors toward land for millennia. (Credits: Mike Davison, Drew Avery.)

Seabirds offer additional clues. Some, like petrels, happily stray hundreds of miles from shore. Others, like noddies and white terns, sleep only on land and rarely stray more than twenty miles out. So if you see noddies or terns zipping off a certain direction at dusk, you can follow them to terra firma. Meanwhile, long-distance migratory birds, like long-tailed cuckoos and golden plovers, can tip navigators off to truly remote islands, far beyond the horizon. Long ago, some clever man or woman likely noticed that those birds always traveled the same routes each spring or fall, year after year. From that, they deduced that land must lie in that direction, however distant.* New Zealand, Hawaii, and Easter Island might have been discovered this way.

Navigators relied on sensory details as well. Tiny flicks of luminescence called *te lapa* can appear a hundred miles offshore. No one knows what causes te lapa, but unlike the bioluminescent algae or plankton that ships sometimes churn up, te lapa flashes on and off quickly and has a definite direction—the lightning bolts always dart away from land. Another clue involves water temperature, which rises near land, especially atolls. Salinity differs as well, allowing some navigators to taste the approach of home. Odor-wise, coral reefs emit sulfides (a cooked cabbage or seafood smell), and vegetation emits odors, too. Sound-wise, all canoes creak and groan as water currents cause the beams and other wooden parts to rub together. Subtle differences in the creaks—in volume, frequency, timbre—can alert a navigator that the direction of the current has shifted, perhaps due to the presence of land.

Other changes due to currents are subtler. When ocean currents encounter an island, some of the water's energy reflects off and bounces back as waves. Meanwhile, water that doesn't strike the island directly will bend around it (refract) and proceed at an angle to its original direction. The result is an intricate interference pattern of bent and reflected

---

* Birds also paved the way for Polynesian expansion by carrying seeds from island to island, ensuring that nutritious fruits were available when people arrived. Birds became direct food sources, too—especially those that "went dodo" and lost the ability to fly. Some species were so blithe about predators that hunters could walk right up and wring their necks. People were less "hunting" the birds than gathering them, the way you would coconuts.

Sadly, Polynesian people killed off untold numbers of bird species as they migrated from island to island, sometimes as food, other times by clearing forests or introducing rats that poached their eggs. One biologist called this avian slaughter the "largest vertebrate extinction event ever detected." It's a touchy topic today. Many Polynesians think of their ancestors as "guardians of Paradise," but the archaeological record refutes that simplistic idea.

waves, and as canoes move through this water, the waves will strike the hulls at slightly different angles each hour. As Loa did, navigators can sense these shifts by lying down, feeling the lap of waves against the hull, and back-calculating where islands must be—essentially using the body's vestibular (balance) system as a sort of compass to detect them. Some blind sailors developed this and other senses to such a degree that they became master navigators despite their handicap; although they couldn't read the stars, they could draw on smells and sounds and vestibular clues to cross unfathomable fathoms.

The amount of knowledge that master navigators could summon was astounding—and that knowledge afforded them great power. If you wanted to travel to distant islands to trade or make war, you needed a top navigator at the helm. And once onboard, your life rested in their hands. Traditionally, navigators guarded this power by restricting who could learn it, often passing it down only to initiates. (While this was largely a male vocation, some women became master navigators as well.)

On voyages, navigators helmed their ships all day and all night for several days running, leaving them with bloodshot eyes. And while many were no doubt humble, some were rather cavalier about danger. I talked to a few modern navigators who recalled, with shrugs and chuckles, situations that would have made me wet myself: violent storms with thirty-five-foot swells, leaks sprung mid-voyage that required days of sustained bailing.

As the effective captains of their canoes, navigators also had to manage friction among crewmembers and passengers. Especially when ships were adrift, food and water had to be rationed, leaving people hungry, thirsty, and testy. Even on routine voyages, people got on each other's nerves in all sorts of ways: by snoring, by shirking duties, by smelling bad. Worst of all, people couldn't escape each other. Tensions would smolder and smolder until the pressure and frustration built too high—at which point, the only outlet was an explosion...

Eponi helps Loa climb up on deck, where she announces that she's detected land. She expects delight, wonder, cheers. Instead, she's met with skeptical stares. Quietly, Eponi asks how she can be sure. Loa explains what she felt—the currents bending around the islands and the subtle lifting of the bow—but they look confused. She has them lie on deck to feel the waves themselves. Mala-Wife-Of refuses to, saying it's beneath her. The others humor Loa, but admit they can't feel anything.

Loa finally appeals to their desperation.

—We can't wait any longer. Help me. Dip your paddle in.

Without a word, Eponi rises from the deck and walks over to Loa's side. Others follow. Mala's nostrils flare at the betrayal, and Loa tenses for a fight. But Mala merely turns and stomps into the hut.

Loa exhales hard. Then she turns to instruct the crew—*her* crew, now that she's navigating.

First thing, they need to mend the sail. Loa assigns each woman a task. Some cut rope with the shark-tooth knife and unravel it to make thread. Others bore holes into the mahi-mahi bones, fashioning them into needles. Still others use teeth from the shark's lower jaw as awls to puncture the leathery sail. They mend the tears with patches from their sleeping mats. The finished sail looks terrible, but Kako chirps *good girl* at her in her grandfather's voice, and Loa feels proud as they hoist it.

With the wind low, they only creep along at first, the sails limp. Still, for the first time in days, the other women don't seem listless. They have purpose now: when Loa gives an order, they snap to. As if channeling their excitement, Kako keeps fluttering up to the end of his tether, straining on it before returning to Loa's arm. Even Mala-Wife-Of skulks out of the hut to watch, chewing a royal fingernail.

An hour before sunset, the soft breeze turns chilly. Loa senses there will be a storm by morning, which could spell disaster. Worse, between the haze and lack of light, their visibility is dropping fast. The bending of a current can point a navigator in the right direction, but it's a rough guide only. Making landfall requires actually seeing land, and if it's too dark, they risk sailing past.

Loa considers their options. They could lower the sail and heave-to—essentially tread water until morning. But if a storm is coming, she fears their battered canoe won't survive. Besides, her crew is looking exhausted again, despite their initial enthusiasm. They need to land. She rubs Kako's feathers and tries to think.

With no other options, she finally calls out to everyone to keep an eye out for birds. Surely a few have ventured out to eat. If they can just spot one returning to land, they'll know which way to go.

Over the next twenty minutes, there are several false alarms—claims of seeing birds that aren't there. Loa finally loses her temper.

—Do any of you know what the hell a bird looks like?

—How dare you.

She turns to see Mala-Wife-Of glaring at her.

—I lend you my attendants, who work themselves ragged for you—and you insult them?

—They need to look harder. All we need is one bird.

—Then sacrifice yours.

She points at Kako. Loa feels a shiver of fear.

—What do you mean?

Mala-Wife-Of strides forward, drawn up regally in her tapa dress.

—We need a bird. Send yours.

Loa points out how absurd the idea is. Kako is a pet, not a wild bird who's used to flying long miles at a stretch. She fears he'd never reach shore alive.

—He doesn't need to, Mala says. He can rise much higher than us. He'll see land and take off. That's all we need, isn't it?

A volcano erupts inside Loa.

—I'd sooner sacrifice you! He's worth ten of you—twenty!

She's shaking with rage. But Mala-Wife-Of just smiles. Loa realizes why. The crew—her crew—are abandoning their posts to gather around Mala. Their faces plead with her. Eponi finally speaks up, her voice hoarse.

—There's only a few minutes of daylight left. Please.

Loa's world suddenly gets very quiet. The sea, her thirst, her aching hand—the volume on everything falls sharply. She can barely speak. But she says yes.

She withdraws to say goodbye to Little Kako, stroking his chest. Then she reprimands herself—it's not goodbye. She promises they'll reunite on the island; she'll hear her grandfather's voice again. After a quiet moment, she grabs the shark-tooth knife she made and severs his tether.

Had Little Kako simply shot off and never looked back, it would have been easier. Instead he darts up and does a few acrobatics in the sky. Then he makes a wide circle around the canoe and, finding nowhere to land, returns to Loa's arm.

She bucks him off, and he flutters up again—only to circle in the sky and return. This time she throws him off, ordering him to fly.

When he returns yet again, Loa pulls her arms in to deny him a perch, ignoring his squawks. Kako finally seizes her shoulder, digging his talons in.

—Good girl.

When Loa tears him off, she can feel the blood. She flings him up, hard, and when Kako tries to land again, she slaps him away, screaming at him to go already. He's a blur of green wings behind her tears.

A confused Kako circles once, and hesitates in the air. Then he takes off—a green dart toward the horizon. Loa wipes her eyes and orders the crew to follow him to land. She just prays they make it there before the storm.

Since the 1970s, there's been a swell of enthusiasm for reviving Polynesian voyaging, including the construction of several authentic(-ish) canoes in Hawaii and elsewhere. Collectively, those canoes have visited dozens of islands, made trips of tens of thousands of miles, and inspired millions of people—there's just something stirring about a big ship under sail. Polynesian vessels are every bit the cultural icons that the pyramids of Egypt are, or the Great Wall of China.

Still, there's an awkward secret at the heart of this revival—that most "Polynesian" navigation nowadays has its roots outside of Polynesia.

Like many aspects of Polynesian culture, a great deal of knowledge about traditional voyaging was lost over the past few centuries. One reason was the devastation wrought by colonization, including both the denigration of traditional skills and a forced shift to Western lifestyles. But even before colonization, many Polynesians had already forgotten how to navigate the deep ocean—especially those who lived on so-called high islands. High islands are the peaks of underwater volcanoes; they're wide and spacious, with lush forests and room to grow crops. Because of this abundance, people on high islands had little need to travel long distances. As a result, navigation techniques atrophied and even went extinct in such places.

But this ancient knowledge lived on in pockets of Micronesia in the western Pacific. Micronesia contains high islands as well, and similar to Polynesia, traditional navigation techniques died out in those places. However, Micronesia also contains low-lying coral atolls. Atolls have fewer terrestrial resources, and people there rely more on the sea for sustenance. Atolls also get wiped out by storms sometimes, forcing inhabitants to pick up and start over elsewhere. As a result, these were the places where navigation techniques survived. The recent revival of "Polynesian" navigation, then, was really just the spread of Micronesian knowledge to Hawaii and other islands. The fact that Micronesia rarely receives credit for this today has left some people there bitter.

To learn more about traditional navigation and sailing, I visited the island of Palau, the same place where I processed coconuts. In addition to its long history of navigators, Palau also boasts rare examples of authentic working canoes built by native people. The canoes belong to an American expat there named Ron Leidich, who commissioned them after his daughter, who's half Palauan, watched the Disney movie *Moana* and asked Daddy for a canoe of her own. Leidich couldn't deny his little girl, and ordered two canoes from local builders. "They have no signs of modernity," Leidich says. "The fibers, the glue, the wood—it was all from the forest." When I asked if he could tell me how much the vessels cost, Leidich groaned. "Without stomach pains? I wish my daughter asked me for a pony instead." Regardless, $150,000 later, they got built: *Moana I* and *Miluu's Pride*.

I arrive at Leidich's dock one September morning, where a man-bunned guide named Mac Sasao—the one who summons sharks by crinkling water bottles—shows me the tools of the canoe trade. First are the paddles for rowing out to sea; they're painted red and have surprisingly sharp tips, because they doubled as spears during raids. (Similarly, the giant wooden scoops for bailing water doubled as shields.) Next, to teach me a bit about canoe construction, Sasao lets me hack at a stump with an adze whose blade is fashioned from a giant clam shell. Frankly, I'm horrendous at this task. Instead of striking one spot repeatedly, my blows veer drunkenly all over. It's nevertheless satisfying to hear the *chunk, chunk, chunk* of a solid tool.

*Moana I* is a modest eleven-footer with an outrigger. The hull is two feet wide with planks to sit on, and is at least a yard deep; when I sit, my feet dangle in air. The mast, outrigger, and tiny deck are to my right. Behind me sits the pilot, Ismael, a skinny, smiley fellow with severely bloodshot eyes.

As we paddle out, the rich indigo water ripples like Murano glass. I'm practically bouncing with excitement. When I first proposed this book, sailing around Oceania on an authentic canoe was the one activity I was most looking forward to. I can barely fathom the courage it must have taken to point the prow at an empty horizon and just take off. To be sure, I'm not fooling myself here. I'm taking a mere day trip, and I know I won't pick up even a hundredth of the clues a master navigator would. I just want a taste of the adventure—a glimpse of what ancient Micronesians and Polynesians would have seen and smelled and felt in their hearts upon setting out.

Alas. Although the forecast promised healthy breezes that week, every trace of wind dies mysteriously on the morning we set out. The weather then proceeds to taunt us for hours. Most of the time we just sit there on the water, staring glumly at nothing. Then the wind stirs for a moment; the sail crinkles. Roused, we scan the treetops in the distance, searching for any sustained ripples...only to watch the wind dissolve again, no more substantial than a ghost. It happens over and over.

There is one moment of excitement. The stupid, lying forecast predicted winds that day from the west, but the one sustained gust arrives from the opposite direction. We scramble to take advantage. Sasao chirps, "Sam, pass the sail to Ismael." I turn, expecting Sasao to hand me a rope or some other means of winching the sail about. Instead, Sasao actually hands me the sail—drops the boom right in my lap. It turns out the sail is not fixed in place. The tip of the boom—the wooden arm along the sail's bottom—merely rests in a groove at the end of the ship. Sasao has pulled the boom out of one groove and handed it to me. Unsteadily, I lift the boom over my head and hand it to Ismael, who jams the tip into the groove behind him. The sail is now pointing the other direction, and the front of the ship has become the back. This maneuver to switch the sail is called shunting, and it's common among vessels in Oceania. Shunting allows sailors to make progress against contrary winds, similar to tacking. It also allows vessels to change direction quickly to take advantage of shifting winds.

In this case, the wind takes advantage of us. Mocking our efforts, it dies as soon as we've shunted, leaving us no better off than before. To say I'm crushed would not be accurate; I'm pulverized. I keep fretting about all the time and energy and money it took to reach Palau, and how my hopes for adventure on the high seas have deteriorated into this—me just sitting there, drifting idly on the water, screaming silently at the wind to please, please do something.

It never does. There's no wind my entire week in Palau.

My only consolation is that the trip isn't a total loss. After all, I get to snorkel with sharks and whales. I also learn navigation tips. As we float on the water, Sasao points out the chocolate noddies streaking by. The birds have silhouettes like fighter jets, and their habit of returning home each night at dusk has been steering sailors toward land for centuries. We study clouds, too, and watch them grow shinier as they glide over reefs and beaches. Some look strangely flat on the bottom, as if lopped off. Others have black mushroom stems of rain streaming into the water below.

Perhaps most importantly for this chapter, my thwarted sailing adventure allows me to truly feel what Loa did—the sheer impotent rage of wasting hour after hour in a stationary ship, praying for the wind to pick up and knowing in your heart it won't. Sometimes, the gods refuse to listen when you need them most...

Loa's crew sails all night in the direction Kako flew. Loa spends most of it begging the sluggish winds to pick up and speed the canoe along. She also scans the darkness for Kako, feeling equal parts hope and fear—sick with worry that he won't return, yet knowing that if he does, it means he never found land.

Given the clouds, daybreak arrives slowly. But an hour later, she finally spies something. It's no grand vision of paradise, more an irregular black urchin on the horizon. But it's land—real, solid ground—and Loa's throat tightens.

She calls out to the somnambulant crew. A dozen heads whip around—then the deck erupts in cheers. Even Mala-Wife-Of smiles. Whether it's their new home island or not, Loa doesn't know. At this point, it doesn't matter.

But as the sun rises farther, Loa sees angry black clouds looming ahead. A few minutes later, rain hits. It feels refreshingly cool at first, washing the crusty salt off her skin. It's a short-lived relief. The tapa pennants start snapping—the wind at last—but now it's driving them into the storm.

They need to prepare. Loa grabs the shark-tooth knife and severs a spare rope into a dozen lengths, each roughly five times her height. She jams the knife between two deck planks for safekeeping, then hands a length to each woman onboard, with instructions to tie herself fast to something.

Everyone does as she's told. Except Mala-Wife-Of. Mala binds the rope around her waist, beneath her bulging belly. But she insists on tying the other end to the hut, so she can stay inside and protect her tapa dress. Loa shakes her head no.

—The hut's too flimsy. You have to tie up to something sturdy.

—A queen carrying a chief inside her cannot go ashore in a ragged tapa.

Loa doesn't even argue. Despite her throbbing hand, she snatches the free end of the line and binds it to a nearby cleat. It's a ragged knot, but it'll have to do. There's a grim pleasure in knowing that at least the storm will ruin Mala's dress.

Before long the rain is lashing them sideways, cold and stinging, like stepping on an urchin. Waves begin bucking the ship, splashing water around Loa's ankles. The bow keeps rising and sinking, rising and sinking—the view pure sky one moment, pure sea the next. It's dizzying.

Most islands have circles of reefs surrounding them, and Loa spots the telltale sign of a reef here: a line of churning froth. She searches anxiously for a break in the line—a smooth patch that indicates deeper water and safe passage. In calm weather, it would be easy to see, but here the entire sea is boiling.

She finally spots something, and hollers at Eponi and another stout woman in the back to turn the massive steering paddle. But the largest swell yet arrives at that very moment. Lurid green water leaps onto the deck and kick-sweeps Loa's feet from under her. She lands on her ribs, hard. Several other women go down, too. Had they not been tied fast, they likely would have been dragged overboard.

—Help!

In a storm so violent, Loa would have thought it impossible for any single sound to break through. But this voice does. Loa turns to see the hut collapsed, and the water from the swell dragging the planks overboard. Along with Mala-Wife-Of.

Loa stares dumbly. Did her knot come loose? Or did Mala retie the line to the hut, to save her dress? Loa scrambles to the edge of the deck, straining against her rope to peer over. She can just see Mala floating there, face down amid seaweed and wooden flotsam.

*Once overboard, always overboard.* She pushes the thought away and gropes toward the mast. There, she finds the remnants of Little Kako's long tether, which she fastens around her waist after unlashing the rope. Three running steps, and she plunges in.

However cold the rain, it cannot match the sea. Loa comes up gasping, the saltwater biting the wounds on her hand. She bumps into something heavy, and recoils to see a shark. Then she sees the gap-tooth grin and skin flopping open. It's the carcass of yesterday's catch, washed overboard.

She splashes in a circle, unable to find Mala. But with the rise of the next wave, she spies a clot of limbs and hair and seaweed, and starts thrashing toward it.

First thing, Loa gets Mala on her back so she can breathe. Then she starts disentangling her from the seaweed. But the weight of Mala's pregnant belly keeps flipping her over. Loa finally just starts tearing at everything, ripping Mala's dress along with the weeds. When she's finally free, Loa locks an arm under her shoulder and starts reeling in the tether toward the canoe.

Eponi and the other oarswoman hoist them up to the deck, and set about reviving Mala. Loa just lies there panting.

She's forgotten about the reef until they strike it. The canoe is riding high on the crest of a wave as they approach, and when it releases them, they slam down hard on the coral. The deck—already creaky from the previous storm—explodes into splinters, ejecting everyone and everything on it into the air. Loa finds herself tumbling like a flying fish. She watches the world go through a complete revolution, upside-down, then lands on her back in the water, knocking every ounce of breath from her body.

Before she knows what's happening, something drags her under, yanking her by the waist. It's the tether, still attached to some hunk of the ship. She claws at the knot until she's free and shoots upward to the surface. She comes up sputtering and rakes her hair from her eyes. She's bobbing in a sea of wreckage just past the reef. Mala is next to her. Loa grabs hold and starts swimming toward shore.

Loa's first steps on the sand leave her reeling. She nevertheless drags Mala onto the beach and confirms she's breathing. Then Loa wobbles over to a nearby grove of palm trees and collapses.

She has no idea how long she lies there. Long enough for the storm to die and the staccato of rain on the palms to stop. The sun even emerges. Loa scrunches the warming sand between her toes, and drinks in the familiar scent of flowers and fruit. She can't believe she's alive.

She finally sits up when she hears dogs barking. She's dizzy, and her back and neck ache, but she makes herself remain upright until her vision steadies. She sees the dogs crash into the surf, where one of them snags something plump and green in his teeth. It's too bright to be seaweed, too puffy for a palm leaf. The other dog grabs hold, too, and they fall into a tug-of-war.

Loa is running before she realizes it. When she grabs for the green body, one dog bites her injured hand. But she pries Little Kako loose, cradling the bird against her while the dogs bark and snap.

She tries talking to him, begging him to speak, say *good girl,* anything, even shaking him when he fails to respond. She just wants to hear his voice one last time.

But it's no use. He's really gone now. While the dogs howl, she squeezes Kako so hard that water begins streaming out of him like a sponge, coursing down her chest and legs. It pools into a dark puddle of misery at her feet.

She's so devastated that it takes her a long minute to fully register the scene around her, and grasp how strange it is. Why are there dogs here? The other canoes they set out with both jettisoned their animals during the first storm.

Then she hears men shouting. A small band emerges through the trees onto the beach. They're dressed familiarly, in tapa loincloths with blocks of solid black tattoos on their legs, chests, and faces.

The men race along the beach, and begin herding the women from

Loa's canoe together, dragging the ones too weak to walk. There are eight total, a few of them naked. Loa realizes Eponi is not among them.

When they're all together a tall man with long black hair steps forward to ask what happened. Loa starts to answer before Mala interrupts. She drapes the ragged remnant of her dress around her and points an accusatory finger at Loa.

—I am Mala, wife of Chief Sioni. Arrest this woman.

—On what grounds?

—Her grandfather abandoned our canoe and nearly killed us. Then she tried to finish the job. She pulled a vicious shark onboard and drove us onto the reef. It ruined my dress.

Loa is near hysterical. Cradling Kako, she turns to the men to defend herself—until she sees them smiling. The tall one, who she realizes now has bloodshot eyes, spreads his hands.

—Do you realize where you are?

Loa glances around. Her island was a hundred miles in circumference, and Sioni's people were not welcome on most of it. But the trees and flowers here seem familiar. And when she scans the mountainscape above, she recognizes a few peaks. The man with the bloodshot eyes continues.

—Sioni is a traitor. He attacked us, and stole canoes. By law, his wife and children are traitors, too.

Several men step forward to grab Mala-Wife-Of. She screams, thrashing her arms and bloodying one man's nose. Another finally slaps her, hard, and they drag her off, her feet trailing in the sand. The other men herd her attendants behind her.

Dazed, Loa watches them leave. The only man remaining is the one with bloodshot eyes.

—Are you Kako's granddaughter?

—Yes.

—I apprenticed with him. I remember you hanging around the beach, spying on us. He said once that you would be a great navigator someday, if you were patient. How much did he teach you?

She cradles Kako's broken body.

*More than I imagined,* she thinks. And enough to get home.

# Rome — AD 100s

〜〜

Many of the trends we saw emerging in ancient Egypt were magnified in the Roman Empire—the specialization of labor, the monumental architecture, the massive population centers. Indeed, ancient Rome was the most populous city in the world at its peak, with well over a million people—due in no small part to the fine roads leading in and out, which allowed for efficient transportation of food and goods.

Another trend was social stratification, including widespread slavery. It's hard to grasp nowadays just how pervasive slavery was in ancient times, and how it permeated every aspect of life. Depending on the estimate, one-third to one-half of the population in Roman cities was enslaved. At one point a Roman senator proposed a law forcing enslaved people to wear distinct clothing as a mark of infamy and to distinguish them from citizens. Other senators quashed the idea: they feared that if the enslaved realized just how many of them there were, they'd quickly start a rebellion.

Unlike the chattel slavery that thrived in the Americas from the 1600s to 1800s, slavery in Roman times was not based on race; most of the enslaved were prisoners of war, sailors seized by pirates, or children taken captive abroad. (Some people in ancient times even viewed slavery as a humane practice, since it meant no longer slaughtering war captives.) In Rome, those who worked hard and obeyed orders could and often did win freedom. But of course, not all slaveholders were so benevolent. Many people under their control suffered beatings, whippings, and brandings, and those who dared run away and were caught were often fit with snug metal collars inscribed with the name of their owner.

Some inscriptions also insulted the wearers as troublemakers, "slutty prostitutes," or worse. That ancient dogs wore similar collars escaped no one's attention.

This chapter is loosely based on the volcanic eruption that destroyed Pompeii and other nearby towns in AD 79. The volcano's initial burst of gas and smoke rose twenty miles into the sky, then flattened horizontally like a tree canopy. As the column cooled, hot ash and pumice began raining down, a blizzard of fiery, stinging snow that accumulated at five inches per hour and eventually buried Pompeii under twenty feet of cinders. "Many besought the aid of the gods," one witness recalled, "but still more imagined there were no gods left and that the universe was plunged into eternal darkness."

Things got even worse a few hours later, when the pyroclastic flows started—avalanches of scorching gas, molten rock, and rubble. Some people swallowed up by the flows famously left behind hollow body casts in the ash, which recorded their final moments in incredible detail. You can see their singed eyebrows, the rings and keys they're clutching, the way they're hitching up their tunics or shielding their faces. Others met more gruesome ends: their body fat ignited like tallow candles, and their brains possibly boiled and exploded through their skulls. Anyone left in the city when the flows began was a goner. But for those who'd fled at the right time, they still had a chance to live, however slim…

Zyrmina wakes to the very ground shaking beneath her. She grabs the stone floor, clutching the earth until the quaking stops. She expects shouts, panic, alarum. But the town remains still. As a foreigner, a Greek slave, she cannot understand how the Romans remain so blasé about such dangers.

It's just before dawn. She's lying outside her matron's bedroom, and drapes her thick black hair over her eyes to resume sleeping, but it's no use. She has a rash from cooking with rue yesterday, and her arms and face itch something fierce. It's worst beneath the metal collar riveted around her neck. The collar's not terribly heavy, being only a fingernail or two thick. But it's two inches tall, and no matter how her fingers dig, she can't scratch all the skin beneath.

At last, Zyrmina hears clapping from the bedroom. She enters to find the thick, squat Marcella standing there in her underwear, ordering her to hurry. As Zyrmina dresses her, she notices how loose her matron's hair is. Marcella has the feast tonight with the priestesses, and her entire mountain of raven-colored braids is near collapse. Zyrmina considers offering to fix it, but Marcella is already rushing out the door.

Outside, no shops are open yet, and the air smells of sulfur. As they hurry toward the temple, Zyrmina indulges her usual habit of calculating escape routes. If she took off *right now,* how far could she get? Given the empty streets, she's sure she could reach the nearby volcano this morning. Then work her way up the terraced vineyards, steal a horse and…

A flare of itching beneath her collar kills the fantasy. After their abduction from Greece, she and her mother ended up here, as slaves in a cloth fullery. Her headstrong mother ran away and was caught—then ran again and was whipped as punishment, as well as fitted with an iron collar inscribed "SERVUS SUM." *I'm a slave.* The creepy overseer at the fullery threatened to collar Zyrmina, too, if she tried again, thinking no parent would risk that. He didn't know Zyrmina's mother. A month later she disappeared for good—perhaps to freedom, perhaps death. But the overseer kept his promise. Now, even if Zyrmina escaped, someone would see the collar and arrest her. It's far safer, then, to *earn* her freedom by currying favor with Marcella.

An iron collar used to punish and track slaves in ancient Rome. The collars came in different sizes and thicknesses. (Credit: Rabax63, Wikimedia Commons.)

Still, the fantasy of running persists.

Their destination this morning lies in a field a mile outside town—a half-built temple to Fortuna, the goddess of luck and Marcella's deity of choice. Marcella has been funding it for years, with an eye toward donating it to the public right before an election—a sort of municipal bribe. After the death of a public priestess last month, she saw her chance, and there's been a mad scramble to finish ever since. As they approach, Marcella hollers at the slumbering construction crew to get working.

The temple is handsome enough, a white block surrounded by columns. But the columns are only half-complete, like a smile missing teeth, and the adornments strike Zyrmina as gaudy: cheap gold trim and poorly painted statues. One team of sculptors is even recycling statues taken from around town—decapitating old emperors with chisels and bolting new heads on, some comically large or small. She wishes Marcella could afford better; it's embarrassing.

The road leading to the temple is worse still. In Zyrmina's experience, Romans are oddly passionate about roads: they brag about them, even compose odes to them. But Marcella hired sloppy surveyors and bought cut-rate material, and the temple road ended up crooked, with a surface as uneven as a stormy sea. She had no choice but to rip up several sections, which delayed the temple's construction even more: instead of using carts, workers have to carry material by hand and drag statues in with ropes. The tension and pressure on-site feel thicker every day: Marcella's already a long shot, but she'll never win election as public priestess if she fails to deliver a temple on time.

A Renaissance engraving depicting the Roman goddess Fortuna (*left*), holding her wheel of fortune. A temple (*right*) dedicated to Fortuna. (Credits: Wikimedia Commons.)

As Marcella curses, the construction workers grumble and climb down into a pit. They're stripped to the waist and streaked with yesterday's mud. Two start swinging picks, breaking the shoddy roadbed into chunks, which others heave onto the roadside. A half-dozen masons, all with hacking coughs, mix fresh concrete nearby. Meanwhile, the hammering of blacksmiths echoes from the temple. For the next hour, Marcella barks orders while Zyrmina stands there scratching at her neck.

She's studying another escape route when someone shouts her name.

She turns to see her fellow slave Hortensia and, behind her, a stalled carriage. It belongs to Valeria, Hortensia's matron and Marcella's electoral foe. She's also Marcella's sister.

Hortensia comes running over. She's a foot taller than Zyrmina and twice as wide—as big as an ox and equally simple. She stumbles on the uneven road, skinning her knee, but pops right up. She then bear-hugs Zyrmina, lifting her off the ground.

—Hi, Zee! We just got into town.

—I see that. Shouldn't you help Valeria?

But Valeria doesn't need help. She emerges through the carriage's silk curtains in a gold-threaded tunic. Then, graceful as an ibis, the richest widow in Rome picks her way across the broken road, waving off the eager offers of help from the construction workers. Zyrmina is struck once again by how different the sisters are. Marcella, as solid as a tree stump; Valeria, so light you'd swear she can lift right up and fly.

Valeria hugs Zyrmina, who can feel the scowl on Marcella's face without even looking. Valeria then runs a hand through Zyrmina's hair, fluffing her thick raven curls.

—Do you wake up looking this beautiful? Meanwhile, my hair hangs as limp as a horse's.

Zyrmina tries to smile. The truth is, people tell her daily, hourly, how wonderful her hair looks. She feels like a walking wig. The fact that her mother had the exact same hair makes things even more fraught—a daily reminder of a woman she'd rather forget. Still, Zyrmina does love her mane; it hides her slave collar, if nothing else. And frankly, she understands Valeria's jealousy. If Valeria has any flaws, it's her flaccid hair—one of the few imperfections she shares with her sister.

—Stop molesting my slave, Valeria.

—Lovely to see you, too, Marcella. That's how you greet me after all this time?

—You're right. It's been, what, three years since you've condescended to visit home? And now you're running for office here.

—Don't be vulgar and talk politics. Let's see this road that nearly killed my slave. You men there, hello!

Valeria glides her way over to the workers. Marcella's face tightens like cracked plaster as she stomps after.

Zyrmina is left talking with Hortensia—although it's more a scattered monologue. First, Hortensia shows off her bleeding knee. Then starts babbling about some glass beads she found to ward off the evil eye. Bored, and with an eye toward currying favor with Marcella, Zyrmina tries steering the conversation toward matters that might affect the election—like who in town Valeria has been exchanging letters with. But Hortensia shrugs and keeps chattering nonsense. She's oblivious to Valeria's plans.

Valeria's late, elderly husband owned a holiday villa in town, near where the sisters grew up. A decade ago, Valeria seduced him and escaped to the high life of Rome. And however infrequently she returns now, her husband's ties to the emperor make her a leading woman here. So when the public priestess died last month, Marcella wrote to her sister with a request—that Valeria host a feast with the other priestesses, and push them to endorse Marcella for the office.

Valeria replied that the feast sounded splendid. Arrangements proceeded. Then, just last week, Valeria wrote to announce her own candidacy for public priestess. She suggested they turn the dinner into a little campaign event instead, to let the other priestesses compare them side by side.

Marcella smashed half her crockery in fury. She also tried getting the feast canceled, and reversed course only when Zyrmina pointed out that it would likely happen with or without her, and that not attending would look petty. Better to show up, act gracious, and outshine her sister; she could even announce the gift of the temple then. In truth, Marcella outcharming Valeria seemed as likely as the Moon outshining the Sun, but the plan bucked Marcella up and won Zyrmina credit in her eyes—another deposit toward her freedom.

Zyrmina finally gives up with Hortensia, and they wander over to where Valeria is holding court. The men keep working—Marcella's right there—but every eye is fixed on her sister.

—Another temple in town? I can't imagine it, Valeria says. When's the last time *you* saw the inside of a temple, you rascal?

She points at a bearded mason who's mixing a batch of cement. He blushes when he admits it's been a while.

—Ah, you're more a bathhouse man.

There's laughter. The bathhouses here are notoriously licentious.

—Well, if I win election as public priestess, I'll build a new bathhouse. Tell all your wives to support Valeria. Your mistresses, too.

The men roar. But only for a moment. There's a sudden sizzling sound, like hot iron plunged into liquid. Zyrmina sees the bearded mason gaping down in horror; distracted by Valeria, he's just dropped a basket of crushed quicklime into a basin of water.

The resulting reaction is quick indeed. The basin erupts like a volcano, spewing chunks of scorching rock. Hortensia screams, and a stampede follows.

Zyrmina gets elbowed in the scrum but keeps her feet. Marcella isn't so fortunate. She ends up sprawled on the ground, her tunic muddy and her mountainous hairdo in tatters. Zyrmina rushes up to help, but Marcella's already on her feet, shaking with fury.

—Who did this? I'll have you whipped!

A long minute of cowering follows. But the construction workers stay mum, even when Marcella threatens to whip them all. Valeria finally defuses the tension.

—Men, I'll leave you to work. My sister's a good soul, even if she has a temper.

Then, *sotto voce* to Marcella:

—A temper that might cost her an election. See you tonight, sister.

Zyrmina watches with a pang as Valeria and Hortensia mount their carriage and clatter off. If Fortuna's wheel had spun another way, she might have been in Hortensia's place. Instead, she listens to Marcella tell the workers they aren't allowed to sleep until they finish. Then she turns and trudges home, with Zyrmina scurrying after.

The home is the same one Marcella and Valeria grew up in, a modest residence brought low—sagging roof, peeling blue paint—by Marcella dumping all her money into the temple. As the guards creak open the door, Zyrmina claims she has work in the kitchen and tries to slip away.

Marcella stops her, saying she has an errand. But instead of explaining, Marcella turns toward a small, gilded statue of Fortuna standing in the atrium. The goddess is holding a cornucopia in one hand and her notorious wheel of fate in the other. Marcella intends to place it in the temple as the centerpiece, and for once Zyrmina approves: it's beautifully rendered—real ivory and gold—and unlike every other statue there, it will do the temple credit. Marcella stands in the atrium now running a finger around the wheel's rim, as if spinning it herself.

Finally, she turns to Zyrmina. She orders her to visit Hortensia at the villa and pump her for more information. As a pretext, Zyrmina can drop off some items from the market—melons, honey, garum—and deliver the mint-rue sauce she made yesterday. Zyrmina nods and withdraws, suppressing a sigh. In the kitchen, she loads the two corked jars of sauce into a leather bag, then sets out into the city.

Walking through town is always surreal to Zyrmina. Most everything about Roman culture was plagiarized from her native Greece— the food, the gods, the architecture—all echoes of home. But the Romans never could capture the elegance of Greece, the refinement. The streets underfoot are filthy, a mess of vegetable peels, leaf litter, and squashed donkey dung, all buzzing with flies. And the raised sidewalks are a cacophony of shoving and yelling, people barging in and out of shops—butchers, bakers, barbers, tailors, knife-grinders, fruitmongers, and more. Despite the similarities to home, she never quite feels at ease here.

Her stomach growling, she purchases some sheep-and-snail stew with her allowance, plus a gourd of cheap wine. It tastes smoky from the pitch used to seal the neck. She eats in the shade of an awning, since sunlight makes her rue rash worse; she can already feel the blisters rising.

Rather than pass by the fullery, Zyrmina takes the long way around, veering left over a series of raised crosswalk stones above the dirty street. She has to use the toilet anyway. While standing in line, she studies the graffiti on the walls. Some relate to the upcoming election, including a stanza comparing Marcella to a donkey. Another scrawl warns of "Sodom and Gomorrah"—one of those annoying Christians.* Then there's the phalluses. For some reason, Romans consider penises good luck, and slap them everywhere—on paving stones, doors, ovens, jars. Here, there's a painting of a giant winged penis that itself has a smaller penis, plus a penis "tail."

Finally, a spot opens up on the toilet bench. Zyrmina sits down amid five other women, the marble warm on her thighs. Upon finishing, she grabs the communal sponge to wipe; it's tied to a stick and sitting in a jug of vinegar. Refreshed and clean, she blinks as she heads back into the bright day.

 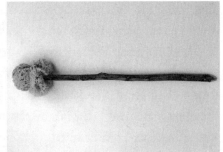

A Roman public toilet (*left*). There were no stalls or barriers between holes. Afterward, people wiped with a communal sponge (*right*). (Credits: Fubar Obfusco, Wikimedia Commons, and Herdemerten Dickson, Wikimedia Commons.)

Per Marcella's instructions, she buys honey for the dormice tonight. The melon shop is one street down. Finally the garum—fermented fish sauce. The jars of discarded mackerel sludge in front of the store always make Zyrmina queasy, but she picks up two top-shelf bottles. Then it's straight to Valeria's villa.

Zyrmina is glad to be dropping everything off. Her shoulder bag is

---

* Oddly enough, pagan Romans often accused Christians of being atheists. After all, Christians denied that dozens of gods in the Roman pantheon existed—and who but an atheist would deny the existence of gods?

getting heavy, and she wants to rid herself of the mint-rue sauce. Zyrmina is allergic to rue, but didn't say anything when Marcella ordered her to chop some up yesterday. At Marcella's request, she even muddled it along with the mint to extract more oils and flavor. On top of that—outrage upon outrage—Marcella made her pick all the chopped herbs out again after finishing. Marcella explained that Valeria claims to hate mint, even though she happily eats dishes when she doesn't know it's in there. But another servant told Zyrmina a different story—that Marcella fears the green bits will get stuck in her teeth, or even stain them after her recent whitening. Zyrmina doesn't know what to believe. It was simply another ridiculous request of Marcella's.

The villa has high yellow walls and a red-tiled roof. As luck would have it, Hortensia is coming back from her own errands; she has a bag of goods and the carcass of a small suckling pig slung over one shoulder. Zyrmina transfers the melons, honey, and garum into Hortensia's bag, then hands her the jars of sauce.

Zyrmina tries to engage Hortensia in election talk, but it's useless. Hortensia picks up chattering again right where she left off, about evil-eye beads. Zyrmina is trying to break in, and failing, when Hortensia suddenly stops.

—What's this?

She's staring down at the jars of mint-rue sauce. Cradling one in her elbow, she picks a green fleck off the other with her thick fingers, and examines it in the sunlight. Zyrmina is about to explain when Hortensia starts chortling.

—Oh, it's mint! It looked like rue. I thought I'd have to scold you, Zee.

Zyrmina can see she's mistaken—it is actually rue—but she's too weary from her errands to bother correcting her. She just asks why.

Hortensia flicks the speck to the ground.

—Because Valeria, she's deathly allergic to rue. Ate some as a girl once and turned red as wine, she says. Throat closed up, too. She couldn't breathe!

She sneezes and wipes her nose with her fingers.

—Well, *some* of us have work to do. You've kept me long enough, you blabbermouth.

Hortensia and the pig turn to go—leaving a speechless Zyrmina standing in the street, staring down at the deadly, discarded speck of green.

Few civilizations could match the engineering genius of Rome. Hundreds of magnificent Roman bridges and monuments still stand today after two thousand years. Roman aqueducts often provided more water than modern plumbing does, and ancient Rome's sewer system was superior to any other European city's before the 1700s. One bathhouse was so lavish that, upon uncovering it, archaeologists initially mistook it for a temple. Many of these wonders were built by rich elites for the public good, and Roman emperors, however cruel and oblivious, believed passionately in public works. To some Romans, even the pyramids of Egypt were contemptible. What purpose did they serve? What did they *do*?

Above all, Romans built roads—53,000 miles of them, 6,000 more than the entire US interstate system. These roads stretched up to twenty feet wide and typically consisted of several layers, starting with a bed of compact sand four feet down. Above that came crushed rock, then layers of concrete and cemented sand. Stone slabs on top provided a smooth surface for carriages and pedestrians. They were so durable that some are still in use today.

To get a better feel for ancient road construction, I visited a Roman technology class taught by Nathalie Roy, a short-haired, high-energy classics teacher at Glasgow Middle School in Baton Rouge, Louisiana. Growing up, Roy says, she "desperately wanted to be a paleontologist. I dug up anything and everything, . . . had a collection of rocks, and kept a human femur in my bedroom (the heirloom of a distant relative)." Sadly, her mother squashed Roy's dreams by telling her that scientists had "already dug up all the dinosaurs." Crushed, Roy settled on studying extinct languages like Latin and eventually became a teacher.

## Rome—AD 100s

Over the years, her Roman Tech classes have spun wool clothing, pieced together mosaic floors, constructed sundials, and fired mini Roman catapults; her classroom also boasts a life-sized chariot with bicycle wheels. A few lessons have earned Roy something of a reputation. One involved writing on papyrus with squid ink, which reeks like fish. "My principal was walking down the hallway," Roy remembers, "and she said, 'What is that smell? It has to be Miss Roy.'"* Another vivid lesson involves sea sponges, which the Romans used in lieu of toilet paper in communal bathrooms. In her classroom, Roy hands me a dry yellow sponge. "Students always want to know, 'How did they wipe their booties with this?'" Holding one, I must say the students have a point. The surface feels coarse, and a bit prickly. Bad bumf. But after we soak the sponge in water, it transforms; the texture turns silky, even slightly creamy. I wouldn't mind wiping my booty with this—provided a thousand other people hadn't already.

Roy's big student project in 2023 involved building a modified Roman road in the school's courtyard. Work began that January, and Roy aimed to finish by Mardi Gras on February 21. The students started by surveying the road with plumb bobs and laying down seventy yards of bowling ball–sized boulders to mark the left and right edges. During my visit, they're filling the space between with fist-sized chunks of crushed limestone. I've volunteered to help haul rock, and before we start, I survey my fellow workers. Like with most junior highs, the range in the kids' sizes looks like a sight gag. Some stand taller than me; others I could smuggle into a movie theater in my coat.

We each grab an orange five-gallon bucket (the Romans used baskets) and start scooping in rocks from a five-foot-high pile near the playground—part of the forty-two tons of rock the road will require. My hands, pants, hair, and lungs are quickly coated in dust. (Roman masons probably had horrible coughs.) Then we lug our loads over to the road site a hundred yards away. It's a slog. Near-record rains this month have

---

* To be sure, Roy's reputation is mostly exemplary. She was the 2021 Louisiana Teacher of the Year.

transformed much of the courtyard into a bayou of mud puddles, complete with crawfish. Even the dump truck that hauled the pile of rocks to the school got stuck in the morass. This doesn't deter the students, though; quite the opposite: on the walk to the roadbed, few can resist the temptation to muck around in the thickest, squishiest ooze. (Even Roy got stuck in the mud one morning, and had to be pulled free by a hefty eighth-grader.) After a shift hauling rock, it's easy to tell which students in the hallways at Glasgow are taking Roy's class—the ones whose pants are spackled with mud. The back of her classroom is one big pile of caked-over boots.

Stupidly, at the start of my shift, I insist on lugging two buckets at once, one in either hand. They're only thirty pounds, after all, a small dumbbell. At first I feel peppy and zoom past the pint-sizers, some of whom are too short to carry a bucket by the handle and have to cradle it instead. Predictably, I get my comeuppance. After the twentieth round trip, my arms are drooping, and I'm sweating in the early-February (!) heat. But of course I can't downgrade to carrying just one bucket; God forbid I look wimpy in front of some sixth-graders. It doesn't help my morale that each bucketful, when dumped onto the roadbed, covers a depressingly small area. All that sweat and arm-ache for a measly square foot.

Overall, during my three-day visit, the sixty (admittedly distractible) students finish a mere thirty yards of road. And that's just one pass; each foot needs several layers of rocks piled up. If Roman roads have lasted for millennia, there's a reason—the workers earned it.

Given the slow pace, I have to leave Baton Rouge before the class completes the road. I follow their progress online instead, and slowly but surely, the students finish the crushed rock layers and fill the gaps between the stones with dark sand. Then they mix up the key ingredient in Roman roads—and in fact, the key ingredient in most Roman construction: concrete.

Concrete is artificial sedimentary rock—stones glued together with cement. (Some Roman concrete was pink from the rosy terra-cotta rubble they used.) Nowadays we associate concrete with brutalist architecture, but the glorious dome of the Pantheon was made of concrete, too, and Roman concrete in general lasted far longer than modern varieties. For centuries, scholars assumed that the secret lay in the fine volcanic ash that Roman workers mined near Naples and mixed into their cement. (Adding fine ash produces stronger, denser cement-glue. In her classes, Roy has bags of white volcanic ash that are as fine as powdered sugar.) But while good ash helps, experimental archaeologists have recently discovered the true key to Roman concrete. Hot-mixing.

Before the Roman Empire, cement production involved mixing small chunks of quicklime into water to form a paste. Patience was paramount because quicklime and water react vigorously, and a slow pace gives the heat from the reaction enough time to dissipate before the water gets dangerously hot (temperatures can spike to 400°F). In contrast, hot-mixing involves dumping loads of quicklime into water all at once. If done carelessly, the quicklime will sizzle and explode, as the bearded mason at Marcella's temple learned. But if done properly, hot-mixing has advantages. Most importantly, the final concrete contains nuggets of unreacted quicklime called clasts. Upon seeing these clasts in old structures, many armchair archaeologists assumed that Roman builders hadn't mixed their cement properly—the equivalent of an inexperienced baker leaving lumps of flour in dough. In truth, clasts strengthen concrete, because they prevent cracking.

Experimental archaeologists at the Massachusetts Institute of Technology proved this with a recent project. The team made several squat cylinders, some of modern concrete, some of Roman concrete, and deliberately fractured them. Then they poured a continuous stream of water over. Normally water ruins concrete, either causing cracks or widening existing cracks until the structure fails. But something magical happened in the Roman concrete: As the water trickled into the fractures, it reacted with the undigested clasts of quicklime ($CaO$) to produce calcium hydroxide ($Ca(OH)_2$), a big molecule that swelled and plugged the crack automatically. Unlike modern concrete, then, Roman concrete is self-healing. This explains why Roman concrete has lasted for thousands of years, while modern stuff crumbles in decades. They call Rome the Eternal City for good reason.

Roy's classes* focus on the technical aspects of Roman construction, but during the days of empire, municipal works had political dimensions as well. Again, Roman elites often built temples, aqueducts, and bathhouses using their own fortunes, both to defuse class tension and gin up votes during elections.

In Pompeii, archaeologists have excavated a building financed by Eumachia, a patron of the fullers' guild. Fullers processed wool and laundered wool clothing—a despised trade, since making and cleaning wool required splashing around in aged urine, the cleanser of choice back then. (Cities sometimes collected the raw material in giant public piss pots.) That said, Roman wool was a valuable material and fullery owners could grow rich selling it. Eumachia even parlayed her patronship of the fuller guild into the prestigious job of public priestess. In this role, she carried out and advised on religious ceremonies, and given how superstitious ancient Romans were, no family would dream of building a home or marrying off a child without consulting someone like her. This was one of the few routes to political influence that Roman women had.

In building a temple to Fortuna, Marcella was trying to duplicate Eumachia's rise in a time-honored way. But when things got desperate, Marcella was not above resorting to other, extralegal means of securing power...

---

* Through superhuman effort, Roy's ragtag student crew did get the concrete poured just before Mardi Gras break (a thing down there), and went home feeling triumphant. Sadly, though, some drunken reveler rode a bicycle across before it set and left fat tire scars, ruining the road's surface. Roy reported that the sight left the students heartbroken and angry—a sharp lesson in how mean-spirited the world can be. Fortuna is fickle.

## Rome — AD 100s

Zyrmina wanders home in a daze. Braying donkeys, the shouts of shopkeepers, even another small earthquake — none of it reaches her. Upon arriving back, she finds herself staring dumbly at the statue of Fortuna in the atrium. Does Marcella really intend to poison her sister?

She debates the question for several minutes, until something finally pierces her thoughts — Marcella screaming.

Zyrmina darts across the peristyle garden, weaving around the trees toward her matron's snug, red-walled bedroom. She finds Marcella inside sitting on a bed, with her ancient hairdresser Bellona standing over her. Everything looks normal. Then Marcella turns, and Zyrmina sees blood trickling down her neck.

—You blind old goat!

Quick as an adder, Marcella slaps Bellona, knocking out her ivory teeth.

—Get out of my house.

When Zyrmina asks what happened, Marcella points to the sharp, bloody bodkin in Bellona's hands. Bellona was apparently trying to salvage Marcella's mountain of hair when her hand slipped and she stabbed Marcella.

Zyrmina runs to the kitchen for a rag to staunch the bleeding. She returns to find two guards reluctantly leading the simpering hairdresser away. As Zyrmina dabs her matron's neck, Marcella asks whether any blood has dripped onto her stola, her fancy woolen dress gown with the golden collar. Zyrmina admits yes, which starts Marcella cursing again. It's her only formal stola, and she planned to wear it to the feast tonight.

Zyrmina unwraps the garment and lays it on the bed for Marcella to examine. She fears Marcella will ask her to take it to the fullery. Instead, Marcella does something worse. She asks about her errand to the villa.

—What did Hortensia say? Did you deliver the sauce?

A flare of itchiness erupts under Zyrmina's iron collar.

—She, uh, couldn't talk.

—Why are you scratching like that? Is that a rash?

—It's the rue. I'm allerg...

Hortensia's words flash through her mind. *She's deathly allergic.*

Zyrmina feels her face flush red. Marcella notices, and cocks her head. Then her face hardens.

—Why are you acting guilty?

—I'm not.

—Did you mention your rue allergy at the villa?

—No.

—But something happened. Is it about my sister?

Zyrmina hesitates, then nods.

—So you know. Did you tell them anything?

Zyrmina swears no, but Marcella presses her.

—Did you mention the rue in the sauce?

When Zyrmina says no, Marcella exhales in relief.

—So it's still a secret. You probably think I'm a monster.

—No.

—Don't lie to me. But ask yourself this. How do you think I feel, watching her come back? I've lived in this city my whole life. I've invested in it. I *care*. The roads, the fountains, that blasted temple. Meanwhile, she's gallivanting around Rome. Then she sweeps back here to steal my office. So, yes, I gave her some rue. I hope she drinks the whole goddamn jar.

Her face has twisted into a gargoyle of hate. Zyrmina drops her eyes, frightened. But she makes herself speak.

—You can't poison her.

—What did you say?

Zyrmina knows she's risking every ounce of goodwill she's curried with Marcella over six years. She could be demoted to laundry girl, or even sold. But she can't keep quiet.

—You can't poison your sister.

Marcella takes so long to answer that Zyrmina glances up. She's shocked to see Marcella laughing.

—Poison her? It's just an allergy. I want to *embarrass* her. She'll puff up like a toad and break out in hives and feel ugly for once. Who told you it would kill her?

—Hortensia.

—The girl's a simpleton.

Zyrmina suddenly feels stupid. Hortensia does say silly things. She tries apologizing, but Marcella cuts her off.

—I'll overlook it. *If* you help me prepare for tonight. You style your own hair, right?

Zyrmina sighs. Her hair again. But in truth, her hair is what first brought her to Marcella's attention. They met in the cloth fullery that Marcella's family owns. Zyrmina was eleven, her mother already a memory. Marcella commented on her hair, and Zyrmina answered that it was the only gift her mother ever gave her, besides the iron collar. Zyrmina didn't see what was so funny—the creepy overseer said that all the time. But Marcella laughed and laughed. A week later she plucked the girl out of the fullery to work in her kitchen, then as servant-at-large. Zyrmina envies Hortensia's position sometimes, but the spin of Fortuna's wheel also rescued her from a life of filth and drudgery. She's loyal to Marcella for that reason if no other.

Zyrmina admits that yes, she does her own hair. Marcella grips her arm.

—I want your hair. I want something incredible, like that tower of braids the empress wears. I admit I was willing to make Valeria ill with the rue. But if I can beat her at her own game, it would be even sweeter. Help me do that. I won't forget this.

Zyrmina studies Marcella's droopy hair. How can she possibly style that into something grand? But what else can she say? She nods.

—Excellent. Here's some money for supplies. And while you're out…

She hands her some coins, then grabs the blood-spattered stola from the bed.

—Take this to the fullery to clean it. I will not forget this, Zyrmina.

Zyrmina can smell the stale piss a block away. As she approaches the fullery, in fact, she sees two municipal slaves lugging inside one of the giant ceramic pots that sit outside taverns, to collect urine for laundering. She can also hear the guild songs, the ones the fullers mutter to get through their days.

At the battered door, she reviews her plan—duck in, hand the stola off, and escape. She'll pay extra from her allowance to have the garment delivered, too, to avoid having to return. And at all costs, avoid Marcus.

It's dim inside, but Zyrmina could never forget the layout. To her right are troughs filled with urine, plus chunks of pig dung for extra cleansing power. Beneath the songs, she hears the women splashing, treading the clothes to loosen grease and dirt. After the trough wash, two fat men carry the dripping garments to wooden racks in back, dousing them with water piped from the aqueduct and twisting them in corkscrew presses to wring them. The garments dry on more racks to the left, draped over burning sulfur to bleach them. Old women then nap the surface with thistles or hedgehog pelts and sprinkle white clay on to restore the wool's gleam.

A Roman fullery. Fullers stamped garments with their feet in vats full of aged urine or other cleaning agents. (Photo courtesy of Miko Flohr.)

With each step she takes, memories gag her. Slimy dung squishing between her toes. Slipping in a trough of urine and soaking her tunic. The smell clinging to her hair at night. Indeed, as her eyes adjust, she sees that several older women have chopped their gray hair off, making them look thin and haggard. None stop splashing or singing, but their eyes never leave her.

—Zyrmina, my love!

She recoils to hear Marcus. He's as tall and scrawny as ever, with a handsome face but even fewer teeth than before. He makes passes at anything that moves, man, woman, or child. As he sidles up to Zyrmina, he slides his hand down her back.

—Why do you never wave back in the street?

—Because you're a filthy reptile. Does Marcella know you entertain soldiers here at night?

The remark hits home, and his hand withdraws. Zyrmina holds up the stola.

—I need this cleaned.

—We've no time today.

—It's Marcella's garment.

—Then Marcella's little house slave will have to get her hands dirty for once.

He turns on his heels before Zyrmina can protest. She curses herself. One grope, and she could have dumped the job on him. Now she has to clean the stola herself.

She can't stomach climbing into a trough. Luckily, there's an empty pitcher in the corner. She uses it to skim some urine from the giant tavern pot, then splashes the liquid on the blood-spattered stola. A wave of nausea overwhelms her. When it passes, she grabs some dung from a nearby pile and crumbles it on top, grinding it in with a wooden stylus.

To her relief, the blood lifts quickly, the crimson fading to pink and then disappearing. She rinses the spot with vinegar to neutralize the odor, then hurries to the back of the shop. A dash of white clay, and a little napping with a hedgehog skin, and it looks good as new. She feels oddly proud. Despite Marcus's sneers about house slaves, she can still do scut work. In fact, she parades the garment past Marcus as she leaves, although she refuses to meet his eyes.

Back outside, after several deep breaths, she seeks out hair supplies. She visits the shop of the age-spotted old woman who peddles fish grease, then grabs black wool thread to sew things up. She arrives back home feeling grateful. She despises working for Marcella sometimes, but her visit to the fullery has reminded her, viscerally, how awful things were before.

Still, Marcella doesn't make it easy to like her. When Zyrmina enters the bedroom again, her matron crinkles her nose at the smell that wafts in with her.

—Lay the stola on the bed. Now, let's get started. Here are the scissors.

Zyrmina accepts them, confused.

—I don't understand.

—They're for you. Get going.

—Did you want a trim first?

—How can you be so clever one minute and so stupid the next? Cut your hair off and start weaving it into mine.

Zyrmina feels no less stunned than when Hortensia mentioned the rue allergy.

—But—but you said you wanted me to style your hair.

—I said I *wanted* your hair. I was quite clear.

Zyrmina's collar feels tighter than ever. Her voice comes out strangled.

—It's my hair.

—Stop overreacting. I hate their game of beauty queen, but I need your hair to beat my sister.

When Zyrmina shakes her head no, the gargoyle look returns to Marcella's face. She pulls Zyrmina close and grabs her collar.

—We both want something here. I want your hair. And you want *this* gone. So I'll offer a deal. Your hair for your freedom. As soon as dinner ends tonight, I'll release you.

Marcella lifts Zyrmina's hand with the scissors, and slides the blades into her roots.

Zyrmina stares at her for a moment, then closes her eyes. Before she even realizes she's decided—*snip*—the first beautiful black tangles tumble to her dirty feet.

Just like people today, ancient Romans fretted over beauty. They crash-dieted, cinched in their bulges, padded their shortcomings, and plucked or shaved unsightly hair. They also lengthened their eyelashes, attacked wrinkles with bean meal, and whitened their teeth by rubbing ashes on them. To reach the very height of pulchritude, your eyebrows had to kiss in the center.

Clothing had its fashions, too. Men mostly wore sack-like tunics, along with loincloths beneath. Contrary to popular belief, they did not wear togas on a daily basis. Togas were essentially long blankets (up to twenty-three feet) that were wound around the body. No one could don one without a servant's help, and they were worn only on special occasions—like tuxedos, only more impractical, since you could barely bend over. Indeed, the lack of practicality was the point: wearing a toga proclaimed that you didn't do menial labor.

Roman women also wore tunics, often over underpants or corsets. On formal occasions they dressed like the Statue of Liberty, donning long-sleeved stolas and outer mantles called *pallas*. Although less swaddled than a toga, these garments were no less impractical; imagine trying to work or even walk around in Lady Liberty's getup. Footwear (for men and women) included slippers and sandals, while elite women wore high-laced, open-toed boots like the Greek goddesses they worshipped. And the whiter your garments, the better. Only slaves and laborers wore dun-colored clothing.

One major aspect of the Romans' beauty regime involved hair. They fussed with cowlicks, dyed the gray away, curled straight hair with irons, and covered bald spots with wigs. Rich households employed an ornatrix, a servant who specialized in hairdressing. This was the role Zyrmina was thrust into. So to learn the ins and outs—and risks—of her task, I took a train to Baltimore one January morning to visit the world's leading expert on Roman hair, Janet Stephens.

Stephens has a bubbly demeanor befitting her background in theater. She wears her own locks in loose gray curls highlighted pink and blue. Professionally, she styles hair in a salon four days a week, and spends her free time doing archaeology. "I read *Vogue* and *Bazaar* for work," she says, "and scholarly work for entertainment." She lives in a brick rowhouse, and upon entering, I'm ushered into her hairstyling bunker in the basement.

Beneath some klieg-level lighting, her tools are spread out, dentist-like, on a few tables: bodkins as long as knitting needles, combs of bone and horn. There are also two fingerbowls, one filled with water, the other with beef tallow—the same bright white grease that McDonald's once crisped its French fries in. In Roman times, ornatrices used tallow scented with lavender or myrrh to smooth women's hair. It also, Stephens suspects, suffocated lice.

Next to one table stands what Stephens calls her "severed heads"—two hairstyling mannequins, one blonde, one auburn. "When workmen come down and see the decapitated heads," she says, "they blanche. Now I warn them: I'm a hairdresser, dude. There are heads on stakes down here."*

Stephens's interest in archaeology began one afternoon in 2001. "I was waiting for my kid to get out of a music lesson. It was raining... and I did not want to sit in a coffee shop and get fat and caffeinated." The Walters Art Museum was nearby, so she ducked in and checked out some of the Roman busts on display.

---

* Seeing these bodiless heads, I'm reminded of a scene in the movie *Return to Oz*, when the evil queen strolls through her Versailles-like hall of heads in glass cases. This creeped me out as a child, but Stephens hoots with laughter when I mention it. She loved that scene.

Roman women sculpted their hair into elaborate mountains of curls, using secrets that experimental archaeologists have recently uncovered. (Credit: Statens Museum for Kunst, in Copenhagen.)

Most museums line their busts along a wall, like spectators at a parade. But the Walters had arranged theirs in a circle facing outward. And by peeking between them, Stephens could study the backs of their heads—the usual vantage point for a hairdresser like her. What she saw fascinated her. Several of the female busts had flamboyant hairdos with snaking braids and mountains of curls. One had what looked like "a loaf of bread" atop her head. Then and there, Stephens decided to re-create the styles at home with her mannequins.

She failed. Roman hairdressers used pins and bodkins, but even with modern tools, she simply could not get the hair to stay up. Confused but intrigued, she decided to research the matter. How did Roman women create such fancy 'dos?

Unfortunately, most scholarly papers on the topic were a mess—a hodgepodge of assumptions and projections that showed no real understanding of hair or hairstyling. So Stephens decided to investigate the original sources that the papers were based on—a daunting task given that she failed Latin in school. She nevertheless spent years typing and retyping passages into Google translate, scouring for clues.

The penny finally dropped in 2005, when Stephens read a line in a Latin dictionary that glossed the word *acus* as "the needle that the cloth-mender and hairdresser used." Suddenly, she understood. For centuries, scholars assumed that ornatrices were pinning women's hair up. In reality, they were *sewing* it. That is, after braiding the hair and twisting it into comely shapes, they would bound the braids in place with needle and thread. Armed with this insight, Stephens got to work again and found she could now re-create any Roman hairstyle she chose.

I'm visiting Stephens's bunker to learn her tricks. First thing, she fires up a playlist of mid-century French songs, heavy on Django Reinhardt and Josephine Baker. We work in parallel, I on one severed head, she on another. We start by combing out the snarls, and even though it's a dummy, I wince every time I catch one and silently apologize. Stephens flits back and forth between her dummy and mine, checking my work and encouraging me with a chirpy "Okay!" or "Great!"

Next we start braiding, making a half dozen plaits per head. On Stephens's instructions, I've spent the past week practicing with yarn. Based on this, I believe I'm perfectly cromulent at braiding, but Stephens humbles me. She can braid hair almost as quickly as you zip up a jacket; her fingers fly. My braiding looks clumsy in comparison, as if I lack opposable thumbs. Worse, while her plaits are taut and clean, like a braided leather belt, mine look as frayed as a rope bridge over a gorge in an Indiana Jones movie.

 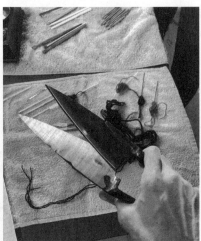

A hairstyling dummy with crooked braids (*left*), courtesy of yours truly. Notice the black thread to bind the hair. A pair of Roman scissors (*right*), known as a forfex. (Copyrights: Sam Kean.)

That's not my only shortcoming. Some strange instinct to look the dummy in the eye keeps pulling me toward the front of the head, instead of staying behind it like a proper hairdresser. As a result, the braids I make—which are supposed to be symmetric around the scalp—are laughably cattywampus, as if the hair went cross-eyed.

Stephens comes over to inspect. "Okay! Great!" She assures me that, however crooked and loose, my two-foot-long braids are serviceable. We can move on to sewing.

First we gather a few different braids and sew them together into one giant mega-braid, using a needle and wool thread. Normally, an ornatrix would match the thread's color to the woman's hair to make the thread invisible, but I'm using black yarn on the auburn dummy so I can see what I'm doing. With everything sewn together, Stephens has me twist the joint braid into a nautilus spiral, a popular style circa AD 250. We then affix the spiral to the scalp with more thread.

Finally, we tie the yarn off and reach for a pair of Roman scissors—a so-called forfex. It looks like giant tweezers with kitchen knives welded on. You cut by squeezing the two sides together, and instead of the smooth snip of a modern scissors, a forfex snaps like toenail clippers. We snip off the excess lengths of thread, and bingo—Roman hair. At this point, an ornatrix would ask her matron to shake her head and see whether things felt sturdy. I take my dummy's silence as tacit approval.

After breaking for a suitably Mediterranean lunch (pasta, chickpeas, toast with cheese and olive oil), we attempt a second, more complicated hairstyle—the tower. We switch dummies, and after undoing blondie's braids and combing out every snarl (sorry!), I part her hair into fourths using her ears and nose as landmarks. Then I twist each quadrant around a bodkin; her head now looks like a pincushion. I braid eight new plaits, using the beef tallow to smooth the hair and fight the static electricity in the dry winter air. My fingers now feel slimy, and by the end of the eighth braid, my hand muscles are tired and twitchy.

Not surprisingly, Stephens has finished her braids a half hour before me, and plays on her phone until I say I'm finally done. ("Okay! Great!") We then sew the braids together and layer them onto the crown of the head. Bit by bit, we build up height in rings, like a woven basket. Soon, we have a tower. After affixing all this to the scalp with more thread, I step back and examine my work.

It probably wouldn't pass a building inspection. And I don't believe Stephens when she coos, "Oh, I like yours better!" But all in all, it's not bad for some French fry grease and thread. Despite my clumsiness, the hair is undeniably stylish.*

Of course, not everyone's hair responded equally well to such attention. The tallest, most fashionable Roman 'dos worked best with voluminous, full-body hair. Women with limp hair had to work much harder, or suffer in silence. Unless, like Marcella, they had a better source of hair to commandeer…

On the walk to the villa, Zyrmina keeps telling herself that it will grow back, that she hasn't lost her mother's hair forever. She thinks about her freedom, too. But the humiliation stings. She looks like a fuller again, and she can sense people staring at her exposed collar. No matter how many times she blinks one away, another tear forces its way out.

Marcella, meanwhile, looks radiant. Her mask of makeup gleams; her stola and palla glisten; her golden urchin earrings sparkle. When another tremor shakes the ground, she barely breaks stride. And above it all shines her new hair. It took Zyrmina an hour to shear her own locks, sew them onto Marcella's scalp, and pile it all up. The result might have been carved on a statue.

Guards lead Marcella and Zyrmina inside the villa, where they change into felt slippers. The forecourt is decorated with ivory and gold leaf, with mosaic tiling underfoot. A doorway leads to a lush central

---

* I later have occasion to test my hairdressing skills, when my mother slips and shatters her shoulder while shoveling snow. She needs surgery, and with her right arm useless, I have to style her hair before the operation, so she looks presentable. I wish I could say that the lessons Stephens taught me translated to a brilliant coif. But my mother's pained smile when I finish shows that I have a ways to go. I tried, Mom.

garden with the town's only private fishpond. Murals of Zodiac animals cover the walls. The guards steer them toward the dining room. At the threshold, Marcella hesitates. But Zyrmina has to credit her. After a deep breath, her matron tosses her shoulders back and strides right in.

The entrance works. Valeria is telling a story inside, but her words stumble when she sees her sister. And all the priestesses crowded around her leave her side to gawk at Marcella's—Zyrmina's—hair. Zyrmina has been dubious about Marcella's chances in the election until now, but she can finally see it. For the first time, Marcella *looks* like a priestess, and in Roman politics, that goes a long way. Fortuna's wheel has turned again.

Zyrmina has never worked in such a kitchen so outlandishly outfitted. There's a grain mill, a wine press, even a brazier full of snow carried down from the nearby volcano to cool the wine cordials. But Zyrmina can't enjoy herself. They're cooking for nine people, a lot of work. She also scans around for the mint-rue sauce, but doesn't see it; hopefully, Hortensia's forgotten.

Hortensia keeps up a steady monologue while working, narrating every task aloud. *Now I'll chop the fat off. Then I'll clean the celery. That needs a splash of garum.* Zyrmina politely asks her to stop. Hortensia laughs, calls herself a birdbrain, then pivots to Zyrmina's hair. Does she miss it? Is her neck cold now? Zyrmina wills her to start talking celery again.

But she has to hand it to Hortensia. However simple, the girl's a marvel in the kitchen. In the time it takes Zyrmina to steam some whitefish and flake it into bowls, Hortensia has basted the pig; wrapped the lamb kidneys in caul fat and started them baking; chopped up a mountain of herbs; beaten a dozen eggs with a whisk of twigs; and spiced the honey for the roasted dormice. Sooner than expected, they're ready to show off the first course.

Zyrmina arranges a circle of honeyed dormice on a platter, and Hortensia plops the small suckling pig in the center. Then they hoist the platter and make their way to the dining room, where they find the nine women reclining on three couches arranged in a U. Low tables sit between them with napkins and fingerbowls of water. Zyrmina glances at Marcella, who looks pleased—things must be going well. When the pig and dormice appear, the women snap their fingers and wave the flaps of their stolas. Hortensia flushes at the applause. She and Zyrmina are just about to return to the kitchen to carve the pig when Valeria calls out.

—Zyrmina, what happened to you?

Zyrmina's knees go wobbly; she's grateful that Hortensia takes the platter.

—I cut my hair.

—Not your hair, dear. We all know what bird's nest *that* ended up in.

Valeria pauses to let everyone titter. Zyrmina sees Marcella squirming as Valeria continues.

—I mean that frightful rash. Come here.

As Zyrmina approaches, Valeria produces a jar of ointment from a fold in her stola. She dabs it on Zyrmina's face.

—This might sting. But it's a marvel. Livia gave me some.

She lists the ingredients: goat marrow, snail ashes, ants, crocodile intestines. She declares it a miracle cream. But Zyrmina suspects she's saying all this less to highlight the balm and more to remind everyone of her connection to Livia—the emperor's wife. Valeria goes on.

—Marcella, I'll do you next. It will clear up those blotches on your forehead. No, I insist. I can get more from Livia. And we don't want your skin distracting from that new hair of yours. You can't put a dollop of cream on a scabby apple.

The tittering of the other women breaks into open laughter. Behind Valeria, Zyrmina can see Marcella's cheeks blaze red through her white-marl makeup. Zyrmina braces for an eruption.

Instead, Marcella forces a smile, then quietly summons Hortensia over. Amid the general chatter that's started, Zyrmina can't hear what Marcella is saying beyond a few words. But they're enough.

—Bring the appetizer... there's a sauce...

As Hortensia leaves, Marcella turns toward Zyrmina, her gaze defiant. And suddenly Zyrmina knows—irrevocably—that Marcella lied to her. Valeria isn't merely allergic to rue. She'll die from it.

Valeria continues dabbing cream.

—You're turning bright red, child. Are you allergic to this?

—I should help Hortensia.

—Nonsense, she's fine. I still need to do your neck anyway. It's terrible beneath that collar.

Before Zyrmina can break off, Hortensia returns with a tray of bread and a familiar clay jar. Zyrmina's throat clenches as each priestess dips a crust into the dark liquid, then moans with pleasure at the flavor. Her rue sauce is a hit.

Finally, Hortensia approaches Valeria.

—I can't. My fingers are covered in cream.

The other women protest.

—One bite! You have to!

A plump brunette priestess jumps up to serve her. Valeria coyly opens her mouth. The priestess dunks the bread until it's dripping. Zyrmina uses the little splashes as a mental countdown. Three, two, one...

—Oh, that stings!

She's not lying; the cream does sting her neck. But not enough to make her yelp. And certainly not enough to make her fling her arms out and topple the tray. She does both anyway.

The mess ends up on poor Hortensia's tunic; she seems ready to cry. Zyrmina apologizes as she scoops up the spilled bread and pottery shards, then hurries to the kitchen. She tosses everything down and looks about wildly for the second jar of sauce. *There*. It's sitting out now.

Before she can grab it, someone grabs her, yanking her backward by her shorn hair. Marcella hisses into her ear.

—Get back out there with a jar of sauce, or you can kiss your freedom goodbye.

—I'll tell people. Ouch!

—One word, and it's back to the fullery.

Then her matron shoves her and stomps out.

Alone now, Zyrmina rubs her stinging scalp. The pain fades, but Marcella's threat keeps echoing in her mind. *The fullery*. Piss-soaked clothing, Marcus's busy hands. Who is Valeria to her anyway? One small deed, and Zyrmina will be free.

But she can't. She walks over to the second jar, and cups it in her hands. Then she turns toward the window, gazing at the volcano. *Where would I run right now?* She indulges the fantasy one last time. Before she's ready, before she'll ever be ready, she releases her grip, and watches the jar drop toward the floor.

It makes the loudest bang she's ever heard. Or *felt*, rather — a blow to the chest that staggers her, knocking her into the wall. In her stupor, it takes several seconds to realize that a fallen jar could never make a noise that loud.

She turns back to the window and freezes. Outside, it's the same old familiar view — the coves of trees, the terraced vineyards — except for one awful difference.

The top of the volcano is missing. And in its place, a giant plume of smoke is spewing toward the sky.

When people think of Italian food today, they think tomatoes and pasta. Ancient Romans ate neither. Tomatoes originated in South America, so no European ever tried one before about 1500. And while the Romans did fry up sheets of flour paste, pasta as we know it didn't exist until the late Middle Ages. Despite their dominance today, then, tomatoes and pasta are not traditional foods on The Boot.

To taste authentic Roman cuisine, I had but one choice, and it wasn't even in Italy: the Buckland Club of Birmingham, England. It's a dining society named for the outrageous father-son duo of William and Frank Buckland, nineteenth-century British biologists known for their passionate zoophagy.* The Buckland Club has kept their spirit alive since

---

* The eating of animals. Buckland *père* once commented, "The taste of mole was the most repulsive I knew, until I tasted a bluebottle [fly]." Frank had a standing agreement with the London Zoo that he got a shank of whatever animals died there.

1952 by hosting black-tie dinners to sample outré food. Past meals have included badger ham (Romani cuisine), kangaroo-tail soup (Australian), snakes and locusts (biblical), and squirrel-hazelnut bon-bons (Elvis, baby). Some meals proved more successful than others. Many people hated 2013's horse d'oeuvres.

The club's Roman feast took place at a stately, oak-paneled clubhouse attached to a golf course. When I arrive, men in homely sweaters are sitting around in plush chairs, sipping warm beer. The evening's food adviser is Sally Grainger, a short, lively chef and food historian with a head of curls. She rolls in at four o'clock bouncing with excitement and sits me down to explain her path to Roman cuisine.

In her early thirties, Grainger was working as a pastry chef at some private clubs in London and studying classical history at uni. She heard about some classmates throwing "toga dinners" that likely resembled *Animal House* more than *I, Claudius*. "And I thought, 'Well, let's do it right,'" she remembers. She researched some ancient recipes, and although she was essentially winging it, "The dinner came off nicely."

Eventually Grainger became a Roman reenactor, schlepping a portable kitchen to fairs and festivals. She'd bang a drum to summon people for food, which she spiced with cumin, coriander, and pepper, a blend not unlike curry. (Romans also used cloves, cinnamon, mace, and ginger.) She found the work fun, but acquired a serious purpose, too—to fight the widespread canard that Roman food was disgusting. Even classicists share this prejudice; they're constantly banging on about how bland or gross or vile every morsel was back then. But to Grainger, this accusation made no sense. "The Romans were sophisticated in so many areas," she says. "Why would they eat rubbish?" She finally tracked the slander down to some historians who'd re-created Roman dishes and found them revolting. But reading between the lines as a chef, Grainger realized the scholars had no idea how to cook. Most surviving Roman recipes don't include quantities, just ingredients, and she sensed that the historians hadn't properly balanced the acid, salt, sweet, and bitter tastes crucial to Roman food.

This ignorance proved especially glaring with the most ubiquitous aspect of Roman cuisine—a fish sauce called garum. Lowbrow television shows and websites, the type of places that reduce Roman culture to orgies and gladiator fights, often describe garum as rotten fish sauce, as if the Romans were guzzling putrid herring juice. Not at all. Garum does consist of decomposing fish, but fish decomposing under controlled conditions. To make garum, Roman chefs layered mackerel, sardines, or anchovies into jars along with gobs of salt to stop the growth of microbes. Over a few months, the fish flesh dissolved into a brown liquid. Admittedly, a crock of garum mid-brew can look gnarly—a jumble of bones, scales, and eyeballs, like something a whale puked up. But the thick liquid garum that collects on the bottom is bursting with umami, the rich protein taste that makes food seem savory. Color-wise, garum runs the gamut from honey-hued to black. Flavor-wise, it resembles Thai fish sauce, albeit less salty. The Romans used garum in pretty much everything they ate. Much like with wine nowadays, you could pay a premium for artisanal varieties or visit the corner store for a jug of cheap plonk.

The infamous Roman fish sauce, garum (*left*). The Romans doused everything they ate in garum, including desserts, and produced the sauce in industrial quantities (*right*). (Left copyright: Armand Maréchal. Right credit: M. Rais, Wikimedia Commons.)

As our interview winds up in the clubhouse, Grainger shows me her dark green bottle of homemade garum, rendered from mackerel she caught herself. Then she whisks me into the clubhouse kitchen, which is already thrumming for tonight's feast.

It's a typical restaurant kitchen with stainless-steel counters, walk-in fridges, and racks of knives and utensils. The crew faces a significant challenge tonight. Eighty people are dining, and there are ten courses total. And while Grainger is the food adviser, the man in charge of actually making the meal is Chris Haynes, a tall, round-faced chef with glasses and spiky hair. He's worked at Michelin-star restaurants in Italy, but he looks harried tonight, and no wonder: it's his tenth straight day on the job. Moreover, all his fancy experience means squat right now, given how little Roman cuisine resembled the dishes he's spent his life crafting. Indeed, as he's flaking steamed bass into some soufflé ramekins, I ask Haynes whether he's ever made anything like tonight's menu. "Nothing," he admits.

Grainger pipes up, "You have to do well. This is your only shot. It reflects badly on you otherwise."

Haynes doesn't seem to want this truth spoken aloud. His face cringes into a smile. Tomorrow happens to be his fortieth birthday, and he already looks ready for his first drink.

Over the next two hours, I shadow Grainger as she helps prep some dishes. First up is *patina*—flaked sea bass in a batter of eggs, wine, garum, olive oil, and fish "liquor" (drippings from the bass). If done properly, the bass sits at the bottom, and the batter coalesces on top like savory custard—a sort of fish brûlée.

Worryingly for Haynes, Grainger can't recall the exact number of eggs needed. Thirty? Fifty? This isn't a minor detail. Too few eggs, and the savory custard will turn into soup; too many, and it will bake into a dry omelet. Grainger plunges ahead anyway. She dumps the ingredients into a tub, and starts beating them with the biggest whisk I've ever seen; its head is the size of a cabbage. As she works, a smile lights up her face—she misses big kitchens. To test its flavor, she spoons up some batter, and the taste makes her hit an opera note of ecstasy: *Aaaah*. I try a spoonful, too. It's so rich I start coughing. But it is delicious, like savory hollandaise. We decant some into a ramekin with flaked bass and pop it into the oven to see how well it sets.

Not well. It comes out soupy. Grainger grins impishly at me: "Sally's blown it!" Haynes looks dyspeptic, and disappears to find more eggs. The only ones on hand are premixed in a carton. Grainger glugs some in, then adds more garum. She thinks it's right now. We'll see.

Next Grainger pulls me into the kitchen's back room to work on appetizers. The room smells deliciously of tonight's dessert, poached peaches and plums in mulled wine; a chef in the corner tends the pot. Meanwhile, Grainger chops up rue for an olive tapenade. It's a sweet-bitter herb that the Romans imported from modern Croatia. As I pop some into my mouth, Grainger mentions that it causes near-fatal allergic reactions in a friend of hers; the woman cannot even be in the same room as the stuff. I swallow and pray.

The rue chopped, Grainger switches dishes, dicing up some mint to add to an appetizer of wine-garum sauce with pepper and honey; the Romans dipped melons in. Grainger keeps muttering that the sauce needs more kick, but no matter how much garum she adds, it falls flat. She finally raises a puckish eyebrow and dumps the last few inches from her bottle into the bowl. It's simply got to have enough kick.

At this point Haynes materializes. "Is that the last garum?"

Grainger nods. "We don't need more, do we?"

Haynes swallows. "My leeks?" He's referring to a cabbage-leek-olive salad on the menu. The dessert chef calls over as well. "Don't the poached peaches need garum?" Yes, they do; the Romans used fish flavoring even in desserts.

Grainger looks at me wide-eyed and laughs. "Fish sauce!" she declares, and grabs some from the pantry. She explains to the chefs that they need to be "gentle" and dilute it, since fish sauce is saltier than garum and could ruin the dishes. Haynes nods wearily. I sincerely hope he makes it to forty.

At 6:45 p.m., I duck into the clubhouse bathroom to don my tuxedo. My shirt looks crisp and my bowtie snappy; I'm hoping they'll distract from the fact that I somehow packed one black and one navy sock. I emerge to meet the gathering diners. They skew geriatric, although I do spy a few thirtysomethings. I was told to expect costumes, but while a handful of folks are wearing wreaths of laurel and bay leaves, the lack of togas disappoints. This would be just the occasion.

Our predinner cocktail is a wine cordial with saffron and honey that's Riesling-sweet. The olive tapenade with rue is circulating; I pray no one's allergic. Then I spy the mint-pepper-garum sauce, on a platter with green and orange melon balls. I spear a cantaloupe one with a toothpick and splash it in. It tastes fantastic. The salty tang of the garum blends wonderfully with the bright melon and mint. And there's plenty of kick.

At 7:30 p.m. we stream into a white-walled ballroom for the meal proper. Sadly, the clubhouse has refused to let us recline as we eat, à la the Romans; we sit at tables instead, each with a centerpiece celebrating a Roman god—coins for Pluto, lightning for Jupiter, a papier-mâché volcano for Vulcan. When we're seated, someone says grace in Latin.

Soon the first course arrives—lamb kidneys wrapped in caul fat and stuffed with coriander, pine nuts, and fennel seeds. It's a dish for Roman aristocrats, Grainger says. "Offal is fundamentally elite in Roman society. Boring old lean meat is not." (Roman elites also scarfed down gonads, livers, and uteruses.) As Bucklanders, my fellow diners relish organ meat. I, the nominal vegetarian, do not. But Grainger claims to have "cured" people's fear of kidneys with this dish, so I dig in.

I'll note here that the popular notion of Roman diners gorging themselves with food, then vomiting to make room for more, is a myth. The Romans certainly ate multiple courses at fancy meals, but small portions of each. Which is good news for me, since I can get away with merely nibbling the kidneys. They have a urine-y tang, and the texture isn't helping: some bites are as chewy as erasers, some crumble like old rubber. I don't fear kidneys, but we're not on the best terms.

Twenty minutes later\* comes the fish brûlée that Grainger couldn't

---

\* To save space, I'm skipping over a few dishes, like the *sala cattabia*, a hillock of savory sourdough garnished with pecorino cheese and onions and drizzled with a sauce recall the recipe for. It's served in ramekins, with a fried scallop atop. At a glance the custard looks perfect. As for the taste, well—it's more than perfect. The bass is delicate and flavorful, and the custard is one of the most delicious things I've ever eaten. It liquefies in my mouth into a scrumptious umami goo, and I find myself taking smaller and smaller bites so it lasts longer. In general, Grainger says, the less you chewed a dish in Rome, the higher its status; coarse, tough food was for peasants. By this measure, fish brûlée must have been the summa of Roman dining.

The next three dishes include the cabbage-leek-olive salad; pucks of mashed peas with ginger and coriander; and the roast loin of a wild boar that someone in the club spent three days shivering in a blind to shoot. I adore the pea puck—the ginger adds a wonderful zip—and while I find the wild boar tough, I savor the thick, grainy cumin gravy it's slathered in.

While we dine, a few Bucklanders stand and give speeches peppered with wisecracks. ("I hope you don't feel that one boar is being followed by another.") These include remarks on the wine. Unlike the Egyptians, most Romans despised beer as swill and upheld wine as the drink of the sophisticated. They flavored their wine with fenugreek, irises, vinegar, and even seawater, and sealed wine vessels with pitch, which sometimes leeched inside and imparted a smoky, resinous flavor. We also know that Romans diluted their wine 1:2 with water.

The Buckland Club does *not* follow this last precept; the wine flows freely. And as Grainger gets a bit tight, she launches into a story about an international smuggling ring she's set up. Perhaps the most prized seasoning in Roman times was the white resin of an unidentified plant called silphium, which apparently tasted like rotten garlic. (*De gustibus…*) High demand in ancient times led to the plant's extinction—or so people thought. Botanists have recently discovered, possibly, some lingering silphium in Turkey. Six months before the Buckland dinner, Grainger flew there to cook some Roman dishes with it, including lamb

---

of ginger, coriander, mint, and stewed raisins. You can see a full menu on my website.

stew with prunes. The work left her hungry to experiment more. But the Turkish government has clamped down on exporting silphium. So, as Grainger explains to the table, she's working with a contact there to smuggle some out in boxes of Turkish Delight. There's a very Cold War feel to the caper.*

Dessert arrives around ten o'clock, the peaches and plums with mascarpone in mulled wine—and fish sauce. As we dig in, there's a hue and cry for the chefs, and a haggard but happy Haynes emerges from the kitchen to take a bow, along with his assistants. Everyone agrees this meal will rank high in the annals of the Buckland Club.

To my delight, as things break up for the night, Grainger's husband Chris produces a toga and demonstrates how to dress someone. The process is every bit as intricate and ridiculous as I've heard—over and under and do-si-do, like wrapping a mummy. As with Marcella's Statue of Liberty garb, togas were garments for leisure, for idle lounging. Not for running through the streets during a disaster...

Zyrmina rues not fleeing when she had the chance. She's spent years fantasizing about escape, calculating routes. Collar or not, she should have at least tried—perhaps ducked into a blacksmith shop to remove it in all the confusion. Instead, she just stares dumbly as the blue sky fills

---

* I later ask Grainger about her dream dishes—what Roman food she'd cook if ingredients and money were no obstacle. She mentions one recipe with cow udders. But topping her list is dormice, an aristocratic delicacy. These aren't the critters from the tea party in *Alice in Wonderland* but a bigger, furrier, continental species that the Romans fattened in special jars. Grainger actually has four dormice at home in her freezer, and is waiting for the right dinner party to whip them up.

Because dormice are a protected species, the Mafia sometimes traffics them. Indeed, according to the BBC, "godfathers...make their most important decisions in front of a plate of dormice," and in 2021, police in southern Italy seized 235 illegally trapped ones. The family involved is best known for smuggling cocaine.

People in other parts of the world dined on mice, too. There's even a restaurant in China today that serves mouse steak, mouse bacon, and other cuts. Given how much time and precision it takes to butcher a mouse, it's expensive fare.

with smoke. And when Marcella barges into the kitchen and orders Zyrmina to run home with her, she nods meekly and follows. She hates to think how deeply the slave mentality has infected her.

The streets on the dash back are mobbed with people, most of them fleeing town with bundles in their arms. Marcella, however, is determined to protect her home against looters, and she and Zyrmina fight their way against the tide. The smell of sulfur and sweaty bodies is sickening.

An hour after the explosion, the ringing in her ears finally dulls. She's huddled in Marcella's forecourt along with her ranting matron.

—Did you see Valeria flee? Meanwhile, who stayed to protect the town? Me.

It's getting dark outside from smoke, and six inches of white ash have accumulated like snow. The ash is interspersed with streaks of black rain—bits of molten rock—as well as a hail of pumice. The two guards out front—Marcella promised them freedom if they stayed, collars if they fled—are huddled in the doorway to shield themselves. Marcella keeps pacing in front of the gilded statue of Fortuna, lost to everything but her own obsession.

—There's no way she can win the election now. I'll crush her.

A noise interrupts her. A fist-sized chunk of pumice crashes through the roof. Rotten tiles crumble around it, and hundreds of pounds of ash pour in like hot flour. Then more tiles crash down, and several ceiling columns begin wobbling like drunkards. The one thought running through Zyrmina's mind is that she can't die with her collar on, still a slave. She turns to run, but notices Marcella trying to lift the statue of Fortuna. Zyrmina screams to leave it. But the determined Marcella heaves it up and staggers out.

Within seconds of their escape, the forecourt collapses into rubble. The guards are nowhere in sight. Standing exposed in the street, Zyrmina feels the sting of molten rock pelting her skin. Several homes nearby are on fire. She runs a few steps one direction, then another, desperate to find shelter.

Then she sees it—Fortuna's temple, looming over the rooftops. She considers just taking off. But the thought of being alone right now frightens her; better Marcella than no one. She might need her help reaching the temple anyway. So against her better judgment, she points. Marcella nods, and gestures to the statue.

—Help me carry this to the temple. Her luck can protect us.
—No.
—Help me, or it's back to the fullery.
—There's not going to *be* any more fullery. The whole town's burning. Let's go.

To Zyrmina's exasperation, Marcella bends down and picks it up anyway. Zyrmina can hardly believe her stupidity, but won't waste time arguing.

They weave through the ash-filled streets, several of which are dead-ended by fire. The hail of hot rock has intensified, sending sparks up wherever it lands. There are no other people around, but Zyrmina sees ovens with bread still baking, half-eaten drumsticks, abandoned dice games. The burning fullery smells more pungent than ever.

Swathed in her stola, her arms full, Marcella can barely keep up. Zyrmina keeps stopping and waiting, and hating herself for waiting. But she does.

As they turn one corner, they both trip on a crosswalk stone concealed in ash. Zyrmina catches herself, but Marcella stumbles and drops the statue. It shatters into a hundred pieces. She begins crawling on her hands and knees, searching for fragments. Zyrmina tries yanking her to her feet.

—Leave it!

Just then, she hears an ominous groan. She barely has time to glance up before the flaming wall of a nearby storefront buckles and comes crashing down.

They're showered with fire. Several embers get lodged beneath Zyrmina's collar, searing her skin. She claws at her neck like a dog until they drop out. Then she hears screaming, and looks up to see Marcella's mountain of hair—*her* hair—roaring in a blaze.

Marcella is swatting desperately at it. But fish grease burns hot, and she succeeds only in dislodging the wig. It slumps down and swallows her face like a hood of flame.

Zyrmina tries. She grabs a half gourd of wine lying on the ground and douses her matron. When that fails, she beats the flames with her own hands. But the grease keeps sizzling and crackling, until Marcella stops moving. All that remains of her head is a beehive of black slag. Her body is already half-buried in ash.

Zyrmina doesn't remember the sprint to the temple. But upon arriving she notices that the facade has already been stripped of gold; Marcella wasn't wrong about looters. She takes the front steps two at a time, and hurls herself into the farthest corner.

As she catches her breath, her eyes adjust to the darkness. She realizes she's sitting in the abandoned workspace of a blacksmith. This time, she doesn't hesitate. She crawls over and starts going through the baskets of tools, tossing aside the hammers and chisels until she finds what she's looking for.

She jams one blade of the thick forfex shears beneath her collar and squeezes. The pain leaves her gasping. The chafing rash, the burn wound, the choking pressure against her windpipe are too much. But she braces herself and tries again.

It takes several attempts. Her lips bleed from biting down. But at last, the twisted metal collar clangs onto the floor. She takes several deep gulps of air despite the smoke, and runs her fingers along her neck. The burned and blistered—but bare—skin feels wonderful.

With her hair shorn, with the collar gone, no one will recognize her or realize she's a slave. All she has to do now is escape the fury of the volcano. She hurries to the entrance of the temple and studies the smoky sky.

Time for one more spin of Fortuna's wheel. She rubs her neck again, takes a deep breath, and runs.

# CALIFORNIA — AD 500S

~~~

Life in ancient Rome would only grow more tumultuous over the next few centuries, until the empire ultimately collapsed. But the rest of the world took little notice. Kingdoms in Africa, India, East Asia, and the Middle East continued to flourish, and people in Polynesia (after a long, mysterious pause in their explorations) began pushing outward to even more remote islands. Cultures in the Americas thrived as well, in part due to the spread of technologies like the bow and arrow. In short, there was a lot going on across the globe, and the remaining chapters of this book will map parallel trends on different continents and build toward the looming clash between Old World and New World civilizations.

This chapter focuses on life among the natives of Northern California, especially tribes like the Maidu who cultivated acorns as a staple food. Given that acorns grow on trees, people there didn't farm in the traditional Eurasian sense of sowing seeds and plowing. Still, it's incorrect to say they didn't practice agriculture, since they used methods like pruning and clearing land (often with fire) to fertilize soil and give favored species a boost. Indeed, there was such an abundance of food that some California tribes, although usually classified by scholars as hunter-gatherers, had population densities approaching traditional agricultural societies, nearly three people per square mile.

Given their investment in acorn crops, many tribes in California staked claims on the most productive oak groves — a fact that surprises some people, who grew up hearing that Native Americans never owned any land and perhaps didn't even understand the concept of land ownership. Not true. Their ideas differed from European and Asian notions, but as one anthropologist noted, "There appears to have been considerable ownership of oak groves and/or individual oak trees in California." Many societies also punished thieves in various ways—slicing their earlobes off, thrashing them with thorny branches—which makes sense only if some form of property rights existed.

Keep in mind, too, that mutilations and whippings were the mild punishments, intended for people from your own village or local band. If an outsider attempted to poach or pilfer, the consequences could be far more deadly...

Nadu awakens to a stabbing pain—then jumps, wide-eyed, to see his wife Hembem hovering over him. She's holding a three-inch-long bone needle, her fingers smeared red.

She smiles slyly at his fright.

—Who did you think I was? Rattler?

He chuckles and drops his head back onto a goose-feather blanket, letting his heart rate wind down to normal. Dawn light is streaming between the pine branches that they propped up for shelter here on their journey. Hembem is giving him a tattoo on his chest. She's been working all night to finish, and he must have drifted off. Only a prick on his scar woke him.

He examines her progress. She's tattooing a series of flying geese, using a sharpened bird bone to jab holes in his skin, then rubbing the red and black ink in with her fingers. The red comes from ochre, the black from roasted wild nutmeg—he could smell it in his dreams.

Both of them, of course, already have several tattoos: She, seven vertical lines on her chin, plus diagonal stripes on her cheeks and breasts; he, a simple, thick black line between his eyes, like an extra lock of his long black hair. But Hembem's clan adores elaborate tattoos, and she's hoping this one will help Nadu fit in when they arrive. He's not optimistic.

He's certainly earning this tattoo, especially around the scar. It's a relic from a tussle with Rattler, and is quite sensitive. Once or twice when he and Hembem were making love, she rubbed a finger across it. He wondered what she was thinking at such moments—whether she has regrets—but she never spoke, never even raised her eyes.

She brushes aside his bear-claw necklace and keeps working. She's wearing a fetching basket hat and lovely tule skirt, but she looks exhausted, with black raccoon rings beneath her eyes. It isn't just the long nights tattooing. Her clan denies pregnant women meat and fish, believing it weakens the child. He's tried to get her to eat some anyway, but she refuses.

He studies her belly now, imagining what the mound will look like. The thought both thrills and terrifies him. Per custom, they were married in his village but are moving in with her parents for the first year; it's where they're journeying now. But her band will soon know, by simple arithmetic, that she got pregnant months before marriage—and as the outsider, Nadu will shoulder the shame. He doubts a few tattoos will make up for that.

Still, impulsive as ever, Nadu's attention soon shifts. Hembem's lying across him now, her bare breasts brushing his skin. And however exhausted she looks, she still inflames him. It takes only a minute. He loosens his buckskin breechcloth and tugs it aside, then arches his hips and starts jabbing. She frowns and turns to see what's poking her—then rolls her eyes and laughs.

—Again? You can't be serious.

—Can you blame me, with you? Give me a kiss.

—I've seen birds less flighty than you.

He loves even her scolding. She tries to finish a goose wing she's outlined, but he's already pulling her on top of him. She does stop him then, seriously; she needs to wrap her needle in its leather case. But with that chore complete, she gives in, and they greet the day as one.

Hembem's snoring, like a little mosquito, always delights Nadu. He snuggles into her and reaches for some breakfast—a brown acorn loaf blended with fat. He rips off a hunk with his teeth and watches the day take shape outside.

The year with her family has been looming over him for months. They'd lost Hembem's brother, a brilliant hunter, to an unlucky snakebite two years ago, and according to custom, they're no longer allowed to speak his name; it's taboo. But they haven't stopped trying to replace him—nor have they stopped reminding Nadu that he falls short. Partly because Nadu *is* short, shorter than Hembem. He's also a poor hunter, too impulsive to track well and too undisciplined to master the new technology of bows and arrows. And while everyone admits that Nadu has his charms—he's a tender lover and can spin spellbinding

## California — AD 500s

tales—that's hardly compensation for a lost hunter, a lost provider. Sometimes, Nadu honestly thinks they would have preferred that Hembem marry Rattler, despite what he had done.

True to his nature, Nadu had fallen for Hembem in a flash. Several villages were gathering for a spring flower festival, with bear-dances and hours of ecstatic music—drums, rattles, flutes. But Nadu barely heard a note, fixated as he was on the young woman with the bright green feather belt, her black braids twirling as she spun. When she passed near, he swooned to smell the spicy wormwood leaves she'd rubbed onto her skin for perfume. As the dance ended, he marched up and declared his interest. Before she could answer—she looked surprised—her father cut in and declared her too young. But over the next few years, at every gathering, Nadu sought her out, and thrilled to see her laughing at his stories. He began tailoring them to highlight parts she liked—the skittish squirrels, the lumbering oafish bears.

At last, her father declared her old enough to marry, and began soliciting suitors. After the disaster with Rattler—word of the rape spread fast—several other young men dropped out, leaving only Nadu. Even then, his hopes were not high; she could still reject him. But he'd never forgive himself for not trying.

Hembem's band has unusual nuptial rites, and the next step fell to her. One warm spring week, Nadu visited her village, and they both slept outdoors on opposite ends of a clearing. Nadu marked exactly how far away she lay down that first night. The next night, he checked the distance again—and his pulse raced to see that she'd moved closer. Still, she hadn't moved much closer, which worried him. If Hembem suddenly shifted farther away, that would signal her rejection of his courtship. And in fact, she did not move closer the third evening, a sign of wavering. Nadu stayed awake all night staring at the stars, gnawed by doubt—doubts made worse by the whispers he could hear from huts around the clearing, people peeking out and tittering. He dragged himself through the next day, and settled down to sleep in defeat. Only to see Hembem striding toward him and dropping her blanket by his side. Nadu didn't sleep that night, either—for now it was his turn to act.

When the first streaks of dawn appeared, he reached out and touched her shoulder. With that, they were engaged.

They shouldn't have gone sneaking off into the woods before their marriage; now they'll pay the price in shame. And it's not just the pregnancy that's on his mind. For the next year, he'll also be required to hunt with her father, and he can already hear the old man scolding him.

Still, with his little mosquito in his arms this morning, he's feeling braver. And he suddenly gets an idea — a gift to charm her family.

As he unwraps himself from her, she snorts and murmurs in protest; she's scared to be alone in the wild nowadays. He fibs and says he's merely heading off to pee. But once outside the hut, he waits for her to crash asleep again — it takes just seconds — then pads off.

On the way in last night, he noticed that the acorns here were already dropping — the coveted first fruits, the sweetest of the season. He's decided to gather some for Hembem's family.

To do that, he needs some baskets, so he heads to a nearby marsh to collect tule. These tall skinny bulrushes can reach ten feet in height, and aside from acorns, no other crop is so vital to Nadu's people. They make ropes and rafts and hats of tule, even decoy ducks to lure waterfowl. Above all, Nadu's people make tule baskets; they're the greatest basket-weavers in the land, fashioning everything from cup-sized trinket-holders to yard-wide behemoths to cook acorns for feasts. Most are woven tightly enough to boil water in, and they're often decorated with colored grasses, woodpecker or quail feathers, and shell beads.

In the marsh, Nadu locates a patch of yellow, leathery, sunbaked tule and uproots it. The task smarts a bit, his chest still sore from the new tattoos. Then he squats down and lets his fingers dance. While he works, he munches some fresh tule stalks as a snack; they taste slightly sweet with a soft crunch, a cross between lettuce and celery. Twenty minutes later, he places the basket underwater, anchored with a rock. Now for a second basket. While he's pulling up more stalks, he finds some sparkling aphid excrement — delicious honeydew — and greedily licks the sweet sap. Only afterward does he frown, wishing he'd saved some for Hembem.

Native Californians wove elegant baskets of all sizes (*left*). They often used the baskets to store acorns (*right*), which they ground into flour. (Credits: Library of Congress, US National Park Service, Wikimedia Commons.)

When he finishes the second basket, he checks the one underwater. The dry, thirsty fibers have swelled and expanded, plugging all the tiny holes in the weave; it's now watertight. He fills it up with marsh water and hurries off to the acorn grove.

It's situated on a little mesa, up a winding path. As he draws near, he smiles to hear the clatter of falling acorns. Still, he remains on guard. This grove belongs to a rival tribe, which tends the oaks here carefully, pruning branches and starting fires to fertilize the soil and wipe out competing fir trees. They guard this investment jealously, and have killed poachers in the past.

Nadu pauses and listens for danger. He hears only songbirds and the hammering of woodpeckers. It seems safe.

The oak boughs are sagging under the acorns' weight. He finds a long stick and swats at them, again feeling the soreness in his chest. He laughs when several drum onto his head. After sweeping them together with a leafy branch, he tests each handful by plopping them into the basket of water. Wholesome acorns sink; he discards the rotten floaters and culls the ones that are misshapen or missing their caps. This gift of first fruits has to be perfect.

When the second basket's halfway full, an animal scream tears across the mesa. Nadu freezes, his hands dripping with water. Is someone hunting nearby? If so, have they seen him?

When the second scream rings out, his confusion disappears, replaced by something darker: that's not an animal scream, it's human.

And it sounds for all the world like Hembem.

The study of ancient tattoos can be frustrating. There's overwhelming historical evidence for tattooing in nearly every culture on Earth, from the poles to the tropics. Archaeologically, though, very little remains to investigate. Ink doesn't last, and the tools for applying it—bone needles, cactus thorns, stone awls, obsidian flakes—could have been used for other purposes, so researchers have to be careful about drawing conclusions. Even worse, the medium for tattoos, human skin, rarely survives the vicissitudes of decay.

Still, there are opportunities. Many Egyptian mummies have tattoos (hieroglyphs, baboons, the evil eye), as does Ötzi, the 5,300-year-old "Iceman" who was discovered encased in a glacier in the Alps. The five dozen lines and crosses inked onto his skin are the oldest-known tattoos in history, and archaeologists had long assumed that they were applied by slicing his skin with obsidian and rubbing soot into the wounds, a primitive method befitting a primitive practice. But recent experiments have cast serious doubt on that conclusion.

The experiments were led by Aaron Deter-Wolf, an archaeologist in Tennessee who resembles a young Elvis Costello and is one of the world's leading experts on ancient tattoos. Fittingly, he sports plenty of ink himself, and has even tattooed his own arm with bone needles, which he crafted by hand from a deer skeleton he found in the woods.\*

Based on his experience, Deter-Wolf wondered whether he could look at the microscopic details of Ötzi's tattoos and determine how they

---

\* In a different experiment, Deter-Wolf explored how well different potential needles from ancient times could apply ink. The best material was bone. Feathers stunk, as did sharpened shards of mastodon tusks. Incidentally, ancient tattoo needles weren't the dainty little sewing implements you might be imagining. They were more like stilettos: they'd easily do for a murder weapon.

were applied. First, however, he needed a reference set of tattoos to compare them to. So he reached out to Daniel Riday, a soft-spoken, dreadlocked tattoo artist in New Zealand who practices the art of hand-poked tattoos—applying ink manually with needles.

To create the reference set, Riday gave himself eight tattoos using eight different tools and methods, including obsidian slice-and-rub, hand-poking with various needles (copper, steel, bird bone), and a subdermal "sewing" method from the Arctic. The design for each tattoo was the same, a series of stacked triangles over a small trunk, like a child's drawing of a Christmas tree. Riday applied all eight to his left thigh, partly because that area was easy to reach and partly because he didn't have enough room elsewhere on his body, given how heavily tatted he is. "I was running out of space," he laughs. "I was glad to find a big enough spot." The top two triangles on each tattoo were inked in completely, to test how well each method filled an area. Riday also tracked how long it took to finish each tattoo, how long they took to heal, and the amount of pain involved.

Applying all eight tattoos took twelve hours—a full, grueling day. But the time and misery were not distributed equally over all eight. Riday finished the tattoo with a modern steel needle in fifteen minutes, and barely flinched. The hand-poked copper and bone needles didn't hurt much, either. Meanwhile, the obsidian slice-and-rub method left his leg looking like he'd been attacked with a tiny vegetable grater: the cuts oozed blood, and the skin flared an angry red. Microscopic bits of obsidian also seemingly flaked off and got embedded in the wound, which dragged the healing time out for weeks. Even more excruciating was the Arctic sewing method, which he undertook with the guidance of Maya Sialuk Jacobsen, a tattoo artist from Greenland. That tattoo took three long hours of jabbing his skin over and over with a fat needle, then jerking an ink-soaked thread through. "It's more mentally and emotionally damaging" than other styles, Riday says, "because your body just really wants you to stop doing that." He described his skin at the end as "Swiss cheese."

Riday's torment paid off handsomely in scientific insight, however.

Under a microscope, the different tattoo methods produced notably different ink patterns on his skin—sometimes continuous line segments, sometimes a series of overlapping dots. Most importantly for Deter-Wolf, the line segments in the obsidian slice-and-rub tattoos showed a characteristic narrowing at the ends, where they tapered off to invisible wisps. In contrast, the lines of hand-poked tattoos remained thick throughout, with fat, blunted ends. Armed with this data, Deter-Wolf examined close-up photographs of Ötzi's line and cross tattoos. The ends were rounded, not tapered.

Deter-Wolf admits the experiment wasn't perfect: "We can't take [Riday] and encase him in ice for five thousand years," to see how weathering affects ink and skin over time. But despite the claims of traditional archaeologists, all current evidence indicates that Ötzi's tattoos were applied in the same general way as Nadu's, with the point of a hand-poked needle.

However cool these insights, merely hearing about ancient tattoos felt flat to me; I also wondered what getting one looked and especially felt like. The problem is, I'm not a tattoo guy—I've never even considered getting one. Still, given how universal tattooing was in prehistory, I realized I'd always have a gap in my understanding of life unless I sucked it up and got a hand-poked tattoo myself.

That's how I found myself in a strip mall five miles west of Disneyland. Spiritual Journey in Orange County is one of the few shops around that still does hand-poked tattoos. When I step inside, nineties hip-hop music is playing—Puff Daddy, Naughty by Nature. The walls inside are decorated with Filipino tribal masks, reed mats, and axes with carved handles.

My tattoo artist is Joseph Ash, a young man with long black hair and Filipino, Mexican, and Irish roots. He's wearing a burgundy shirt and black baseball cap, along with ink-spattered khakis. Feeling journalistic, I ask him how long he's been tattooing. "It's my first day," he says—startling me before I realize he's pulling my chain. After initially pursuing a career in welding, Ash took a job at Spiritual Journey

## CALIFORNIA — AD 500s

cleaning up the shop, and worked his way up to tattoo artist, a job he's held for five years now. He started off using a modern, electric gun, but something about hand-poking tattoos called to him, and for a while he did hand-poking exclusively.

He asks me what tattoos I have. When I confess zero, he nods. "Cool. I get to pop your tattoo cherry." To select a design, he grabs a laminated booklet and starts flipping through; he has three or four ideas in mind.

As dumb as it sounds, until this moment, I've barely given any thought to what tattoo I want. I'm more interested in the technique and the feel, not the final outcome; I'd be fine with a few lines or dots on my thigh, something casual. That's not Ash's style. He points to several options in the booklet — a frigate bird, some elaborate stars. He's especially excited about a Filipino shaman, a shrunken little wizard man. He ties all these designs back to my work as a writer. Shamans, he says, enlighten people through words, much as I (strive to) do. He spins similar tales about birds and stars, which act as wayfarers for people at sea, guiding those who are lost back to shore.

It's all very poetic. But I can feel pinpricks of fear on my skin, because I realize suddenly that I'm in way deeper than I expected. Ash gives me an especially hard sell on the shaman — which is not only far larger than I want, it's a frickin' shaman. How the hell would I explain that to anyone? I start stumbling over my words and flip back to the page of stars. Ash taps a flamboyant one with loops at the end, like a starfish wearing earrings. He says we'll plan on the shaman, but keep this one in mind as a backup. I agree to that, and scurry off to wait on a red leather couch while he preps his tools. I need a moment to think.

It turns out that bone needles are illegal in California,* so Ash has to use a metal one. He tapes the needle to a pair of chopsticks and wraps everything in adhesive tape for grip. Then he starts arranging the ink

---

* Bone needles have pores that trap blood, making them less hygienic. Riday told me that bone and metal needles feel nearly identical on the receiving end; the artist just has to poke harder with bone, since the tip is less sharp.

and other supplies. This takes a minute, which is a good thing—because I'm feeling increasingly panicked. I'm a pasty European mutt; I can't have a shaman on my leg. Unfortunately, I'm also one of those people who can't even tell a barber when I don't like a haircut. I'm having those same flustered thoughts now, except ink is a lot more permanent than a crappy hairdo. My mind starts racing. *Can I say no? Should I just bolt for the door?*

Ash finally beckons me over. I sit down in the dentist-style chair and summon all my courage. I first fake-laugh and say—haha!—that maybe I should have thought things through before coming in. But I veto the shaman. To justify this, I say that I'd feel uncomfortable appropriating Filipino culture. Ash rolls his eyes, and the shop's owner, who's been eavesdropping, chimes in: "Only white people care about appropriation." He says that as long as the design isn't tied to a specific clan or family—and the shaman isn't—no one will care.

*Shit.* If I don't think fast, I'm going to have a shriveled wizard man staring up at me every time I shower for the rest of my life. I just start talking.

"I can't do a shaman. I can't. And I'm not so hot on the star, either. It's too big and loopy. But in the booklet there, there were other stars, right? Small ones. One looked like an asterisk, like punctuation." I suddenly spy an opening. "And I use asterisks in my books all the time. For my footnotes. I wander a lot when I write, haha, and I have all these little asides that I can't bear letting go of. But with footnotes, that lets me keep those bits around. So yeah, let's go with an asterisk, because an asterisk tattoo would be really meaningful for me, you know?"

Reader, I do love me some footnotes, but this is the biggest load of steaming horse manure I've ever uttered. Ash looks dubious, not to mention disappointed: he put serious thought into my tattoo, and I'm blowing his ideas off. But for once, I stand firm. I want the teeny asterisk. Ash just nods: he's a pro, and doesn't protest. He turns his baseball cap around and begins prepping me.

First, he shaves a patch of thigh above my kneecap with a razor.

Then he draws the asterisk in pink marker and swabs the skin with alcohol and Vaseline. Now for the needle. For a hand-poked tattoo, Ash will be depositing a series of overlapping ink blots beneath my skin, similar to an old dot-matrix printer. He spreads the area with his fingers, dabs the needle tip with pigment, and digs in.

The first jab hurts, but no worse than pricking your finger with a sewing needle. I relax and settle in, thinking I've got this made. I don't. With every subsequent prick, the pain ratchets up; I start sweating, and I can barely take notes for all my squirming in the chair. I wouldn't call the experience torture — it's not *that* bad — but the natural human reaction to pain is to flinch or recoil, to pull away. I can't do that in the chair. The needle just kept digging, digging, digging into my skin, and I have to bite my lip and just take it, take it, take it. Ouch.

To distract myself, I ask Ash to describe his technique. He explains that he's using a method called flicking, in contrast to another method called stabbing. Stabbing is fairly crude: you jab the needle in straight up-and-down, at 90 degrees, and force ink beneath the skin. The problem is, the ink often gets pushed too deep, at which point it "blows out" and blurs the edges of the image. (Prison tattoos, often given by people without proper training, can look fuzzy for this reason.) Flicking is a more complicated motion, but it's easier to get the depth right. In flicking, the needle breaks the skin at a 45-degree angle. Then, instead of pulling straight back out, you flick the tip upward as you withdraw. This causes the skin to tent up a bit, like a single goose pimple. It also deposits the ink roughly an eighth of an inch down, the proper depth to prevent blowing out and keep the image sharp.

Eventually, as I'm sitting there, the pain in my thigh dulls. But every so often, one stray prick jolts me, and the writing in my notebook turns into a seismograph. Thankfully, Ash works quickly; in just ten minutes I have my asterisk. The shop's owner wanders over and nods his approval: It's crisp and sharp, he says, "as good as a machine."

Which reminds me. I'm curious how the sensation of hand-poking compares to the machines used for most tattoos. So I ask Ash to run the electric gun over my thigh with no ink, just to see what it feels like. Secretly, I'm hoping my hand-poked tattoo hurts more—that I suffered for something authentic.

I'm wrong. The tattoo gun hurts way worse. I grab my thigh and stifle a yelp; it feels like a mini lawn mower chewing my skin. Ash explains that electric guns have multipronged needles that jitter up and down and tear your skin up far more than simple pokes. That's also why modern tattoos require you to bandage them afterward and rub lotion on, to help the raw wound heal. In contrast, my hand-poked tattoo requires no aftercare. As soon as I hop out of the chair, I'm good to go.

But I'm not ready to hop out of the chair. Feeling reckless, I ask Ash, "Can I tattoo myself?" He says sure, so I don some gloves. Once again, I don't know what design I want, so Ash grabs the pink marker and draws an insipid smiley face: two dots and a curved grin. I wonder if he's mocking me—a tattoo for simpletons. "How about just the eyes," I say. For one eye, I decide to use the 45-degree flicking method, for the other the 90-degree stabbing.

The flicking method is easy to abide. Two or three jabs and I'm done. The stabbing method is another story. I keep flinching and stopping short, unable to dig the needle in far enough. I finally take a breath and just go for it. My leg jerks. I've succeeded, but just like Ash predicted, the ink went in too deep and begins bleeding sloppily outward. I'm suddenly glad I didn't opt for the full smiley face: he would have had one normal eye and one psychotically blown pupil.

By now, I'm ready to head home, happy with my tiny tattoos. But Ash looks me dead in the eye. "Want to tattoo me?"

I'm not sure I've heard him right. When he assures me I have, I feel an ethical obligation to warn him how crappy my artistic skills are; I can't even draw convincing stick figures. He shrugs. "It'll be fine." Then he motions for us to switch spots.

Ash hikes up his ink-spattered khakis. His socks have kitschy Virgin Marys on them, like votive candles. I'll be tattooing his shin, which already sports several Y-shaped insignias as well as a dozen inch-long lines, reminiscent of Ötzi. I ask him what design he'd like. "Anything you want," he says. Then he grabs my hand. "No dicks." That settled, I decide to add another line to the array.*

At the risk of sounding obvious, giving a tattoo is much more pleasurable than getting one. I feel calmer and more detached, better able to concentrate on aesthetics. I get into a rhythm of jabbing the needle and flicking the skin—snap, snap, snap. Halfway through, I recall tanning the deer hide, and have more dopey stoner thoughts about what an amazing material skin is—supple yet hard to tear, more elastic than rubber. But there's an intimacy here that tanning lacks. You're adorning someone's body, marking it permanently. Snap.

In the end, it's not a bad line. I have to fill in some gaps, and it's a touch crooked at one point. But I'm happy overall, and thank Ash profusely. I have to be one of the few people in history who's ever gotten and given their first tattoo on the same day.

Beyond the mechanics of applying tattoos, archaeologists also study what the markings *meant* to ancient people. Among Native Americans, tattoos often had metaphysical power: they warded off evil spirits, initiated people into cults, or announced a transition into a new phase of life—adulthood, motherhood, marriage. Other tattoos documented feats of bravery in war, or branded someone a slave or prostitute. Given how they clustered around his joints, Ötzi the Iceman's crosses and lines were likely medical-shamanic treatments for arthritis or another ailment. Above all, tattoos proclaimed people's identities, announcing, *We belong to this group, not that one.* They were permanent marks of belonging, marks that nothing on earth could possibly erase, save death itself...

---

* When I ask Ash about this string of tattoos, he tells me something remarkable. Tattoos were an integral part of native Alaskan culture, especially face tattoos, before missionaries and colonialists all but stamped the practice out. Recently, some native Alaskan women have tried to revive it, so Spiritual Journey hosted a seminar to teach them the ins and outs of inking. And to give the women some practice, Ash gamely hiked up his khakis and let the women use his leg as a doodle-pad. Some of the lines look pretty shaky, honestly, but Ash considers this a small price to pay for helping these women restore their heritage.

Nadu kicks over the basket of acorns as he takes off running. He skids straight down the mesa bluffs, nearly tearing his breechcloth off. Then it's a mile sprint to where he left Hemben ninety minutes ago.

Just before reaching the hut, he notices bear prints in the mud. When he turns the corner, devastation: the hut collapsed, food scattered everywhere, a circle of more prints. And inside them, a battered Hembem. Her arms are still clutching her stomach in protection.

Nadu throws himself to his knees and gathers her head. Blood has soaked through her basket-hat; it's sticky as he presses it to his chest, begging her to respond. She remains lifeless. His new tattoos are now smeared with blood; she'll never get to finish the last one.

A thunderstorm of grief wracks his body. He rocks back and forth on his knees, clasping her, as the soreness of his chest gives way to deeper, sharper pain—like someone driving a thick bone needle deep into his breast. Not knowing what else to do, he keeps calling her name, calling it out—then stops cold. He can no longer say *Hembem*. Her name is now taboo.

Still clutching her, he looks around, trying to make sense of what happened. Where did the bear come from? Why did it attack? His people identify with bears spiritually, consider them near kin. And however fearsome, bears aren't vicious—they rarely attack unprovoked. But he can see the bear prints, as well as claw marks raking her arms and breasts. They're proof enough.

Or are they? Even amid his grief, something feels wrong. The scattered food, for instance. A bear would never leave rich, fatty acorn bread behind, nor salty venison. Stranger still, he sees that his quiver of arrows is missing, along with his best bow.

Wiping his eyes, he sees the scene anew. Then he takes a closer look at his wife's wounds, steeling himself as he peels off the basket-hat and gently probes her skull. There's a sticky crater there. He's heard his uncles tell stories from the old days of the acorn wars, the raids and ambushes. One detail always stood out to him: that warriors back then often dispatched people with war clubs. Some still do, as they consider it a more personal revenge. A rogue bear could certainly have killed Hembem. But staved in her skull?

After a few more seconds, he's certain. He's not looking at a bear attack. He's looking at a murder scene.

Amid the bear tracks around the hut, Nadu can now pick out human prints. But they could be his own from yesterday. So he circles wider, searching for other signs of disturbance. He finally spots a torn spider web between two small trees, in a direction he hasn't gone. He drops to his knees and puts his cheek nearly on the dirt. The sun from that angle provides a strong contrast, and a moment later he sees it: a left footprint.

He shifts a few feet past it, drops low again, and spots a second print, a right one. A yard farther, in some grass, another print amid the morning dew. Rising to his feet, Nadu picks out a zigzag of subtle dark marks on an otherwise unbroken lawn. The trail's as clear as a stone skipped across water. He rushes back to the hut to retrieve a few broken arrows and a bow with a busted string; he'll mend it on the way. Then, with a final caress of his wife's face—*I'll return as soon as I can*, he promises—he takes off in pursuit.

So far, he's been assuming the obvious: that a war party killed her in revenge for him poaching acorns. But the farther he follows the track, the more uneasily that conclusion sits. For one thing, there's just a single set of prints—one person. And why would the other clan, however angry about poaching, kill her and not him? Like bears, they are fearsome, but they're rarely vicious.

A quarter mile later, he smells a carcass. He enters a clearing to find a dead bear. It looks scrawny, with maggots crawling out of its nose; most of its teeth have rotted out, which means it probably starved. But a few anomalous details seize Nadu's attention.

First, a missing paw—severed, no doubt, to plant false tracks at the hut. Second, the exposed leg bones and missing muscle; a close look reveals butcher marks. In Nadu's culture, certain animals are taboo to eat—coyotes, buzzards, lizards, frogs. And bears. Only someone desperate would eat one, especially gamy, rotten bear. In tandem, the clues add up to a single suspect. Nadu cannot say the man's name, not after his exile. But he thinks of him as Rattler.

Rattler grew up with Hembem. He stood a head taller than everyone else and was always the cleverest boy in the village. But he had a hole where his conscience should have been. A favorite pastime of his involved tricking animals into running off cliffs, just to watch them break bones. He earned his nickname at fifteen, when he tried baiting a rattlesnake to bite a dog. The snake bit him instead, on the face. He lost most of one cheek to the toxin, and his sunken profile made him look like a serpent.

Rattler had been pursuing Hembem since childhood, refusing to accept that she despised him. Her father nevertheless gave Rattler, an excellent hunter, permission to try marrying her. During the initial courtship night in the clearing, she slept rather close to him, surprising everyone. She moved even closer the next night—a shocking turn of events.

On the third evening, however, Hembem purposely brushed past him. *I'd rather kiss the snake who bit you,* she spat. Then she walked a mile into the woods to sleep as far away from him as possible. The whole village laughed and laughed, deepening Rattler's humiliation.

The next time she wandered off alone, to weave baskets by the marsh, he stalked and raped her.

When the crime emerged, Rattler assuaged the village elders by paying a heavy fine of deer- and bearskins and fasting for a whole month. He also claimed that her unprecedented cruelty in rejecting him had inflamed him to the point of insanity; his anger, he said, was now sated. This was a lie. After Nadu and Hembem became engaged, he trailed them into the forest during one of their dalliances and ended up stabbing Nadu with a flint blade; Hembem fended him off by clawing his facial scars with her fingernails and screaming for help. The elders banished him in absentia, and there's been no sign of him since. Until now.

In the clearing, Nadu remembers his wife begging him not to leave her alone. In a burst of rage, he kicks the dead bear in the face, watching its jaw unhinge. It's a profane thing to do, but he doesn't care. He finds Rattler's prints again and tears off down the trail. His fury focuses his mind, and he tracks like he never has before.

A mile later, the prints disappear into a grove of pine trees in a ravine; its walls pinch together in a vee. A hundred yards beyond, at the notch of the vee, he sees a black crevice high in the wall. An elevated cave. And there's a glint inside it, movement. Someone's up there.

*Thwack.*

Something streaks down and punches Nadu in the breast. He staggers back a step, gasping at the pain. He thinks he saw a feather, and wonders fleetingly if a bird attacked him—until he glances down, and sees one of his own arrows buried in the unfinished tattoo on his chest.

People worldwide have been eating acorns for thousands of years, albeit with varying degrees of enthusiasm. Some cultures regarded acorns as hog feed or famine fodder, a last resort. Others enjoyed acorns as seasonal treats only. Native North Americans, meanwhile, recognized what a nutritional bounty acorns are—crammed with vitamins, fats, and starches—and wolfed them down year-round.

Curious to try some acorn dishes myself, I began gathering them around my neighborhood, scooping them up and filling my pockets on walks. Despite the occasional side-eye from a neighbor, I enjoy the task, and find that it shifts my perspective on local trees. Instead of just a vague green canopy overhead, I begin seeing them as individuals, and moreover as resources—potential food. I also need rocks to process the acorns, so I head to a park. Finding a hammerstone is easy; I just needed something potato-sized that I can grip in one hand. Finding the anvil is trickier, since I need a specific shape: flat on one side, so it won't wobble on a countertop, but with a hollow on the other side to hold the acorn in place. After much rooting around, I excavate a suitable stone that looks like a flat, dented kidney bean.

Unlike with flintknapping, there's no precision involved in cracking acorns: you just nestle one into the anvil and whack it. But I wish I'd known beforehand about Nadu's trick for separating out rancid acorns by floating them in water, because I end up opening plenty of duds. The cause of the rot is usually acorn weevils. Adult acorn weevils actually look pretty cute, like furry, owl-eyed Snuffleupagus beetles. Unfortunately, the adults don't dine on acorns. That would be their loathsome maggot children—quarter-inch larvae with pebbly white skin and alarmingly blood-red mouths. When exposed to air, the larvae writhe like someone being electrocuted in a sleeping bag. Worse still, the maggots eat most of the meat inside the acorns they inhabit, leaving behind what's politely called "frass" but is really just stringy maggot turds.*

Eventually I learn to differentiate good acorns from bad ones by sound. Rotten acorns make a dull noise when smashed, like a soggy drum. Good acorns snap cleanly, and open to reveal a firm nut that's the same waxy yellow as a peanut. I crack roughly five hundred overall, and after discarding the foul ones, end up with three cups of acorn meat.

Next I leach the meat to remove the tannins. Tannins are common in many plants, and provide a subtle note of bitterness in tea and wine. But there's nothing subtle about the tannins in acorns. Out of curiosity, I bite into a raw one, and wow. There's a slight green apple taste, with a touch of citrus. But any discernible flavors are quickly overwhelmed by

---

* Reportedly, Northern California natives sometimes plucked the nutritious, protein-rich grubs out of their acorn flour to roast and eat them. Children were even told that nibbling them would keep their hair shiny and black (an ancient echo of my mother telling me as a child that eating spinach would put hair on my chest). After seeing dozens of these grubs wriggling inside my own acorns, I couldn't shake the thought, and God help me, after a few beers one night during the pandemic (we all did things we're not proud of during that time, right?), I cracked open a few acorns, found a maggot, and bit down. It *popped* between my teeth, and squirted juice like a tiny berry. But the flavor was surprisingly bland.

the insurrection in my mouth. My cheeks cinch up, my throat goes dry, and my tongue shrivels into a pimply pickle. It's like gnawing on sour wood pulp.

To leach those nasties out, I wrap the acorns in an old T-shirt and let the bundle steep in a kettle for several days, changing the water whenever it looks like beef broth. Nadu's people soaked them in earthen pits, or else placed baskets of acorns in streams for continuous water flow. Some people nowadays soak acorns in toilet tanks, since every flush brings a fresh rinse—a hack that strikes me as both bold and lazy.

To grind the meat into flour, I grab a mortar and pestle for making guacamole that I received as a gift once and never used. It's slow, dreary work—far less caveman-satisfying than smashing the acorns open. Worse, it reduces my three cups of meat into one measly cup of flour—and calling it "flour" is generous; it looks like bran meal. More than anything else in this book, seeing that single cup—after all the hours I'd spent collecting, cracking, sifting, soaking, and grinding acorns—drilled home just how bloody hard it was to make food way back when. I've heard archaeologists claim that the prehistoric shift from foraging to farming—from plucking food to processing it—was the dumbest mistake human beings ever made. The statement seemed like heresy the first time I heard it. Not anymore. The prospect of such Sisyphean drudgery, of having to grind your food like that day after day after day, has wised me up.

At least I've arrived at the reward stage now—cooking and eating. I attempt tiny acorn muffins, with sugar, eggs, milk, and butter. After baking an hour, they puff up beautifully—rich brown tuffets with springy tops. I bite in eagerly, expecting something like a tea-flavored bran muffin. What I get is the tannin equivalent of a ghost pepper: a shrieking klaxon of bitterness. The main problem is that, unlike with tea or wine, no other flavors rise up to balance the tannins. I choke down two muffins out of a sense of duty, but soon wish I hadn't. They leave my guts roiling, the same icky feeling you get after drinking wine on an empty stomach. I end up throwing the rest of the batch out. While staring into the garbage that night, toting up the many hours wasted, I seriously considered subtitling this book—not for the first time—*All That Work for Nothing.**

Given the importance of acorns to Nadu's people, oak groves were a frequent cause of war. In most cases, warriors attacked each other with bows and arrows. But sometimes they resorted to blunt-force "shock weapons" like war clubs.

Clubs were almost certainly the first weapons human beings ever used. Despite this, archaeologists have largely ignored them in favor of flashy projectile weapons like arrows and atlatls. But not all archaeologists are so seduced. Among them is Joe Curran, a private-sector archaeologist who was a graduate student at the University of Nevada at Las Vegas when I met him. He's a hearty, giant Viking of a man—six-foot-five with a sweep of blonde hair and ruddy beard. He also has a warm smile, and in fact did not stop grinning the entire time we chatted, even when showing me how to bash someone's brains in.

Much of the archaeological evidence for shock weapons comes from staved-in skulls, which usually show damage on the left side—exactly where a right-handed assailant would land a blow. (These skulls come largely from Southern California and the desert Southwest, although clubs found use in Northern California, too.) Curran wanted to know the shape and size of the weapons that would leave such marks, so he started experimenting with three classes of cudgels: stick clubs, sometimes called "skull-crushers," which look like sawed-off baseball bats; mallet clubs, which look like stocky bass-drum mallets; and ball clubs, which look similar to mallet clubs except with rounded ends. Note that these were crafted tools, not just knobby sticks that people picked up off the ground.

---

* The sheer yuckiness of these muffins got me wondering: Had I screwed up somehow? Or did people back then simply tolerate bitter flavors better? To answer this question, I later cooked some acorn dishes with the wilderness gourmands at the Moose Ridge School in Maine. It turns out the fault lay entirely with me. Fresh acorn pancakes cooked on a hot stone griddle are delicious.

Replicas of Native American warclubs, alongside the watermelon we bashed. (Copyright: Sam Kean.)

Curran's initial experiments involved bludgeoning a crash-test dummy that had accelerometers implanted in its noggin. His team dressed the dummy in a black T-shirt but no pants—a Winnie the Pooh getup that left its surprisingly humanoid buttocks flapping in the breeze. Despite the modest size of the clubs Curran tested—the heaviest weighed less than four pounds—the blows produced some wicked accelerations inside the dummy's head, up to 437g (i.e., 437 times the force of gravity). Most people pass out around 5g, and fighter pilots rarely pull 10. The highest recorded g-force that a person ever survived was sustained in a race car accident at 214. Curran doubled that with a blow from a stick.*

Sadly, I missed Curran's crash-test dummy experiments, so when I meet him at UNLV, we improvise. He walks me over to a tool bench in a concrete courtyard behind the archaeology building, where he unpacks

---

* For comparison's sake, Curran bashed the dummy's head with a rock a few times. It proved a much poorer weapon than a club, harder to grip and prone to stinging his hand.

several wooden truncheons from his bag. Then he holds up a straining grocery sack and grins: "We'll get set up here with some clubs, then you're going to beat some watermelons in."

Curran carved the clubs himself from mesquite logs. Each one took about six hours, and the heaviest one on hand today weighs eight pounds—although considering their intended use, they seem menacingly heavier in my hands.

To set up, we nestle the first watermelon into a bed of sand in a yellow litterbox, then set the box at chest height on the tool bench. This catches the attention of two small boys present, the sons of a UNLV student who's working on a pottery project nearby. The boys spent the previous hour bored, puttering around the courtyard on scooters and giggling about how the wet potter's clay looked like dog turds. (It really did.) But the war clubs seize their attention. The younger one—a moppy-haired lad with gaps between every tooth in his smile—asks if we're really going to smash watermelons. Curran assures him we are: "We're channeling our inner Gallagher." I'm not sure the joke lands with grade-schoolers, but they look impressed.

Unlike Gallagher, we don't swing the war clubs wildly. Curran explains that Native warriors attacked in a choreographed, two-step sequence. First, they butted their opponents in the face with the blunt end of the club to stun them. Then they grabbed them by the hair and crushed the side of the skull with a blow.

He has me start by practicing the butt. Feeling a bit silly—a dozen people around the courtyard are watching—I pick up the mallet club and mime a few strikes. Then, as ready as I'll ever be, I thrust like a fencer with a stumpy épée. The blow lands harder than I expect, producing a squishy *thud* that cracks the rind. Curran smiles and calls it a textbook butting injury. On a skull, it would have left a hairline fracture.

Because watermelons lack hair, we can't replicate the second step of the attack exactly. Instead, Curran suggests I just swing the club down as hard as I can. He demonstrates the proper form first, and his grace and balance impress me. (I later learn that he teaches martial arts in his spare time.) Then he hands the mallet to me. I back up, take a few warmup swings, and let fly.

I end up pulling my punch a bit. Yes, it's just a watermelon, but it's still the size and shape of a human head, and something inside me flinches. The blow nevertheless lands squarely, with an even thicker *squash*. Curran examines it forensically and points out several cracks spreading across the surface like lightning bolts. With another grin, he declares it an "excellent radiating fracture." Pink juice is now pooling in the sand beneath.

After a few more blows to another melon-skull, the young boys and I feast on watermelon brains in the desert sun. It's delicious. Curran then asks the lads if they want to try the clubs on the last watermelon. Boy, do they. They drop their scooters with a clatter and scamper over.

Curran places the watermelon in the litterbox, then places the box on a chair. The moppy-haired brother, Julian, grips the mallet with both hands and cocks it over his shoulder. Unlike me, he has no hesitations; he absolutely slobber-knocks it.

He looks up giddy, his wide eyes begging to know whether he can go again. Curran chuckles and says sure.

What unfolds next is basically Gallagher meets *Goodfellas*. In short, Julian beats the ever-living piss out of the melon. Eight or nine blows fall in rapid succession, until juice is splashing his clothes, spattering his face. *Splat, splat, splat, splat*. Curran—whose smile has disappeared—finally yells out, "Whoa, whoa, whoa! It's dead, Julian. It's dead."

Julian stops with a snap, looking dazed. Then he blinks, drops the club to his side, and shrugs. "It's called a war club for a reason, right?" He has us there.

As we wrap up, another student asks Curran why Native warriors bothered attacking each other with clubs after the invention of bows and arrows. After all, why risk getting your head smashed in when you can pick someone off from afar? The answer is twofold. First, arrows appeared rather late in North American history, by which point war clubs were firmly established. Second, while some tribes did use arrows to harass their enemies, killing someone with a club, in close quarters, took more courage and therefore won the warrior more honor. There's just something primal about looking your enemies in the eye before you dispatch them.

All that said, if a warrior got desperate, there was no limit to the depths he might sink for revenge...

Nadu wakes to find the sun blazing overhead. He's lying in the marsh, covered in mud, and has never been so thirsty in his life. Every breath throbs in his chest, and he can feel the arrow there rising and sinking in rhythm. It takes two attempts to even open his eyes.

After the arrow hit, Nadu escaped to the marsh here, splashing into the water to conceal his tracks and staggering onto a small, tule-covered islet to pass out. He half-expected to never wake up again; Rattler no doubt expected the same. But, having woken up, Nadu now has to attend to himself. First thing first, he grips the arrow in both hands and pulls.

A heave of blinding pain follows. He takes a few minutes simply to breathe. Then he rocks his body like a turtle, flipping onto his hands and knees, and crawls to the water's edge to drink and drink. Finally sated, he tugs his breechcloth aside and pees like an animal, letting it splash down around his legs. To staunch his oozing wound, he grabs several long stalks of tule and cinches a tuft of grass to his chest.

He assesses his situation. His desire for revenge has not diminished, but it has changed character. Before it was hot, agitated, fierce. Now it feels hard and dense—a bitter acorn of hatred in his heart. He's got to be more tactical, even devious.

He starts the process of standing, using his bow to brace himself and struggling to his feet. Being upright leaves him swaying, nauseous, but he manages to avoid pitching over. This victory secured, he uproots some more tule and weaves the crudest basket he's ever made—more of a sling. Then he grabs his bow and the arrow he freed from his chest, and limps off to find the bear carcass.

Rotten liver is one of the two ingredients in the poison Nadu's people make. The bear's liver feels slimy as he digs some out and plops it into the sling. The purple-red chunk smells sour and barely holds its shape.

Now for the other ingredient. Nadu hobbles to a nearby rock formation, which is baking in the afternoon sun. He probes a few shady nooks with his bow. Nothing. He's about to leave and try elsewhere when he feels something squish. Followed by a rattle.

Grimacing, he kneels down and sees the glint of two cat-slit eyes. It's a dusty yellow snake with fat brown patches—just the species he's looking for. He slides the bow beneath its coils and gingerly drags it out. He knows that snakes are surprisingly patient; they don't strike unless threatened. But every so often you get an ornery one.

Exposed to the sun, the snake tries to slither off. But Nadu holds it down with the bow, then rolls the tip up the length of its back, until it's pinned just below the head. This, the snake hates—it hisses several times. Maidu whispers to it, soothing it. Then, quick as a hawk, he snatches it up.

He's gripping it right below the skull, which is good and bad. The snake can't bite his hand from this position. But it can whip its body back and forth, slapping his legs and chest. His arm feels shaky, and he almost drops it twice.

With his free hand, he gropes for the sling with the bear liver. He swings it back and forth, bopping the snake on its snout. The snake flares its mouth, baring its fangs. Nadu keeps smacking it, riling it up more. Then he dangles the offending liver right between its eyes.

It happens so fast that Nadu nearly drops the sling. The snake lunges and locks on to the liver. There's a lightning pulse behind its eyes—a gush from the venom gland. A split-second later it unlocks and hisses again.

Nadu now feels dizzy in the hot sun. But one pulse of venom might not be enough. So he taunts the snake further, smacking its nose until it latches on again. After the third bite, he can see venom dripping from the holes in the liver chunk like thick, deadly honeydew.

Regrettably, he cannot let the snake go—in his weak state, it might whip around and bite him. He's got plans for the body anyway. He whispers a short prayer of forgiveness, then crushes its head with a rock.

Laying the body aside, he empties the envenomed liver onto the ground and mashes it into a paste with the head of the arrow from his chest. It needs more moisture, so he spits into it. The thick red clot in his saliva is a bad sign. But it heartens him to know that his blood will also be part of his revenge.

Bows and arrows were the single most complicated piece of technology in prehistoric times, incorporating nearly every material used by our ancestors—wood, stone, sinew, antler, resin, rope, feathers. And while atlatls long predated them, bows and arrows eventually replaced atlatls on nearly every continent. There are several reasons why.

The big advantage of atlatls is that their heavy darts pack quite a punch; you can really wallop game. But as megafauna went extinct on continent after continent, the importance of landing a big blow waned in tandem. Hunting smaller game requires stealth and precision, and bows and arrows allow you to hide in a blind and snipe at game instead of scaring them off with the big clumsy movement of an atlatl toss. You also stand still while shooting them, and can sight down an arrow and take aim, something that's impossible with the side-armed atlatl. Arrows offer a superior rate of fire, too. With atlatls, you usually get one throw before an animal flees, but it's possible to fire several arrows in quick succession. (Some Plains Indians could keep eight in the air at once.) All in all, after the decline of the megafauna, arrows proved superior in most hunting scenarios.

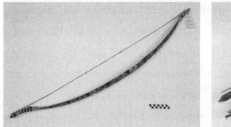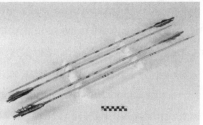

A bow and arrows used by native Californians. (Copyrights: National Museum of the American Indian, Smithsonian Institution (08/7467, 14/7440). Photo by NMAI photo services.)

But as with all technological advances, the switch to bows and arrows was accompanied by social upheaval. For one thing, bows and arrows seemingly favored individuals over groups. When everyone used big, slow atlatls and got just one shot at game, hunting in groups was necessary to hedge bets. In contrast, the precision and stealth of bows and arrows encouraged solitary hunting. The group became less important.

Arrows might also have upended the relations between the sexes. Recall from chapter 2 that women often throw atlatls better than men. Bows and arrows, however, tend to favor males. That's partly because men are generally taller and generally have more upper body strength, both of which provide an advantage when shooting bows.* (Arm length and arm strength allow an archer to use a stiffer bow and pull the string back farther, generating more snap.) That said, it wasn't all biology; cultural factors favored males as well. However clumsy atlatls seem at first, people could master them reasonably quickly; children as young as seven can take down deer with them. Proficiency with arrows takes more practice: few children can reliably kill game with arrows until their mid-teens. And for whatever reason, most cultures in the ancient world—in Africa, in Asia, in Europe, in the Americas—denied young females

---

* It's worth italicizing the words *tend to* and *generally* here. Some women are of course stronger and taller than the majority of males. But *generally*, they *tend to* be shorter and less muscular. And while there's a lot more to hunting than being tall or strong, it does help with bows.

the chance to develop this skill, shunting them off to gather plant food instead. As a result, bows and arrows became a male-dominated weapon.

Historically, hunters often combined arrows with poison. There's indirect evidence that humans have been using poison arrows for tens of thousands of years: archaeologists have found arrows in Africa so slender that, unless they were used to hunt butterflies or something, they simply would not have been effective without poison.

For mastery of poison, no one could top the hunters of the Americas. They extracted toxins from frogs, ants, scorpions, centipedes, tarantulas, and more. Nadu's people hunted with an unusual poison that combined putrid, festering animal liver with venom from the Northern Pacific rattlesnake. The symptoms of the Northern Pacific's bite can vary, but immediately afterward, you might feel tingling and burning pain. By fifteen minutes, you'll feel woozy, with chills and mild sweating. By thirty minutes you'll be gushing sweat, and the chills will flip into a 104°F fever. Soon, you won't be able to walk without help, and ninety minutes in, you'll be shaking and doubled over vomiting. After this point, the pain will only magnify, including a torturous thirst. Then again, after ninety minutes, you might feel no pain at all. You might already be dead.

The Northern Pacific rattlesnake. The native people of California smeared its venom on poisoned arrows. (Copyright: Chad Lane.)

If snake venom is so lethal, you might be wondering how people ate the game they hunted with poison arrows. Wouldn't they slump over dead after a few bites? Fortunately, no. Venoms (snake and otherwise) are cocktails of several different molecules, mostly proteins. To kill you, those proteins need to reach your bloodstream; even the ones that attack your nerves need to circulate within the blood to reach their targets. But if you merely swallow the proteins—as you do when eating poisoned game—your digestive system will treat them like any other protein and break them down with enzymes. In fact, while I don't necessarily recommend it, you can actually drink snake venom and walk away unscathed. To your body, it's just a cobra protein shake.*

All that said, if you're foolhardy enough to drink venom, take care that you don't have any cuts on your lips or inner cheeks. If even a tiny amount seeps into your bloodstream, you're a goner. The same goes for preparing and shooting arrows. Even an inadvertent prick or a tiny scratch will spell doom...

Snake and weapon in hand, Nadu slinks back toward Rattler's cave using trees for cover. It's both painstaking and painful. His wound is still seeping, and he's sinking fast from blood loss. He has to fight the temptation to lie down and sleep; he doubts he'd wake up this time.

He's restrung his bow with the sinew belt from his breechcloth. As a result, he's naked beyond his tattoos. He'll attack with the arrow extracted from his chest. The tip is broken, but it only needs to deliver the poison.

While tracking Rattler this morning, Nadu envisioned confronting him in battle. But the new, tactical Nadu has a different plan. When he

---

* Toxicologists draw a distinction between *poisons* and *venoms*. Poisons contain tiny molecules that resist digestion, so you die after consuming them. Venoms contain bulky molecules that cannot evade digestion, so consuming them isn't necessarily fatal. There's also this saying: "If you bite it and you die, it's poison. But if it bites you and you die, that's venom."

reaches the grove near the ravine, he finds a log with a small crevice and places the tip of a stick inside. Then he starts rubbing his palms back and forth around the stick, as if trying to warm his hands. Out of habit, he stops and spits on his palms to lower the chances of raising a blister — then laughs ruefully. What could a blister matter now?

An experienced fire-starter can get a blaze going in under a minute, but Nadu can work only as fast as his aching chest allows. Gradually, though, the friction of the rubbing wears off small bits of wood inside the hole and turns them into glowing embers. Using the stick, he transfers the embers to a pile of cattail kindling, then gathers some twigs and fallen branches to get a proper fire going. He's heartened to see the thick carpet of pine needles in the grove; they'll burn like mad.

He starts the blaze on the east side of the grove. Orange tongues are soon licking the trunks of trees. With the blaze established, he shuffles over to the grove's west end and slumps down against a tree to wait. By now the sun is sinking. He's glad he'll get to watch one more beautiful fire in his life.

As he waits, he contrasts the day's perfect beginning — seeing his tattoo take shape, gathering the first acorn fruits, making love to Hembem (he'll say her name now, taboo or not) — with this ugly ending. He even had fantasies of parading into her family's village and winning them over. He laughs again. Plans seldom unfold perfectly, but rarely do they go this awry. At least he can salvage some dignity now.

Before long, Nadu catches himself nodding off; the hard acorn of hate has to shake him awake. To pass the time, he watches the resinous gray smoke pour into the air. He starts to cough, and drags himself backward several yards to another tree for better air. But not too far. He still has to take aim at Rattler.

Nadu started the fire on one side of the grove to narrow the possible escape routes and force Rattler to emerge where Nadu is waiting. Sure enough, several raccoons and fat chipmunks scamper past him, even a wolverine.

At last, twenty minutes after the fire started, the outline of a human appears in the haze, doubled over and hacking. Nadu grabs the arrow slathered in venom and putrid liver, and struggles to one knee. A moment later, a familiar, scarred figure emerges from the smoke.

Nadu closes one eye to aim, then whistles, sharp and shrill. When Rattler turns, he fires.

It's a poor shot. Nadu prepared himself mentally for the pain of pulling the drawstring back, but it's fiercer than expected. His arm falters, and the arrow veers left.

But the shot's good enough. It scrapes Rattler's shoulder, opening a gash. Even at thirty yards, Nadu sees the poison spatter upon impact.

Rattler doesn't seem to grasp what's happened. He looks more surprised than anything. When he finally sees Nadu, and realizes that the arrow didn't lodge inside him, he sneers.

— You're pathetic. She deserved a real man.

Nadu tries to answer, but a fit of bloody coughing stops him. He wipes the phlegm away, and hurls the dead rattlesnake in the air. He'd fantasized about saying something biting as he threw it, but can't catch his breath now. It lands a few yards from Rattler, who once again looks confused. But his quick mind soon puts it all together. Nadu never noticed before, but the muscles around Rattler's right eye must have been paralyzed when that snake bit him long ago; he can't open it very far. The other eye, however, flares as wide as the moon.

Rattler begins sucking frantically at his wound, his cheeks heaving in and out. But the putrid liver makes him gag and he stops. He must know it's futile anyway.

Nadu slumps back against the tree, idly wondering which symptom will hit Rattler hardest — the vomiting, the dizziness, the suffocation. He supposes it doesn't matter.

He hears a howl and sees Rattler running for him through the haze. With the background of billowing fire and smoke, he looks like some avenging demon from the depths of hell. But Nadu knows that the revenge is his.

# VIKING EUROPE — AD 900S

~~~

In the past, European historians often described the fall of Rome in the late 400s as the fall of civilization itself—a laughable premise. Civilization did lose a foothold in western Eurasia, but continued to thrive elsewhere, especially in the Middle East, Central Asia, and China.

That said, Europe did suffer gravely after Rome collapsed. Sewers, roads, and aqueducts fell into disrepair. Outside the walls of monasteries, literacy plummeted, and the volume of trade around the Mediterranean dipped so low that it would not recover for 1400 years. Architecture deteriorated, too, as buildings grew short and squat. Especially after Charlemagne died in 814, most of Europe fragmented into miserable villages, with people living much as their ancestors had thousands of years earlier, eking out a living on meager farmland.

The conditions were ripe, then, for thugs to flourish. Today we call them Vikings. The word *Viking* means "pirate" or "raider," and few people in history raided with as much gusto as the Vikings of the ninth to eleventh centuries, terrorizing virtually every coastal settlement in Europe. The Vikings sailed from attack to attack in sleek warships that stretched up to 100 feet and could scoot along at speeds comparable to those of the best Polynesian vessels.

Upon reaching a settlement to pillage, the Vikings preferred blitzkrieg tactics—smash and grab—but laid siege when necessary. (To defend themselves, the besieged would hole up inside castles or stout churches and either fire arrows from the walls, or drop stones, boiling tar, red-hot lead, and scalding sand. Sometimes, though, they could only wait for winter and pray for hunger and disease to deplete the enemy.) Some Viking warriors participated in just one campaign, stealing enough treasure to buy some farmland and settle down. For others, plundering was a passion, a vocation, and they continued until the day they died.

Most of these warriors were men, but Viking poetry did mention female warriors here and there. For typical reasons, scholars long dismissed these women as fuzzy myths, but recent archaeological discoveries suggest otherwise. Most spectacularly, archaeologists unearthed a woman buried in Sweden with a sword, axe, spear, dagger, armor-penetrating arrows, and two horses, plus a board game useful for thinking through tactics on the battlefield. Chapter 2 recounted a woman in the Andes discovered with a hunting toolkit; this unknown Valkyrie had a warrior kit. (I should note, however, that neither she nor any other Viking would have worn a horned helmet. That's a trope from opera.)

The sheer range of Viking raids was astounding. One treasure hoard recently excavated in Sweden included an Irish bishop's staff, a French sword pommel, a silver dish from the Mediterranean, an Egyptian ladle, and a four-inch Buddha statue from Pakistan (possibly obtained through trade). Looking west, the Vikings sailed as far afield as North America, bouncing from Iceland to Greenland and down into Newfoundland, perhaps after observing the travels of geese or other migratory birds in the same way that ancient Polynesians did.

One popular target for raids was Christian monasteries and priories, which often housed ecclesiastical treasures. There may have been psychological reasons for attacking monasteries as well. After gaining many early converts in Southern and Eastern Europe, Christianity took longer to diffuse through the rest of the continent. But however slow, its march was corrosive. Every new Christian meant one less person upholding the venerable old pagan traditions that nourished people's souls for millennia. This loss would have given the Vikings—devotees of Thor, Freya, and Odin—extra reason to hate Christians and extra incentive to destroy them. And this struggle would have played out not just between societies, but in the hearts of individual men and women who found themselves stranded between the old ways and the new...

Ciaran the leech awakes earlier than the junior monks sprawled on the straw around him; he supposes he's less used to flea bites. They're probably also exhausted from the extra prayers they've been saying overnight, for deliverance from the Vikings. Then again, he's twice their age, and has been up all night himself tending to the prior. Their prayers aren't doing much good anyway. It's day thirty-seven of the Viking siege, and there have been no lightning bolts yet from their God.

His bones groan as he rolls his bulk into a sitting position. He feels hungover because he is. Shaggy beards of moss cling to the damp stone walls here in the priory's church where he and the junior monks are sleeping these days, and with the lower-story windows bricked up against the siege, it smells even more like unwashed bodies than normal. Ciaran pulls on his leather boots, wrapping the laces around his shins. Unlike the monks, who wear cowls, he wears brown hose and a belted tunic. He grabs his coarse cloak, and heads off to find some food.

The priory grounds consist of two separate building complexes. The first was, until recently, their living quarters—a dormitory, pilgrim guesthouse, and kitchen; they abandoned these when Viking ships appeared on the horizon. The second is the stout church and its accoutrement, including a courtyard, sacristy, and scriptorium to copy books. As Ciaran enters the cloistered grass courtyard now, the wind flattens his curly hair; it's ash-colored these days, the red mostly a memory. In the makeshift kitchen-shed, the food congealed in the cauldron turns his stomach: plain barley porridge, with the apples already picked out. No honey or butter to liven it up, either. He grabs a cow-horn drinking cup instead and heads over to the woodpile, easing onto his knees to pat around behind it. For a few panicked moments he can't find it. Then he feels the curves of the black ceramic wine jug and relaxes. He'd prefer beer—the priory once brewed a delicious dark amber, flavored with piney juniper branches—but a bloc of pious brothers here objects to anything except wine, which has biblical provenance. It nevertheless does the trick. Ciaran drinks one hornful to slake his thirst, then a second to drive his headache off. The third is to get drunk. Upon finishing, he feels warmer, brighter. After hiding the jug, he ducks back outside to pick some nettles. The greens add an earthy kick to the otherwise inedible porridge.

Although it's half-past seven, the sky remains depressingly dark. Normally he'd chase away a foul mood by exploring the caves along the coast, or trudging off for a look at the bog body. But all such larks have ended with the siege, and he's getting more and more stir-crazy by the day. Of all the annoyances of late, the monks' singing irritates him most — morning, noon, midnight. Such wretched music, too. No drums, no horns, nothing to stir the soul.

He should check on Prior Patrick in the chapel. He recrosses the weed-strewn courtyard, kicking over a few anthills, and turns right inside the church, passing beneath crude wooden carvings of saints. In the corner, he eases open the door of the sacristy, where a half dozen senior monks lie sleeping. The walls here are drier, less mossy. Rather than scratchy bare straw, they sleep on mattresses made of cowls stuffed with soft grass. Every time Ciaran enters, he swears he smells cheese; he's sure they're hoarding food.

He tiptoes in and eases open a handsome wooden cabinet; the brothers decant jugs of wine from barrels in the cellar and store them here for communion. Prior Patrick's medicine requires wine, but Ciaran figures a swig of Jesus-blood for himself won't hurt anyone. He removes the wooden stopper from a jug and tips it back. But the cabinet door groans when he closes it. He cringes to hear rustling.

—Who the devil is that? We're trying to sleep.

He turns to meet two smoldering black eyes. Brother Andrew is even more rotund than Ciaran, with a thick wooden necklace of prayer beads resting on his belly. As a leech, a doctor, Ciaran notices that the rash on his face and arms looks worse — the skin is peeling. Brother Andrew considers the rash a blessing from God, a Job-like test of his fidelity. Ciaran suspects scabies.

—God bless, Brother Andrew.

—You're into the wine early today, Cornelius. Another toothache?
Ciaran grimaces. It's the excuse he used last time.
—Just testing it. I can't use bad wine in Prior Patrick's medicine.
—And how is the dear prior faring?
The truth is, he's backsliding. But Ciaran won't give Brother Andrew the satisfaction.
—Better, now that someone's doing more than just praying for him.
—Do not mock the Lord, Cornelius.
Brother Andrew refuses to call Ciaran anything but his baptismal name, a name Ciaran never even heard until he moved into the priory a few years ago as caretaker. Brother Andrew says the current siege is punishment for Prior Patrick's lax rule. Other senior monks agree. Ciaran tries to ignore priory politics, but it's obvious that if the prior dies, Angry Andrew will assume control—and evicting Ciaran will no doubt be his first order of business.
Contemplating this, Ciaran suddenly feels drunker than he'd like. But it gives him a devil-may-care courage.
—I was not mocking the Lord, Brother Andrew. I was mocking you.
Before Andrew can react, Ciaran grabs the wine and pivots toward the chapel door. He hears Andrew rise and lumber after him, his prayer beads clacking. But the light-footed Ciaran darts through the door and drops the wooden bar in place, locking it behind him. As Andrew pounds his fist, Ciaran grins.
—Brother Andrew, please! People are sleeping.
The pounding grows more furious. But as Ciaran grabs a candle from a nearby nook and heads down the hallway, it fades to impotent echoes.

They call it the chapel, but Ciaran thinks of it as the treasure room. It's crammed with silk tapestries, silver incense burners, pewter candlesticks, a marble altar inlaid with amber, jeweled chalices and reliquaries, gilded croziers. Plus, the cross—nine feet tall, studded with gems, gleaming enough to blind you even in moonlight. There's a reason the

Vikings are laying siege here. One glimpse of the room a few decades ago instantly converted Ciaran's father to Christianity: this god, he declared, must be the most powerful of all. The man had already shed his pagan beliefs by that point, after the old gods failed to save his wife from dying during Ciaran's birth. The treasury then convinced him that the Christian god was far more powerful, and he grew devout enough over time to become the priory's caretaker. After his death, Ciaran took on the duty for more pragmatic reasons. He was tired of practicing medicine and wanted a sinecure—food, shelter, easy access to drink. Still, the chapel never fails to impress him.

But however lovely, the chapel smells foul today, rank with the sickly sweet aroma of human decay. Prior Patrick, barely fifty, lies on a cot. His skin has turned gray and transparent, exposing a delta of veins on his tonsured scalp. He wheezes when he breathes. Most unsettling, his right eye has erupted with infection like a volcano. It's already spreading into the nearby flesh, and it's only a matter of time before it spreads everywhere.

Despite this affliction, Prior Patrick's left eye sparkles a bright blue when Ciaran enters. He points.

—You forgot the plug.

Ciaran glances down at the open jug and flushes red. But Patrick chuckles and waves a hand.

—I'm teasing you, Ciaran. Please, get to work.

Ciaran plunks the wine down on an altar and picks up a brass bowl. Inside is a salve of chopped wild onions and garlic, plus equal parts wine and ox gall. It looks like soggy burgundy stew. He sniffs it and frowns. When fresh, the onion, garlic, and bile make his nostrils burn; now, there's no punch left. He made it three weeks ago, after some bored Viking hurled a rock at the priory. It smashed into an upper-story window, raining shards down on Prior Patrick during Mass; a sliver got into his eye. After infection set in, Patrick agreed—over Brother Andrew's objections—to let Ciaran try this salve, an old pagan cure his grandmother taught him. And at first, it worked: the volcano of Prior

Patrick's eye shrank. But as the salve's grown weaker, the bulge has grown back, angrier than ever.

Ciaran splashes some wine into the bowl and swirls it. They're out of wild onions and garlic to make more salve, so he's trying to stretch the last of it. While he stirs, Prior Patrick clears his throat.

—I'd like us to pray before you apply any.

—I'm not much of a pray-er.

—For me?

Reluctantly, Ciaran kneels on the cold stone floor. He's hoping for something quick: "From the fury of the Northmen, deliver us, O Lord!" But Prior Patrick pulls his prayer beads out. Ciaran groans inwardly. His knees already ache thinking about the next half hour.

As always, his mind wanders. What a strange religion Christianity is—virgin births, holy ghosts, people eating their God's flesh. And yet here he is, muttering paternosters like a full-throated believer. Ciaran's grandmother was furious with his father for converting, and Ciaran himself, though far from devout, always felt more sympathy with her pagan faith—the sacred oak groves, the solstice hymns, the tales of faeries and the old god Crom Cruach. But after converting, his father taught him to despise such ideas, and what played out within his family has played out across the land: belief in the Old Ways grows scarcer every year.

When they mutter the last amen, Ciaran feels Patrick take his hand.

—Thank you. I know you don't believe a word of it, but...

—I do believe.

The prior laughs.

—Ciaran, you're a terrible liar. An extra Our Father for that. But it comforts me that you try.

Ciaran swallows, a lump swelling in his throat. There are times he wishes he did believe; it would make life easier here. Instead, he's stranded between faiths, unable to believe in either the heaven of Jesus or the Otherworld where the pagans spend eternity.

Before he gets emotional, Ciaran stands to play leech. He grabs the brass bowl and has Prior Patrick close his good eye. With his fingernail, Ciaran scrapes the crust off the other, matted eye to tease it open. The eyeball beneath looks bloodshot and yellow, worse than last night. Using an owl feather, Ciaran drips some burgundy-gray salve into the eye. He's concerned when the prior doesn't flinch; the salve has lost its bite.

He feels compelled to speak. He's been weighing a decision for days now, and it's time to say something.

—We need more onion and garlic for another salve.

—There's none growing in the courtyard?

—No. But I know a way out.

Prior Patrick's good eye snaps open, and it's not twinkling with amusement this time. Ciaran explains that the scriptorium in the cellar—where the monks copy and illustrate Bibles to send to remote parishes—was once the monk's brewery. More to the point, the brewery was connected to a tunnel that led to a cave along the shore where the monks malted grain. No one has accessed the tunnel in years, but Ciaran knows from his father's time where the door is, boarded up behind an old mash tun. In the morning gloom, he can slip out before the Vikings see him, then return after sunset.

When Ciaran finishes, Prior Patrick chuckles.

—If it's related to drink, of course you know about it. But why didn't you escape before?

—And go where? The Vikings have laid waste to every farm and village within a hundred miles.

The prior frowns. He concedes the point, but asks Ciaran if he's sure he can slip out unseen. Ciaran nods. Prior Patrick takes his hand.

—Then God be with you.

*Which one?* But he thanks the prior and takes his leave. He sneaks the wine out with him.

Luckily the stairway to the scriptorium is off the hallway, so Ciaran can avoid Brother Andrew in the sacristy. And despite the danger he faces outside, as he descends the damp steps into the cellar, he feels elated at the prospect of escaping the church walls. Even better, with more onion and garlic, he can save the prior, and keep Andrew from taking power and expelling him.

At the same time, Ciaran does feel a pang of conscience. Because despite being teased for his poor lying, Ciaran has just put one over on the prior. He's sneaking outside to gather onions and garlic, certainly. But he's also going to raise a bog body.

From the vantage point of today, Ciaran's salve of wine, onion, garlic, and ox gall seems likely to do more harm than good—a placebo at best, and liable to introduce microbes from dirty plant matter and cow juice. But recent experiments have shown that it not only works but can outperform modern drugs.

The salve appears as one of several remedies in a millennium-old manuscript called *Bald's Leechbook*, with *leech* being ye olde slang for *doctor*.* Historians think the manuscript originated in a monastery. It presents the salve as a cure for styes (infected eyelash follicles), which in the pre-antibiotics era could result in blindness or even death if the infection spread.

We know the salve works thanks to two professors at the University of Nottingham in England, including Freya Harrison, a microbiologist who specializes in chronic infections. She's also a medieval reenactor. Harrison is currently constructing a book from rawhide parchment, and she learned to sword-fight in her youth—a desire born, she says, from

---

* Doctors were called *leeches* because they used to apply the bloodsuckers to patients as a medical treatment. Historians once considered this practice a perfect example of quackery, but leeches have made a comeback in medicine lately. Surgeons sometimes apply them to reattached fingers and skin grafts because leech saliva contains compounds that prevent clotting and dull pain. Leeches also drain off excess blood, a tablespoon every forty minutes.

"having read lots of fantasy literature as a teenager." Upon moving to Nottingham, she tried to join an Old English reading group, only to find that it had been disbanded and replaced with an Old Norse reading group. She attended anyway and met a kindred spirit in Christina Lee, a historian of medicine and associate professor in Viking studies.

Over coffee one day, Lee and Harrison began discussing the monastery leechbook. They were particularly curious whether any of its remedies could actually fight infections. There was reason to be skeptical, given that most "cures" from the era are obvious nonsense. Some required smearing burned bees on a patient's scalp, or leaving a potion beneath a church altar until five Masses were said over it. One eyebrow-raising treatment involved having a virgin gather water "from an eastward-flowing spring"—a sure fix for any cyst "which pains the heart."

Compared to those remedies, the eye salve seemed at least plausible, so Harrison began gathering ingredients. She selected organic onions, garlic, and wine to avoid confounding pesticides. As for the ox gall, it's a standard compound in microbiology labs, used to inhibit the growth of certain bacteria, and was readily available online. (Gall, or bile, is a liquid produced by the liver to digest fat.) The recipe also called for steeping the brew in a brass or bronze vessel, which Harrison simulated by adding small plates of plumbing brass.

Now came the tricky part. From the recipe, Harrison knew she needed equal parts onions and garlic and equal parts wine and ox gall. But she knew little beyond that, since medieval recipes are often vague on amounts. (A typical direction from a leechbook reads, "Take just enough honey.") The list of instructions was equally cryptic. As Lee recalled, "There were lots of discussions about things like what 'pound well' means. How much elbow should you put on that?" But she and Harrison muddled through. Afterward, they let the mixture rest for nine days, at which point they planned to test it on some bacteria.

Inspired, I also make some salve at home. I find onions, garlic, and wine at the supermarket, and in lieu of a brass bowl, I have a bank teller pluck a dozen shiny pennies from a change drawer. (Pennies are made of zinc coated in copper, the two metals that make up brass.) I order ox-gall pills online, capsules with a mottled brown powder inside that looks like sawdust blended with cinnamon. Figuring what the hell, I sprinkle some of the powder onto my tongue as I'm mixing the ingredients. This proves a bad idea. It tastes like a shot of ballpoint-pen ink. That soon fades into a savory oak flavor, but then the ink comes roaring back—only it's now supercharged with the most bitter flavor I've ever tasted in my life. My mouth puckers like a cartoon character's. I spit and spit to no avail.

Despite the awful taste, I dutifully empty several capsules of gall into two bowls of garlic-onion-penny broth—one with red plonk, one with white. I set them in my spare bathroom to age.

Over the next few days, I can practically see the clouds of stench whenever I open the bathroom door—imagine jamming your nose into a vat of garlic and vinegar and taking a big whiff. (Pity my upstairs neighbors.) After a week, the smell grows tamer, but the broths start to look funky. The red salve turns a murky burgundy color, and the chopped onions and garlic in it are now gray and soggy. Meanwhile, all the floaties in the white salve look psychedelic green, a result of copper ions being stripped from the pennies. (For whatever reason, Lee said their white-wine salve did not turn green. Instead, "it looked a bit like wee.")

As luck would have it, I tear a fingernail shortly after making the salves, and the resulting infection gives me a chance to road-test them. I also endure a root canal (it's a long week), and when the dentist tells me to go home and swish some Listerine in my mouth to fight off a potential infection, I decide to use the salve instead. I figure it's the same color.

That night, I start by soaking my infected finger in the bowls, one minute in red, one in white. Upon removing it, I realize that I probably should have swished some in my mouth *before* getting pus in there. I nevertheless slurp up a spoonful of red. It's a doozie. The pungency of the garlic and onion is off the charts, although it does have a sly sweetness that I like. After a palate-cleanser of pretzels, I try the white. It tastes bitter, although not overwhelmingly so, and I get hints of borscht.

Incidentally, if anecdotally, my finger soon heals up and my tooth remains infection-free.*

Harrison and Lee's tests were a tad more scientific. After letting their brew steep, Harrison began dribbling the liquid into some Petri dishes with two nasty bacterial customers inside, staph and gonorrhea. Both have developed strong resistance to modern drugs, rendering them potentially lethal. But to Harrison's delight, the thousand-year-old salve walloped them, wiping these menaces out. That's a big deal considering that drug companies are currently scrambling to find new antibiotics to fight resistant strains: some healthcare experts predict that, if left unchecked, antimicrobial resistance in general could kill ten million people per year by 2050.

Harrison also tested the salve on so-called biofilms, mats of interlocking bacteria that grow inside living things. Because the films secrete a protective slime that repels outside molecules, the bacteria in biofilms can tolerate drug concentrations a hundred times higher than isolated bacteria—concentrations so high that the drug ends up harming the patient more than the microbe. Biofilms are one big reason why so many drugs that show promise in lab tests fail in animal or human trials. Miraculously, however, the medieval salve blasted right through the biofilms in Harrison's experiments and reduced the bacterial population by many orders of magnitude. "When this [result] came out and worked so well, we thought we'd done something wrong," recalls Lee. "It shows that there's a lot more sense in the Middle Ages than I thought before."

Harrison and Lee's work has inaugurated a new field they've dubbed "ancientbiotics"—the search for modern cures among age-old remedies. It's possible the salve will fail in future trials; most treatments do.

---

* Later, when I fish the pennies out, I'm startled to see that they look scorched, as if I've taken a blowtorch to them. They presumably reacted with the acidic broth. Incidentally, Harrison admits to a bit of self-experimentation herself. After her cat gave her ringworm, she dabbed some salve on the ringlet of broken skin. When I mention bathing my finger, she says, "Officially, I have to say, 'No, don't do that.' But unofficially, I think, 'Oh, that's really cool.'"

Maybe, though, this medieval medley will become a vital drug in the modern arsenal.* Which would be humbling, that such a potent weapon has been sitting around for a thousand years in a book, unremarked and unregarded. Harrison and Lee also point out that even some of the ridiculous recipes mentioned above make more sense upon reflection. Sure, the virgin thing was weird, but letting a brew steep under an altar for five Masses might have simply been an easy way to mark the passage of time as the solution matured. We can probably still learn a thing or two from Ciaran and the leeches of yesteryear...

Ciaran emerges, muddy and wheezing, from the cave along the beach; he didn't remember the passages being so narrow or tortuous. A pull of wine steadies him. A hundred yards down the shore, he sees the sleek Viking longships with their dragon figureheads, bobbing at anchor under a gunmetal sky.

He drags his bulk up a slope of scree and seagrass to a small cliff, then peeks over. A half mile away squats the priory. Travelers who rest there often remark that it's the ugliest church they've ever seen, a stolid pile of brown stone. Where are the arches and decorative windows? The florid capitals and spires ascending toward heaven? Prisons have more grace. Ciaran cares little for its architecture himself, but grunts with satisfaction now. Those thick ugly walls have withstood everything the Vikings have thrown at them.

On the grassy plain next to the priory, he sees the Viking camp—a hundred linen tents draped over wooden frames. Above the murmur of

---

* As of this writing, Harrison and Lee's eye salve is moving forward with human trials. If it turns into a blockbuster drug, this wouldn't be the first time an ancient medicine made a big splash. In the 1970s, a Chinese biochemist named Tu Youyou began researching an ancient herbal medicine derived from wormwood. From this work, she discovered the vital malaria drug Artemisinin. She won a Nobel Prize for the discovery in 2015.

the ocean, he can hear hacking coughs from a few tents; ague always runs rampant in armies. Sheep and cattle carcasses lie about, poached from local residents, while cats, a favorite Viking pet, dart among them. Cooking fires are already roaring; Ciaran sniffs jealously and catches hints of beef and mutton. He's also envious to see several Vikings scrubbing themselves with rags, using water from rain barrels. The monks never bathe.

A shout draws his attention. A few warriors are huddled in a circle, rolling dice. The winner snatches the prize off the ground, a chain of some sort. He then takes a hard shove from the loser; everyone laughs as he tumbles into the mud. Ciaran sighs. He too played such games as a young man.

He notes that several of the warriors look Irish. He figures they were kidnapped by raiders in their youth and trained as warriors; he swears he can almost see some familiar faces. In fact, just as he's about to leave for his errand, he notices one of those faces looking his direction — staring right at him.

He drops into a crouch behind the cliff, his heart thumping.

His mind races to reconstruct what he saw. The warrior was blond, with shoulder-length hair. He was standing near the dice game, but not partaking. He had armor on, a padded leather cloak, and Ciaran saw a sword and helmet, too. The thought unnerves him: Why was this warrior dressed for battle? More importantly, did he see Ciaran?

Ciaran has to take another peek. He creeps a quarter mile along the cliff. Shielded by a small bush this time, he raises his eyes over the edge.

His bowels nearly empty in relief. No one's sounding an alarm or even looking his way. The blond soldier is leaning on his sword now, absorbed in the dice game. Ciaran sinks back down and takes another pull of wine. No more spying. Time to get to the bog.

The hourlong walk takes him three. The grassy plain near the priory soon fades into sparse forest, and after thirty-seven days cooped up, he enjoys the simple freedom of wandering. It's a relief to hear new birds sing, too—instead of squawking seagulls, there's the whistles and squeaky drumrolls of starlings.

Out of habit, he avoids passing near any hawthorn trees (his grandmother said that spiteful faeries lived in them), but he does make a pilgrimage to her favorite oak grove, smiling to think how happy she'd be at the good mast year—thousands of acorns crunch underfoot. There's one tree she carved some runes into; he traces a finger over them, wishing he could remember what they meant.

At a birch tree near a toppled bridge, he scrapes off some resin to chew, rolling the sweet, sticky gum in his mouth. At one point he spies an abandoned hut with a thatched roof and wattle-and-daub walls, the occupants no doubt scared off by the Vikings. There's a puddle at the door, hollowed out from the tread of feet in and out. Luckily, there's onion and garlic in a nearby garden. He tucks them into his tunic for the salve.

Near the end of his ramble, he feels the ground underfoot growing spongy. The warble of starlings fades and the squeak of the whinchat and peeps of the meadow pipit emerge. Then he crests a hill and sees it—his bog. The trees are sparse and stumpy. Between them lies a patchwork of mosses and sedges and pools of dark water; it's mostly muted green and black, but dotted with rusty orange and yellow plants in spots. He winds his way down carefully, testing each footfall. Beyond the birds, there's no movement anywhere, not even wind.

In his grandmother's youth, before the Christians came, people believed that bogs were sacred. Bogs supplied iron ore and peat for fires, plus healing herbs if you knew where to look. A few mystics also spoke of bogs as portals into supernatural realms. If so, Ciaran wonders what otherworlds the bog lady has seen.

After some gulps of wine for warmth, he strips down at the water's edge and splashes out to the dead tree that marks the bog lady's position. He found her a few years ago, while cutting peat for the priory—a brown-gray foot sticking grotesquely out of the muck. He stared for several minutes, scarcely breathing. Whenever he misbehaved as a child, his grandmother would scare him with tales of cruel parents throwing naughty boys into bogs. Later, more seriously, she explained that in the old days—back when gods still roamed the earth—human sacrifices were common around bogs. By that point, a man of fifteen, he barely listened to her stories anymore. But here someone was.

He excavated the woman over a few days. When the sun and air began putrefying her—rancid fats were oozing from her bronze skin—he submerged her underwater near the dead tree to preserve her, anchored by stones. He thinks about her almost daily, and checks often to make sure she's still there. But only rarely does he pull her out, when he aspires to connect to something bigger than himself. Indeed, in all the Christmas and Easter and Holy Day Masses he's attended at the priory, he's never felt even a touch of grace or rapture. But the last uncynical scrap of him still hopes he can find such things through her. He's been saving his supplications for a crisis, he supposes. If he's honest with himself, he's also afraid she'll prove powerless.

The water near the body is crotch-high and black; mud squishes between his toes. When he ducks down to search, the cold snatches his breath away. He feels a pulse of panic when he can't find her. Then he grabs hold of something—five leathery fingers. He pats his way up her arm and removes the anchor stones, gently lifting her thirty-pound frame to the surface.

He lays her on some grass, shakes himself as dry as he can, and dons his cloak. Another warming gulp of wine—the last—and he tosses the jug aside. As he rubs his limbs, he looks her over. She's lying flat, her legs stretched out and her face turned aside as if someone slapped her. She has flaming orange hair twisted in a knot. In one hand she's clutching a flute of stag bone. Her breasts look deflated, and her throat is scarred with rope burns and slash marks; dead moss fills the gashes.

There's one feature hidden from sight. His heart beating faster, he parts the twiny hair with his fingers and peels the torn scalp back to expose the hole in her skull—a circular disk the size of an egg. In all the hours he's spent pondering the bog lady, he's spent the most time wondering about this hole. It doesn't look like an injury; it's too neat. His practiced medical eye can see cut marks near the edges anyway, from a tool. Was this ancient surgery? Or were there deeper reasons for it? The monks at the priory shave tonsures into their hair to allow their God easy access to their souls. Had the redheaded woman done something similar? Every time he thinks on it, it seems more mysterious.

The bog's chill soon spurs him, however reluctantly, to do what he came to. Out of habit, he starts to kneel, then stops—his grandmother never did. He remains standing, his palms outspread like hers on their visits to the oak groves. He remembers only snippets of the prayers she recited, but hopes a pure heart will make up for poor form.

A deep breath, and he begins. He prays to the bog woman for several things. For Prior Patrick's health. For deliverance from the Vikings. For the strength to drink less, at least someday. He adds his hopes that the gods won't mind him praying on behalf of Christians, the very people driving pagan beliefs extinct. He mostly just wants peace. He's old and achy, and longs to live out his last days in quiet. He thinks they'd understand.

He tries hard to feel something—aches to. He wonders what people feel when prayers work.

Before long, the cold gets to him. He needs to dress again and walk to get his blood flowing. A little wobbly—he wishes he had more wine—he drops his cloak, raises the plank of the body, and splashes into the water to anchor her down again. After wading back to shore, he rolls in the ochre grass to dry himself, then hurriedly dons his clothes. He's just finished lacing his boots when he hears the unmistakable sound of a sword being drawn from a scabbard.

He looks up to see the blond warrior with the padded leather armor. He's got his helmet atop his head, and not even a wisp of beard. Which would be unusual for a Viking man. Except, as Ciaran now realizes, it's not a man at all.

She's shorter and far slighter than Ciaran but thirty years younger—and that sword looks awfully comfortable in her hand. She's turning it lightly, the blade flashing back and forth.

—I saw you spying on us, old man. From near the caves along the shore.

He's startled to hear her speak Irish. She must be one of the kidnapped ones. Ciaran finally finds his voice.

—Why didn't you sound the alarm? I could have been mustering troops.

—In which case I would have killed you.

—But you haven't.

The sword sags a few inches.

—Do you recall twenty years ago, treating a girl at a nunnery near here? She nearly lost an eye due to infection.

He tears his gaze from her sword. He doesn't remember, but ventures a guess.

—I thought you looked familiar.

—You saved my life. But you also made me pray with you. I thought you were Christian.

—I don't mess around with nuns.

The warrior smirks, then nods toward the bog.

—You were praying over her. Do you follow the Old Ways?

He hesitates, then decides to be honest.

—In my heart. Or at least, I'd like to.

She finally relaxes her sword, returning it to its fish-leather scabbard.

—Then I have no need to dispatch you, old man.

Ciaran's soul lifts. Have his prayers worked so quickly?

—You'll spare us?

Now the woman laughs.

—You, yes. But my brethren were glad you tipped us off.

Ciaran feels a chill even deeper than the cold. He asks what she means, and she smirks again.

—If you got *out* of the priory, there had to be a way in.

Bogs represent one of the last true wildernesses in Europe, and people have been burying things in them for thousands upon thousands of years. This has proved a boon for archaeologists, because the unique chemistry of bogs can preserve organic artifacts to an incredible degree—scarves, leather shoes, wooden cups, wolfskin capes, wagon wheels, floppy felt hats, ancient books, even twenty-pound blocks of butter.* And most spectacularly, bodies.

Over a thousand bog bodies have turned up across Europe, in at least 250 sites from Ireland to the Baltics. The oldest date back 10,000 years. Bogs preserve bodies well for a few reasons. One is the chilly water, which rarely tops 50 degrees. The water also tends to be stagnant and lack oxygen, which inhibits the growth of maggots and putrefying bacteria. Finally, mosses in the bogs produce acidic chemicals that work their way into skin and organs, and vegetable-tan these soft tissues into something durable.

Some bog bodies are graphite-colored, others mahogany or dark red. The acidic water can dissolve bones, too, so some are little more than rubber chickens, with their limbs and torsos twisted like an El Greco saint or the Grünewald Jesus. Many have exactly the sort of ghastly mummy visages you'd expect, but others look quite human: recognizable, fleshed-out individuals, with wrinkles and stubble, chapped lips and braided hair. (Most have red hair, although that might be an artifact of the acidic water bleaching and discoloring the original strands.) Even the remnants of their last meal sometimes survive (fish, barley, mistletoe) as does the occasional tapeworm. As one scientist marveled, you feel like they're about to open their eyes and talk to you.

---

* Burying butter in bogs was surprisingly common—over 400 blocks have been retrieved. The cool, acidic water preserved butter nicely, and because people commonly paid taxes in butter way back when, scofflaws could also cheat their liege of revenue by hiding them this way. A few medievalists have re-created bog butter in modern times. One of them described it as having a "mossy, earthy aroma and a rich umami taste laced with traces of vegetal putrefaction." Bon appétit.

 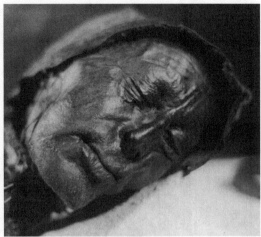

Tollund Man of Denmark, a bog body from circa 400 BC. He was likely ritually sacrificed; notice the noose. (Left photo courtesy of Museum Silkeborg. Right photo credit: Sven Rosborn, Wikimedia Commons.)

A few bog bodies show signs of medical care. In cases where the bones survive, archaeologists have found skulls with holes surrounded by obvious cut marks. The holes also have new bone growth around the rims, proving that people survived the procedures for a substantial time. More commonly, however, bog bodies show signs of violence. These people didn't simply trip and drown in bogs; they were murdered and dumped. Some were disemboweled or dispatched with arrows to the face and chest. One's head was bashed in with an axe. Chillingly, many bodies show signs of multiple traumas. One had his throat slashed, suffered a broken neck, took a blow to the head, and was garroted. With another poor soul, his assailants stabbed him in the chest, decapitated him, slashed his nipples, then sawed him in half for good measure. Archaeologists politely call such deaths "overkill."

However fascinating, it might seem difficult to study such bog bodies experimentally. After all, you can't just chop off a volunteer's nipples or saw their bodies in half. But archaeologists have ways. Tiffany Treadway, a Californian based in Wales, has been fascinated with bog bodies since she first learned about them in college. For her master's thesis, she decided to study their stab wounds—specifically whether the victims were dispatched with spears or daggers. Call it *CSI: Bog Bodies*.

To simulate a human body, Treadway used pig carcasses from a local farm, since pigs are the same approximate size as adult humans and have similar skin. As a first step, to prevent spoilage, she carved out the organs and replaced them with ballistics gel, a standard substitute for organs in forensics work. The gel is usually clear, and feels like blocks of dense, leathery Jell-O. You can order fancy ballistics gel online, but being poor graduate students, Treadway and a friend made their own from boxes of Knox gelatine, mixing the powder with water and pouring the slurry into the hollowed-out pigs. They quickly ran into trouble. "When we were pouring in the gelatin, we didn't realize it would leak out the orifices," she recalls. "So I had my friend help me start sticking cotton balls in certain places." By which she means the snout and, well, where the sun don't shine. But after several minutes of jamming their fingers up the rectums and nostrils, they plugged the leaks. The gelatin set in two days, and when Treadway pressed on the pig bellies afterward, they felt like real bodies—firm but squishy. Time to get to work.

For the experiment, Treadway hung the pigs on timber frames in a forest. To stab them, she recruited a dozen fellow graduate students—an easy sell, she says. "You get a bunch of master's students, and they're all stressed out, and you ask, 'You want to have a whack at a pig?' They're all like, 'Okay!'" The spear and dagger were reproductions of Iron Age weapons. They had roughly similar cross-sectional shapes, although the spear was significantly wider and thicker; each volunteer got to try both. On her go, Treadway remembers feeling surprised at how resistant the skin was; it took force to jab through. But once she hit the gel beneath, she added, "it's smooth sliding." Each wound was then color-coded to the weapon responsible, leaving the pig bellies streaked with red, blue, and green marks.

As the experiment progressed (it took five hours), flies began swarming the carcasses. Treadway lit a fire to drive them off, but the heat remelted the gelatin inside the pigs—at which point it began dribbling past the cotton balls and leaking out certain places again. "It was a bit of a mess," she admits. But the experiment went well overall. The bigger spear tip produced bigger wounds in the pigs—no surprise there. Unexpectedly, though, the spear and dagger tips produced different-looking wounds, despite their similar cross-sectional shapes. In short, dagger wounds gaped and sagged, spreading apart like a hole: 0. Spear wounds, meanwhile, cinched up like a slit: |. (Treadway attributes this difference to the surface area of the wounds. Spear wounds, being larger, have more surface area and therefore more places along the wound for the flaps to stick together. As a result, they don't gape. The opposite holds true for smaller dagger wounds: Without much surface area, the flaps struggle to stick to each other.) Regardless, archaeologists can now use Treadway's findings to examine bog bodies and determine whether victims were speared or stabbed. That's important, because reconstructing how people were killed can hopefully shed light on the biggest mystery of all with bog bodies: *why* so many victims were murdered and dumped.

Some might simply have been robbed, or executed as criminals or prisoners of war. But many show signs of being sacrificed, especially those recovered between 400 BC and AD 400. Again, in some pagan traditions, bogs are sacred places, or even portals into supernatural realms. Ancient Europeans also mined bogs for iron and peat, and to the pagan mind, you can't take something out of nature without giving something back—namely, flesh and blood. Most intriguingly, the overkill deaths call to mind religious ceremonies where people die in ritual reenactments or as scapegoats for a community's sins. Indeed, Celtic and Germanic folklore contains stories about kings, heroes, and gods undergoing a rite of "threefold death" (e.g., stabbing, strangling, bludgeoning) similar to mutilated bog bodies. (In contrast, if you were robbing someone, there'd be no reason to kill them multiple ways.) Treadway's work can help archaeologists sift different possibilities. If thrice-killed victims commonly have gaping dagger wounds, for instance, then daggers were likely a vital part of the ritual.

A few archaeologists even argue that the victims embraced their fates—that they were honored to be sacrificed for the community's benefit. For people of pre-Christian Europe, then, bog bodies might not have evoked horror or fear. They might have been holy. The same could have held for people of later generations, like Ciaran, who knew the Old Ways only dimly, but burned to keep them alive...

They're trudging back to the priory, she in front, him trailing behind. A few miles out, Ciaran hears the piercing call of the lur—a long Viking trumpet made of willow. How many times, while listening to the droning chants of the monks, did he long to hear such crisp clean notes? But he shudders now. The lur is martial music, used to rally warriors.

A hundred yards ahead, the woman Viking—the Irish Valkyrie, as he thinks of her—stops to wait. When he pulls within twenty yards, she forges ahead again. She keeps doing that, never letting him catch up. A few times, she's even disappeared left or right into a thicket, then reappeared from an unexpected direction. It spooks him.

—Get on!, he finally yells. Go.

—I don't want you to get in trouble, old man. Or cause it.

He grumbles. Why does she bother waiting for him? He wishes like hell he had something to drink.

When the forest thins to nothing near the priory, she nods goodbye and walks off towards the ships, leaving him to survey the destruction in the fading daylight. Smoke pours upward from spot fires on the roof. The front door is open, and a parade of Vikings are plodding through, each one lugging a jeweled reliquary or crozier or other treasure—along with several monks, bound in rope and destined to be sacrificed or sold into slavery. Every one of their screams tightens the vise in Ciaran's chest a little harder. He finally has to kneel down on the grass, a hand on the earth to steady himself. A phrase he heard once in their Christian Bible throbs inside his skull. *What hath God wrought?* What pains him most is the sight of a dozen whooping warriors dragging out the giant gold cross. He's sure the defilement of the chapel would have made him laugh before, but right now a lump rises in his throat. Prior Patrick was in there.

After the better part of an hour, the procession of looters stops. Still, not wanting to run into a straggler, Ciaran sneaks down to the coast to reenter the priory. The tunnel stinks—of smoke at first, then something rotten. Where it opens into the scriptorium, he finds two disemboweled brothers, smeared with their own blood and shit. They apparently tried to stop the Vikings from entering and failed.

The rest of the cellar is full of splintered tables and half-destroyed Bibles. The brothers took pains to illustrate their Bibles with pictures, and broken pots of pigment now litter the floor and walls—blotches of pink, yellow, green, and blue, a grotesquely festive look. Especially since some of the blotches are smeared across the bodies of monks. Monks he tolerated, monks he despised. Most dead, a few technically alive. His stomach clenches, and he hurries past. He also can't help but notice, near the stone steps, that dozens of barrels of wine are missing.

Upstairs in the hallway, the door to the sacristy has been smashed and is hanging by one hinge. Two brothers stand examining it. Ciaran asks them how many monks survived. They say around a dozen, barricaded into the choir loft above the church. They seem furious in answering, though, as if they could spit on Ciaran. One looks him up and down.

—Everyone else down here was killed or taken. Where have you been?

—I snuck outside, to gather herbs for Prior Patrick.

The other monk points to the door he came through.

—That's the way the Vikings got in.

Ciaran's conscience pangs him. He tries to apologize, but they hurry off without listening, their eyes livid with betrayal.

Ciaran enters the chapel to find it stripped bare, down to the paving stones—except for Prior Patrick's cot, which has been kicked over and cracked in half. The prior is lying next to it, moaning. That he's alive at all gives Ciaran a surge of hope. The brass salve bowl is propped on his head, a mocking crown—Viking humor. There's onion and garlic in his hair and streaks of red wine on his cheeks and neck. But as Ciaran kneels, his stomach sinks to see that the streaks aren't all wine.

The stye in the prior's eye looks worse, but it's not the source of the blood. Ciaran traces that up to the prior's shiny wet head. There's a crater of depressed skull there—probably where an axe smashed down. Prior Patrick groans at being touched, and tries to say something. His tongue is thick in his mouth, but the pain is clearly excruciating.

—Remove your hands from him, Cornelius.

Ciaran turns to see Brother Andrew, scratching his poxy face. He's flanked by the two brothers from the broken door a moment ago. Ciaran stares coldly back and hisses at Andrew.

—What a miracle of God that you're unscathed. Did you not fight?

—My leadership is the only reason anyone survived this treachery. And as the acting authority of this priory, I order you to remove your hands from him.

—As long as Prior Patrick is breathing, he's in charge.

—He's mortally wounded.

—I can save him.

—God alone can save. We shall pray for the prior's swift recovery. But no more pagan cures. His life is in God's hands now.

The words infuriate Ciaran. Andrew clearly intends to let Prior Patrick languish and die.

—If your god's so powerful, then why didn't he stop the Vikings?

—Our prayers were working—until we were betrayed. What did the heathens pay you to show them that tunnel?

Ciaran freezes. While making his way back to the priory, he of course felt ashamed for accidentally tipping off the Vikings. But he never imagined that the brothers would think he did it on purpose.

He protests his innocence. Brother Andrew simply snaps his fingers. The two monks flanking him step forward and make a menacing sign of the cross.

Ciaran scuttles back into a corner, until he's pressed against the cold stone wall. They step over Prior Patrick's body, looking even angrier than before. The pair come on steadily, squarely, eyes locked on his. And just when they seem ready to lunge—they freeze, arrested by the sound of a sword being drawn from a scabbard.

Gesturing with her blade, the Valkyrie orders the two men to another corner. Then she turns to Brother Andrew.

—You get into the corner as well.

Andrew ignores her, and stares down Ciaran.

—Did she fuck you, Cornelius? You betrayed God for a woman?

The swipe takes Ciaran by surprise. From somewhere unseen, the Valkyrie has produced a dagger and sliced Brother Andrew's necklace of prayer beads; the wooden balls go clattering off the stones. There's also a slit in Andrew's cowl, just above his heart—not unlike a bog body wound. He gasps and clutches his chest. Ciaran sees blood streaming between his fingers.

The Valkyrie turns to the two other brothers.

—It's a superficial wound. The next one won't be. Take him away.

They each get under one of Andrew's arms, and hurry him off. He looks catatonic. When they're gone, she turns to Ciaran.

—I saw you from the ship, old man, stumbling along the coast. You need to be more careful.

When Ciaran doesn't answer, she goes on.

—Is the angry one in charge? With the scabs?

Ciaran's eyes stray to Prior Patrick on the floor, barely breathing.

—No, *he* is. At least for now. But I think I can save him. May I borrow that dagger?

The Valkyrie frowns and hands it over. Ciaran turns it in his hands, thinking of the hole in the bog woman's skull. He wipes his sweaty palms on his tunic.

He has no choice. He'll have to operate on Prior Patrick.

As far back as 10,000 years ago—before cities, before bronze or iron, before written language—human beings were performing neurosurgery, carving up skulls and exposing the raw, naked brain beneath.

Specifically, they were performing an operation called a trepanation, which involved removing a section of the cranium in response to head trauma. If a giant cat mauled you and dented your skull, or if you fell off a cliff and smashed your head in, the damaged bone would often be left pressing down into the brain, a dangerous situation. To relieve that pressure, surgeons would trepan the skull by sawing, drilling, or scraping away the broken bits.

Beyond accidents, the operation found heavy use treating war injuries. Archaeologists suspect this because, much like with injuries due to shock weapons, most trepanation holes appear on the left side of the skull, the very place a right-handed warrior would clobber someone with a weapon. Trepanations also perhaps treated brain bleeds, brain swelling, infections, cysts, and tumors.

Doctors performed trepanations all over the ancient world—France, New Guinea, Peru, Kenya. The first step involved cutting the scalp and peeling it back. To remove the underlying bone, they could have used stone blades, animal teeth, horns, or shells—all of which have performed well in modern experiments. After tossing away the fractured bits, surgeons might treat the wound with medicinal plants or oils. Then they sewed the scalp back up, often leaving the patients with a soft spot in their heads, like a baby's fontanelle. Or several soft spots: multiple trepanations were not uncommon. (One unlucky fellow had seven.) The largest known trepanation holes gaped four inches wide.

Trepanations (or trephinations) allowed doctors in prehistory to remove damaged bits of skull or relieve pressure on the brain. Archaeologists have re-created trepanations in modern times using flint, obsidian, bow drills, and shark teeth. (Credit: Wellcome Trust.)

Given the lack of anesthesia back then, one screamingly obvious question is: Didn't trepanations hurt? Not really. Cutting through the scalp is somewhat painful, but most of the skull lacks pain sensors. The brain can't feel pain, either; someone can slice right through and theoretically, you wouldn't notice a thing. So the discomfort was minimal, and most patients probably stayed awake during the procedure.*

---

* Not every culture left a soft spot. Some South Seas tribes plated over trepanation holes with coconut shells; the Inka used gold. Archaeologists have also speculated that the Inka used coca leaves to anesthetize the pain of cutting through the scalp. In contrast, Kenyan doctors used what might be called anti-anesthesia. During the operation, they jammed stinging nettles into patients' ears. This apparently caused so much agony that they forgot all about having their skull sliced open.

One rural Kenyan tribe, the Kisii, performed trepanations through the 1980s, and could be quite aggressive with the surgery. After hitting his head on a doorway in 1940 and suffering excruciating headaches, a Kisii police officer named Nyachoti had thirty square inches of skull removed—two full palms' worth. Afterward, it

Another big, blaring question is: Didn't people die all the time? Again, not really. Bone regrows after trauma, and archaeologists can therefore examine the edges of trepanation holes for new bone growth and determine how often people survived the operation. They usually did. In fact, traditional trepanations often outperformed modern surgery. One study found that, in the 1870s, 75 percent of patients who had sections of their skulls removed in London hospitals—patients operated on with steel scalpels, by licensed surgeons—died from the operation or its aftermath. During that same rough period, on islands near Papua New Guinea—where doctors often sawed skulls open with shark teeth, washed the wound with coconut milk, and dressed it with banana leaves—missionaries estimated that just 30 percent of patients died. If you had skull trauma in the late nineteenth century, you were far better off in the South Pacific than Europe.

All that said, the procedure was hardly pleasant. If you needed a trepanation, you were probably already nauseous, woozy, and gravely injured, with blood streaming down your face. And while the pain probably wasn't overwhelming, someone was still sawing your fricking head open. Imagine listening to that, or smelling the frass of your own skull dust—like a drilled tooth at the dentist, only ten times worse. There were also lingering side effects: headaches, exhaustion, sensitivity to light and noise.

However gruesome, trepanations have always fascinated me. How did doctors way back when even have the guts to try? For my money, there's no bigger gulf between ancient and modern times than with regard to medicine, and I felt I'd never truly grasp what our ancestors endured unless I tried a trepanation myself.

The problem was, I had no idea how to procure a skull. Should I call a butcher and claim I'm craving head cheese? Trawl local highways for roadkill?

---

looked like someone had blasted the top of his head off with dynamite—the sagging was that extensive. Nyachoti usually wore a hat, and earned the nickname "Hat On/Hat Off" for his habit of doffing it and startling people. You can see a picture of him on my website.

On a whim, I reach out to Laura Wolfer, the proprietor of the Moose Ridge Wilderness School in Maine. I'm already planning to visit there to brain-tan a hide, and given that tanning also involves blood and guts, I figure maybe she has some decapitated animals lying around. I type out a carefully worded email, explaining my long-standing desire to saw open a skull, then immediately feel like a creep for sending it. But Wolfer replies that night—and she's into it. As a former biologist who now works on archaeological digs, she's just as curious as I am about trepanations. And she does indeed have some leftover pig and deer skulls from the butchering classes at Moose Ridge. Just like that, I'm set.

The trepanations begin on a sweltering July afternoon. Wolfer leads me out to a picnic table strewn with stone blades, shards from the Moose Ridge knapping pit. She then fishes around in a garbage bag. Her hand emerges, Judith-like, with an amputated deer head. It's heavier than I expect and, after a long refrigeration, damp and cool to the touch.

To start, I part the hair on the scalp with my fingers—it feels like paintbrush bristles—and pick up an obsidian flake. The edge bites cleanly into the skin and the incision glides along smoothly. So smoothly that I cut too far and slice my finger open. Given how sharp obsidian is, it's the opposite of a paper cut—utterly painless. I don't even realize until I see blood.

After some first-aid, the trepanation continues. I complete a T-shaped incision, then lift the skin flaps and pare back the shiny white fascia beneath that binds scalp to skull. The fascia looks like dense cobwebs; I half expect a spider to dart out. Below it lies the mottled surface of the skull: matte white bone, clingy bits of muscle, looping red sutures where different bones come together.

To open the skull, different cultures around the world used different methods: drilling, sawing, scraping, slicing. I decide to saw. I'm essentially making a Tic-Tac-Toe board of four intersecting lines (#), then removing the center space. My saw is a three-inch piece of serrated gray flint that Wolfer secured from a local knapper. It works beautifully. The sawteeth chew right through the bone, producing little white curlicues of sawdust—skulldust—that I blow away every few seconds, like a carpenter sawing a board. The swift progress thrills me; I imagine myself an ancient medicine man, racing to save a warrior's life.

Good thing this is fantasy, because I soon run into a host of problems.

The first involves flies—they're relentless. Instead of focusing on the surgery, I spend a quarter of my time shooing them away from the open wound, a distraction that modern surgeons don't face. The second problem is the mess. Raw flesh is greasy, and the slicker my hands get, the more the deer head squirts around; this makes it hard to get into a rhythm when sawing. Ancient surgeons probably battled similar issues. Doctors didn't have gloves in the BCs, and given the absence of anesthesia, their patients were probably flinching and fussing. Anesthesia clearly benefits patients nowadays by eliminating pain, but we forget how much it benefits doctors, too, by immobilizing patients. Ancient doctors worked on moving targets.

After nearly half an hour, my flint saw finally breaks through. The white bone gives way to a thin red layer, a vein of sparkling skull ore. Beyond that, a void opens up into the inner sanctum. It's like peering through a crack into a hidden cave. I feel like a real surgeon.

Unfortunately, I still have three cuts to go. At this point, I decide to abandon the Tic-Tac-Toe box and carve a triangle instead. Beyond saving time, this shortcut seems necessary because I'm now running smack into the biggest problem of all with ancient surgery: the blade. Unlike modern saws, my three-inch flint blade isn't hafted to a comfortable handle; I have to pinch hard to hold it steady, and my hand is starting to ache. Even worse, stone blades dull quickly. Ancient surgeons no doubt had multiple saws at the ready. I do not, which means the next two triangle cuts are a war of attrition—my arm versus the skull. Before long, the sawteeth aren't chewing through the bone as much as gumming at it. I need to push down even harder to make progress. But this makes my hand ache more, and also makes the teeth wear down even faster—a vicious cycle.

After another hour—the warrior in my fantasy is barely clinging to life—I finish the second and third cuts. To remove the wedge of bone, I jam a chert chisel under one edge and jimmy it up and down. The triangular chunk is the width of a bottlecap and a quarter-inch thick, roughly the same thickness as a human skull. It comes free reluctantly. As a final step, I grab an obsidian flake and slice through the dura mater, a protective membrane beneath the skull that looks soft but is as tough as canvas. I peel the dura back—and there it is. Brain.

Given the modest trepanation hole, I can't see much—a peek-a-boo of shiny, brown-red folds. To my chagrin, I realize that my first sawcut went too deep and mangled the top layer of neurons, leaving chewed-up bits of pulp. (I cringe for my warrior.) But this mistake can't take away from how profound, and humbling, seeing a real brain *in situ* is. Here was the essence of this deer: its memories, its instincts, its desires, all somehow bound up in flesh. Imagine how much more profound it would be to see a human brain—a *living* human brain, softly pulsating to a heartbeat. It took a ridiculous amount of work to reach this point, nearly ninety sweaty minutes. But it feels worth it. As with scaling mountains, you have to earn a view like this.

Nothing if not foolhardy, I try another trepanation the next day, this time on a pig head. Once again, Wolfer plops it down on the picnic table fresh from the fridge. It's bright pink, with a squishy snout and floppy ears. It's also dripping blood from its mouth—which proves to be an omen, because this trepanation is a debacle from start to finish.

Given all the fat on pig heads, I can't just peel back the scalp after the initial incision; I have to hack at the skin to expose the bone. It's *work*. For the actual trepanation, I try a different technique than yesterday—scraping, which involves scouring the bone away with an obsidian flake; it's like scratching off a very thick, very stubborn lottery ticket. It was probably the most common trepanation technique in ancient times; the resulting hole looked like a shallow bowl.

Unfortunately, there's a *lot* of bone to scrape here: the pig skull is as thick as a triceratops brow. Equally bad, tiny flakes of obsidian keep crumbling off into the wound. Given that I'm cosplaying a surgeon here, I feel obligated to pick these shards out for the sake of my "patient," bit by microscopic bit—a huge waste of time. Overall, the scraping method is an odd mix of brute labor and tedious fine-motor tasks.

About the next seventy-five minutes, the less said the better. By the end my hand has cramped into a lizard claw. And if the flies were annoying yesterday, they're a veritable plague of Moses today. They're tickling my nostrils, dive-bombing my ears, probing beneath my shorts and biting. I spend the last half of the trepanation hopping from foot to foot in a silly jig. I gaze down at one point to see the severed pig head, the swarm of insects, the blood—and realize I've stumbled into *Lord of the Flies*. At no other point in this book did I want to quit an activity so badly.

At long last, I'm close enough to cheat. I find an obsidian flake with a sharp tip, fit it into the hole, and start pounding with a hammerstone. After a few strikes, the tip punches through to the glittering red skull ore. In all, I've scraped through a couple centimeters of pigheaded bone, micron by micron, for a hole the size of a grain of rice. It hardly seems worth it. But as I pop the dura mater, something odd happens: the brain burps, expelling a bubble of liquid. I have no idea why. But if that had been a patient with a brain bleed, the release of pressure might have saved their life. It's something.*

---

* In the 1960s, some counterculture hippies began *trepanning themselves* in order to get high. The practice traces back to a Dutch librarian and LSD enthusiast named Bart Hughes, who developed a (dubious) theory that the confines of the skull restrict blood and nutrient flow into the brain, which somehow squashes our "higher consciousness." So, to recapture that state, Hughes got hold of a hand-borer and drilled open his own forehead. (Again, pictures on my website.)

Hughes also persuaded a few disciples to trepan themselves. (Reportedly, John Lennon was interested.) I managed to track one down, Joe Mellen, from England. Mellen struck me as sweet, charming, funny, and utterly batty. Alarmingly, Mellen failed the first three times he attempted to trepan himself, with a corkscrew drill, because he couldn't get enough leverage. He described the struggle as similar to "trying to uncork a wine bottle from the inside."

Mellen ordered a power drill for the fourth attempt and succeeded. "There was a schlurping sound as I took [the bone plug] out," he recalls, "and what sounded like bubbles." (Even more disgustingly, a woman named Heather Perry told an interviewer that, after her self-trepanation, "I was a bit like a whale: every time I coughed,

Back at my hotel that night, my hands are almost too sore to crack open a beer. As I sip, I reflect. In performing an ancient surgery, I'd expected to learn about blood and guts, and I did. But the pain and frustration were as much a part of the experience as the grease and the smell. Doctors back then had to endure biting flies and throbbing hands every time they operated. And, as with Ciaran and the prior, there was far more at stake than simply the pride of not quitting...

While the Valkyrie keeps watch at the chapel door, Ciaran begins to excavate the crater on Prior Patrick's skull.

He positions the prior's head in his lap and picks out the onion and garlic from his hair. The Valkyrie's dagger makes a sharp incision in the skin, and Ciaran peels the folds of scalp back. He gets only a glimpse of the crushed white skull before a red ooze burbles up and floods the bone. His hands and tunic are soon sticky with crimson.

He plans on making four cuts, and removing a piece three inches square. But however sharp, the dagger blade lacks serrations and is too smooth to make much headway on bone. The first cut takes an hour. By the end of the second cut, his hand throbs. He shifts positions to get a better angle for the next cut and hears Prior Patrick moan.

Halfway through the third cut, Ciaran runs into a wall. He's sweating now, and growing more nauseous by the minute. He doesn't think he's gone this long without a drink in years. Worse, his hands are shaking; instead of smooth, even cuts, he keeps missing the groove in the bone and jostling Prior Patrick. After one such lurch, the prior's right

---

some fluid would come out of the hole in my head.") Mellen still has a soft spot today on his balding pate; during our conversation he gleefully pressed down on it like a deflated blister. He said he could feel his brain pulsating to his heartbeat.

According to Mellen, the trepanation really did get him high. He described it as "a feeling of lightness. I'm quite sure that it was relief from gravity." The sweetness and joy continued to swell for hours, and was still present the next morning. "I've been expanding my consciousness for over fifty years" through drugs and other means, he says. "I like to have as much of it as I possibly can."

arm and leg start seizing; he's having a fit. It's a terrifying moment. Ciaran yells for the Valkyrie's help. They grab hold of Patrick's limbs and pin them down, but it does no good. Ciaran actually finds himself praying, begging any god who'll listen to make this stop.

When the seizure finally passes, he feels a few hot tears escape. He's never felt so impotent as a leech. If the shaking gets any worse, he fears he'll never finish.

As he wraps up the third cut, he turns to the Valkyrie, thinking how to disguise his request.

—I need you to run to the sacristy. It's the room down the hall with…

—I know what a sacristy is. What do you need?

—Some clean vestments for bandages. As casually as he can, he adds, And if you see any jugs of wine, grab them.

He sees her face harden and hurries on.

—I need to make a salve to wash his wounds. I suppose you were spying on me when I collected these?

He digs into his tunic and holds up the wild onions and garlic, now stained red. The Valkyrie leaves without a word.

Ciaran's mind starts churning. He needs a minute alone with the wine: He doesn't want her to watch him drink, to see how desperate he is. But what other errand can he send her on?

Before he can think of anything, she enters with two Easter vestments—and a single, broken, half-empty jug. Ciaran frowns. It's not enough to both take the edge off and finish the surgery.

—We'll need more wine than that.

—There isn't more. We raided the rest.

Ciaran feels his throat catch. He searches her face, hopeful she's just teasing him—but no, she seems serious. His eyes drop to the floor, too ashamed to look at her.

—I don't think I can finish with my hands shaking like this. Are you sure there's no more?

—Yes.

Ciaran nods. Wine will stop his sweating and steady his hands; he can operate more safely, and from that perspective, it's best for him to finish the jug.

But Prior Patrick's salve. He sets it aside untouched.

The final cut takes two hours. His head throbs. Sweat stings his eyes. The jiggling of Prior Patrick's face makes him nauseous, so he focuses on his skull and hums to distract himself—chants, folk songs, anything that flits to mind. But with the blade getting duller, progress is slow. Prior Patrick's breathing is getting dangerously fast and shallow.

At last he's through. His fingers are wrinkled with blood now. He uses the dagger to pry loose the shards of crushed bone; some are still attached by a lip, and he has to work them back and forth to snap them, like green wood. The dura beneath looks as swollen as a pregnant belly. He pierces it, and the hole pukes up a mass of half-clotted blood, like undercooked eggs. But he's encouraged to see the prior's jaw relax when the pressure lifts. His breathing deepens, too. It's a good sign.

Ciaran grabs the jug of wine to flush the wound. The smell makes his mouth water, and despite the jagged shards around the neck, he's tempted to take a gulp. But he pushes the thought aside. In between pours, he can see the pink brain pulsing to the prior's heartbeat; it seems strong. He turns to the Valkyrie.

—I saw some cattle carcasses in your camp this morning. Can you fetch me some ox gall?

She nods and turns, then pauses at the door.

—Do you want me to find the others and swear you didn't betray them?

—Do you really think they'd believe you, swearing to a different god?

When she leaves, he realizes he could take a drink now if he wanted. Instead, he distracts himself, doing anything he can to keep busy. He fills the brass bowl with the remaining wine and steps away to a windowsill to chop the onion and garlic with the dagger. He then eases himself onto the floor and starts cutting the vestments into bandages, trying not to stare at the bowl. He wonders what onion-garlic wine would taste like. It can't make his body feel worse. He wishes the Valkyrie would hurry.

She finally returns, holding three wrinkled gallbladders the color of overcooked peas. He dribbles some greenish-yellow bile into the bowl, then squeezes them to wring more out. He stirs with the dagger. When he's finished, he holds the knife up to her from where he's sitting. She shakes her head no.

—Keep it. But I need to leave you now, old man. The ships are sailing soon.

—Why did you stay with me at all?

The Valkyrie looks toward the ceiling. She takes so long to answer that Ciaran isn't sure she will.

—I hated the nunnery. The songs, the bizarre holidays. They even made me read books. Everything there felt alien. I suppose I was the same as you. A lone believer, trapped among the Christians.

She squats down, eye to eye with him.

—If you want to come with us, I'll vouch for you. We need a leech. And the Norse beliefs are not so different from the Irish ones. You'd feel at home.

For a moment, Ciaran can almost see it. Sailing around, drinking and playing dice, singing the songs of his youth. Then he shakes his head. He's too old, too much of a sot. He drops his eyes again. It's answer enough.

—Will you be okay, old man? What about the rash one?

He smiles at the pun.

—As long as the prior lives, I'll be okay. Andrew's sworn to obedience. And a few of the more scurrilous brothers might even think me a hero for saving him.

She smiles back, and for the first time, dimly, he can remember her as a girl. Or at least he thinks so. But before the memory clicks, she rises and leaves him alone with Prior Patrick.

Only to return a minute later. She's holding two more broken jugs of wine, which she places at his feet.

—There's a dozen more in the sacristy, all half-full. Just remember that you saved him without it.

When she leaves, Ciaran moves the jugs aside. He has priorities. He dips the bandages in salve and wraps Prior Patrick's head. He also dabs his infected eye. His patient's color looks better, and his pulse is strong. There's reason to hope.

He reserves the fuller of the two jugs for the prior's care. Then, with a short prayer of thanks, he hoists the other and takes a long drink. It's every bit as wonderful and terrible as he remembers.

# Northern Alaska — AD 1000s

After their reign of terror in Europe, the Vikings began making excursions across the Atlantic Ocean, to Greenland and North America. There, they encountered the subjects of our next chapter, the people of the Arctic.

The Arctic has been home to several distinct and vibrant cultures, but this chapter focuses on the Thule people who inhabited Northwest Alaska—one of the harshest environments on Earth, where temperatures can drop to eighty degrees below zero. Estimates vary, but the Thule culture first arose around the year 1000 and quickly spread east through Canada and Greenland. The Thule later gave rise to the modern Inuit/Iñupiaq people.*

Unlike many hunter-gatherers, the Thule lived in semi-permanent settlements of up to sixty people. And while they hunted many different animals, their lives revolved around sea mammals, including seals, walruses, and whales, which provided food, fuel, clothing, materials for tools, and more.

The Thule were among the most technologically sophisticated cultures on Earth. (One archaeologist called them the "most gadget-oriented people of prehistory.") This technical wizardry arose directly from the harshness of their environment. In temperate climates, if your house collapsed or your hunting equipment failed, you could stand a night or two outdoors or fall back on gathering plants. Those were not

---

* The umbrella term *Inuit* refers to several distinct groups of people throughout the Arctic. Among them are the Inupiat (singular: Iñupiaq) of Alaska.

options in the Arctic. Shelters had to stand, dogsleds could not break down, hunting equipment had to fell game, or else you died.

Another factor in the success of the Thule was their insistence on sharing and the greater community good. All hunter-gatherers share things generously, to an extent that's sometimes hard for modern, individualistic societies to appreciate, but Arctic people were especially adamant. The only way to survive on the ice was through sacrifice — self-, and otherwise...

The night flashes by so quickly that Amaruq misses it. She's kneeling on the dirt floor behind her young cousin Meriwa, who's squatting over a sealskin mat and gripping two stakes in the ground for leverage. Amaruq urges her to push, *push,* but Meriwa claims she's too exhausted. Then she curses Amaruq, again, for being a bad midwife, for not making the pain disappear. Amaruq swallows her annoyance, and squeezes her cousin's rib cage to help.

The birth has turned the hut into a hothouse; Amaruq is already sweating through her leather pants and parka. The hut belongs to her family; it's a modest structure with an arched ceiling of whale ribs and walrus skins. Black sod covers the outside, and a hole papered over with seal gut provides light during the day. Benches of driftwood line the perimeter, draped with musk-ox blankets. During most nights, the rest of Amaruq's family would be scattered around the benches or lying on the floor, but the males are off hunting seals and the females have given them privacy for the birth.

When it's time, Amaruq shifts around to the front of her cousin and greases her hands with seal oil to help the baby along. Based on all the taboos Meriwa has broken while pregnant—lingering in doorways, walking around with her bootlaces untied—she expects a difficult birth. But to her surprise, the head and torso and legs slide right out onto the sopping sealskin. It takes no more than a minute, the easiest birth she's ever midwifed. When Amaruq sees it's a boy, she wonders whose soul he'll inherit—someone recently dead, or a long-lost ancestor? There's no way of knowing yet.

Meriwa slumps back, moaning. Amaruq ignores her sobs—her sister used to do the same when she wanted attention. The child's good comes first anyway. And right now, the child's not crying—in fact, not even breathing. She picks him up by the right arm first, to ensure he'll be strong, and holds him out to Meriwa.

—His nose and mouth are blocked with mucus. You have to suck it out.

Meriwa recoils, her eyes even bigger than normal. Amaruq says the baby will die, but Meriwa shakes her head no.

Irritated, Amaruq does it herself. She covers his tiny mouth and nose with her own mouth and sucks until a plug of mucus strikes her tongue. It gags her. But she spits and sucks twice more, until the block clears. Then she rubs seal oil on his soft chest with her thumb, warming him until the first cry escapes his lips. It's unbearably feeble. Her womb scar tingles to hear it.

Amaruq grabs her ulu knife and slices the umbilical cord. Then she tells Meriwa to push the afterbirth out. Meriwa swears she can't; she's too depleted. Amaruq explains she must, or it will rot inside her.

Another refusal. Amaruq's patience finally snaps.

—Open your mouth. Open it.

When Meriwa concedes, Amaruq slides a finger down her throat to make her gag. It's an old midwife trick: the contractions force the afterbirth out. It plops down on the crimson sealskin a minute later. While Meriwa curses her as a brute, Amaruq blots the bleeding between her legs with dried tufts of tundra grass.

She tries to teach Meriwa to nurse now, but her cousin won't listen. Amaruq hates losing her temper, but she doesn't know how much longer she can go without slapping her patient. She feels every muscle in her body clench; even her toes curl into fists inside her boots.

But the child, the child. She has to help him.

She just wishes the other women in town were here to witness this. They all defend Meriwa, claiming that Amaruq is jealous—and as much as Amaruq hates to admit it, there's some truth to that. Amaruq has wanted a child her whole life, and it stings to see her silly, frivolous cousin fall so easily into pregnancy at age fifteen—and not even with a proper husband, but after a dalliance with a roaming trader from the inland whom she'll never see again. Meanwhile, Amaruq cannot have children. When she was a child, her silly, frivolous sister Meriwa*—the

---

* Young family members inherit the soul of a late relative and then take that relative's name.

one whose soul was reincarnated in her cousin—was play-hunting with atlatl darts one day and impaled her womb. Amaruq barely lived, and her ability to bear new life vanished. The closest she ever came to children was assuming responsibility for Tuluk, her sister's child, when she died fifteen years ago. Her womb scar tingles during every delivery.

But her frustration with Meriwa isn't just jealousy. She despises how her cousin always puts her own desires above the community good. During meals she mixes land meat like caribou with ocean meat like seal, a crime that offends the gods of both domains. Worse, Meriwa mistreats seals. Before butchering a seal, the Thule always pour water down its throat as a prayer of thanks and forgiveness. But lazy Meriwa skips this step. The Thule believe, in their marrow, that hunters cannot catch game through skill alone; animals must offer themselves up and die willingly. And the spirits of mistreated animals warn the others to stay away and not offer themselves. By denying the seals water, Meriwa threatens the community's survival.

As does her indifference to her newborn son. The village needs children, but Amaruq fears that her cousin won't give this child the care it needs. She actually begs Amaruq to take it away and let her rest. Amaruq hushes her and makes her hold her baby; he needs warmth. There's one last step in the birth rite to perform anyway.

Amaruq unfurls a piece of leather and selects a bird-bone needle for the face tattoo.\* She then picks up the pot of ink she prepared earlier—soot from charred seal oil mixed with sterile urine—and studies her cousin's face.

Men get tattoos for feats of prowess—taking down whales, killing enemies in battle. Women get tattoos to commemorate momentous occasions like births. Meriwa already has the chin lines that all their women do, plus the star Amaruq's late sister had. Usually, for a birth,

---

\* In some Arctic cultures, expectant mothers got elaborate thigh tattoos, so that the first thing the baby saw upon entering the world was something of beauty. Most of these cultures used the painful subdermal-sewing method mentioned earlier.

Amaruq adds a curvy, avian *v* to a new mother's cheek—a new life taking flight.

But as she dabs the needle in ink, the tableau in front of her—a pouting Meriwa, a listless child on her breast—makes her hesitate.

—What's wrong? Meriwa asks.

Amaruq puts the needle down.

—Maybe you're right. You should rest before we do the tattoo.

Suspicious now, her cousin rises on an elbow.

—Why?

—Just until you're stronger. Rest.

Amaruq scoops up the placenta and hurries to exit the hut, crawling out through the low tunnel. Outside, she grabs the scapula of a walrus and drops to her knees. She has to bury the placenta so the village sled dogs don't eat it. But as she digs into the mucky soil, she's not worrying over the dogs. She's worrying over her cousin's child.

She doesn't give tattoos to commemorate stillbirths or children who'll die young. And given how limp the boy looks—plus Meriwa's apathy—Amaruq feels certain he's going to die without ever getting his soul.

Amaruq wakes inside the hut early the next morning—although "morning" loses meaning this time of year. With the solstice coming, nights have been dwindling to nothing, and after last night's fifteen minutes of darkness, now begins a surreal weeks-long "day" of endless sunlight. The prospect does not energize her. Summer is the busiest season of the year for her people, the time they tuck in food for winter, and there's always too much to do. Indeed, despite her fatigue, she should go help the other women now.

But not yet. She takes a moment to wonder after Tuluk. Her beloved nephew is on his first hunt right now, guided by two of her uncles. It's a two-week excursion, the last such trip before the summer sun breaks up the ice. She wonders how he's faring, whether he's proving strong and bold. She fears not—he's always been too gentle. She can picture his crooked half-smile and the white birthmark in his hair, soft as slushy snow. She would have worried over Tuluk all night if not for Meriwa's labor.

Meriwa. Amaruq sits up now and looks her over. Her cousin's sleeping with her mouth open, the baby tucked into her arm. They seem fine—Amaruq finally got her to nurse last night—so she peels off her blanket and rises. It's time to help the others.

Amaruq pads her way over to the exit. The low tunnel leading out forces her to crawl on her hands and knees; it slopes down, then up, a sinuous gooseneck that blocks escaping heat, leaving the inside balmy. Outside, she finds her parka on the whalebone clothing rack. The cold dried it by freezing the water out as ice crystals. She beats the garment to knock the crystals loose—a dozen firm whaps—then pulls it on.

She's heading to the seal racks, which sit on the grassy flatlands a half mile away, along the coast. As she walks, her boots sink into the soggy brown turf. Mist curls up from the ponds along the way and mingles with her frosty breath. A black-and-white snow bunting, plump and proud, warbles to her left. All male snow buntings have a unique song—a territorial display—and this one's sounds intimately familiar, an anthem for her village.

The seal racks, the center of the village's social life, are rough-hewn frames of gray driftwood. On them hang the red ropes of intestines, plus seal flanks drying into black jerky. At her family's rack, she tears off several leathery strips of flesh with her fingers; the underside is crunchy with dried blood crystals. It tastes delicious, like savory fish; the fibers fray between her teeth.

Her belly satisfied, she wanders off to help a family of four women. They've gathered for the most important task of the summer—rendering seal oil. Her people use seal oil for everything imaginable: it's a seasoning and a preservative, an antiseptic and a salve, a lamp fuel and a sealant for kayaks. The oil starts as solid blubber, which they chop with a curved ulu knife before rendering it into liquid. The knife's ivory handle once belonged to Amaruq's grandmother, a strong, resourceful, and prickly woman whose spirit everyone swears she inherited.

Native Alaskans dried whale and seal meat on wooden racks (*top*). They butchered these animals (*left*) using a curved knife called an ulu (*right*). (Top credit: Library of Congress. Left credit: National Park Service. Right credit: Copyright National Museum of the American Indian, Smithsonian Institution (11/3419). Photo by NMAI photo services.)

As they work, the other women ask about Meriwa's newborn. Amaruq admits she declined to do the face tattoo. The women exchange glances.

An hour passes. Amaruq is starting her third seal when, to her shock, Meriwa wanders up. By herself. Amaruq can barely get the words out.

—Where's your child?

—Sleeping.

—What if he falls off the bench?

—I put it on the floor. Is there any seal? I'm starving.

Amaruq scowls as the other women scurry to feed her, actually placing bites of seal in her mouth as if she's an infant. Can't they see how selfish she is?

One woman even asks if she wants to hear a story to soothe her spirits. Meriwa says yes, and requests the poop-knife story. Amaruq objects—it's an idiotic tale—but she's quickly shushed.

Amaruq barely listens. In short, an old man declares that he wants to spend a winter on the ice hunting. His family says no. They're sure he'll die of exposure, so they take away his sled and tools. But the old man is determined—and clever. During a gale one night, he slips outside, drops his breeches, and defecates. When the turd freezes solid, he hones it into a blade. Grimly, he kills a sled dog and guts it. Using its rib cage as a sled, he harnesses his other dogs to it, and races off into the night, cackling as he goes.

Throughout the telling, Meriwa keeps adding details the others miss—until, by tale's end, she's taken over the story. Even in her foulest moods, Amaruq will admit that Meriwa is a gifted raconteur, what with her big eyes, her animated voice, her lively hands. But this morning, Amaruq can't take it. What happened to her being exhausted from labor? Amaruq finally interrupts and announces she's going to check on Meriwa's boy—to make sure he's alive. She wipes her oily ulu in the grass and rises.

She doesn't even make it two steps. A mile beyond the muddy beach, out on the sheet of sea ice, she sees someone. Six days ago, her uncles Siku and Silla packed up two dogsleds and set out with Tuluk on his hunt. They were supposed to be gone two weeks. But out on the ice now, she sees Silla limping back—alone.

She hurries out to help. Through chattering teeth, Silla explains what happened. He and Siku were letting Tuluk practice driving a sled. But he leaned too far around a turn, got tangled in the reins, and fell off; he ended up being dragged behind the dogs. Siku hopped on the other sled in pursuit. After a chase, he caught the runaway dogs and started untangling Tuluk.

Unfortunately, they'd halted over rotten ice. While running up to help, Silla heard it shudder and crack beneath their weight. Siku managed to shove Tuluk onto stronger ice before it gave way, but dropped into the freezing water himself. He clawed his way out, but both sleds and twenty dogs went down. Silla ran up to find his brother already shaking with hypothermia. Tuluk was in even worse shape: The sled had mangled his foot while dragging him along, and he couldn't walk. Silla had been trudging three days and nights to reach the village.

By the time they reach the beach, Amaruq's stomach is coiled so tight with dread that she fears she'll vomit. But she wills herself to stay calm. She leaves her uncle with the other women, then dashes off to prepare a dogsled. No one even debates sending somebody else.

At the clothing rack outside her hut, she switches to a thick parka with wolverine hair lining the hood; wolverine hair doesn't hold moisture and won't grow icicles from her breath. She also grabs thicker boots with stitching made of caribou sinew, which swells when wet to fill the holes and keeps water from seeping inside. She lines the boots with tundra grass to absorb sweat and pulls them on. Then she gathers up an atlatl, darts, shovel, and other gear, including a lamp kit with oil, kindling, and flints.

The dogsled sits around back. Its driftwood frame is bound together with sinew. The whalebone runners are lined with baleen, slick black sheets that snow won't stick to. As Amaruq secures her supplies on the sled, every remaining dog in the village begins yipping, darting back and forth in excitement. They have shaggy black-and-white coats and fat paws—natural snowshoes—and can run a hundred miles a day. At least in theory. These dogs are second-rate; the best ones are out pulling sleds for hunters right now. Amaruq picks the least worst of them and harnesses four pairs to the sled with braided sealskin straps. Two have all-white faces, so she smears soot beneath their eyes to absorb the glare of the sun, which reflects intensely off the ice. To protect her own eyes, she ties on a pair of snow goggles—a mask of antler with an eye slit.

As she steps onto the runners, Amaruq realizes she forgot to check on Meriwa's boy. But saving two hunters means more to the community than a doomed infant. Besides, it's Tuluk.

She hollers at the dogs—*go!*—and feels a jerk as they tear off.

Research for this chapter took me to two diametric, even contradictory, climates: Shishmaref, Alaska, just below the Arctic Circle; and Las

Vegas. I'll start with Alaska, where I sampled arctic cuisine and learned to perform the single most important chore of traditional arctic life: rendering seal blubber into oil.

Shishmaref is a coastal village of five hundred people on a sand spit near the Bering Strait. There's a *Field of Dreams* quality to the place. In most ways, it's a typical modern town, albeit with an arctic flavor: seemingly every home has two or three snowmobiles or ATVs out front. I've never seen so many children playing outside—laughing, shrieking, tossing balls, riding bicycles. (The town must also contain the highest number of trampolines per capita of any city in the world. On my trip there, I count fourteen, every one of them full of bouncing kids, a seeming nod to the famous arctic blanket toss.) Still, when I cross one dirt road on the edge of town, I feel like Kevin Costner—transported to another world. Minus a few elements, this scene could have existed a thousand years ago. Yellow arctic grass stretches to the muddy shore. Shallow ponds give off gothic mists, while a polar bear skin flaps on a rack in the wind. Beyond it, an infinite ice sheet extends to the horizon. Most prominent of all, I see wooden racks draped with seal flesh, a staple of the arctic diet even today.

In the past, most seal meat was eaten uncooked,* for good reason. Human beings need to consume vitamin C or we get scurvy, a fatal condition that sailors at sea often succumbed to. Given the dearth of edible plants up north, arctic diets seemingly lack this nutrient. But there is another source. Marine-mammal organs contain it in abundance. Muscle meat contains some, too—provided you don't cook it, since heat destroys vitamin C. Consuming raw flesh, then, was a necessary adaptation to arctic life.

Historically, too, many Inupiat swore that fully cooked meat tasted

---

* The antiquated name for native Alaskans—Eskimos—supposedly means "eater of raw meat," although most linguists now think that's a canard. Instead, the name probably comes from an Algonquian phrase meaning "netter of snowshoes." Regardless, I've avoided using the term here because it's too vague, lumping together coastal and interior clans from Alaska to Greenland who led widely different lifestyles. It would be like speaking of "Europe culture" instead of distinguishing between the very different lives of Poles, Italians, Britons, and other groups.

less rich to them. If you've ever had perfect sushi, you can see their point. Instead, they prepared meat in other ways. They fermented seal flippers underground to give the flesh a sour tang. They plucked clams from acidic seal stomachs for arctic ceviche. They dried seal meat on racks in the sun to produce a soft jerky. (I tried some in Shishmaref. It tastes like cured salmon—delicious.)

Arctic cuisine also got creative with textures. *Quaq* consists of thin slices of partially defrosted meat; they have a pleasant crunch when you bite down, then melt into savory goo. Packing raw bird eggs into seal intestines and freezing them into sausages achieved a similar effect. Other dishes ran creamy, like *kapuktuk,* a foamy pink fluff of blood and oil, and *akutaq,* an "ice cream" of whipped fat and oil sweetened with berries. Another dessert involved fermenting berries and field greens in oil inside a "seal poke," a waterproof sealskin bag. Spoonfuls of it fizzled on the tongue.

Some dishes combine multiple textures. One day in Shish, my host family generously invites me over for "arctic salad." It's a medley of crisp local greens; rubbery hunks of *muktuk* (whale flesh) with thick black skin and light pink blubber; chopped red rings of seal intestine that are as tough as calamari and taste of hot dogs; and crunchy beige clusters of herring eggs that look like couscous and taste like cauliflower.

Another day, a local couple named Jessica and Archie welcome me over for boiled walrus. It comes in inch-wide chunks made of three layers, each so distinct they look like they were glued together from different animals. On top is the creamy pink skin, smooth as ivory. Next comes the beige blubber, composed of what appears to be hundreds of cooked rice grains fused together. At the bottom are strips of muscle meat that look like scraps of exploded tires. Jessica and Archie serve the walrus on cardboard (to soak up oil) and slather it with French's mustard, whose vinegary flavor provides a nice contrast, they say. I take my walrus unadorned. The muscle meat tastes vaguely of roast beef. The skin reminds me of boiled fat, but with a texture like corkboard. Then there's the blubber. When I bite down, it compresses between my teeth like a sodden rice cake. Oil squirts out from the crushed grains, a gush of fatty goodness. I've never tasted anything like it, melty and chewy all at once. I begin spearing the biggest pieces I can find.

The next day, Archie and his sister Janet teach me how to render blubber into oil. My lesson takes place at their family's meat-drying rack on the muddy flatland near the beach. Four sealskins with thick layers of blubber are lying there. For the most part, the landscape around Shishmaref lacks much color; it's white ice, brown mud, yellow grass. In comparison, the blubbery sealskins look hallucinatorily bright—sherbet orange and bloodred, like fleshy lava flows.

The first step involves separating the blubber from the skin with an ulu, a Native Alaskan knife. It's a wide, semicircular blade with a short handle. You don't slice with it as much as jab. For butchering, ulus have an advantage over regular knives because you can get your weight behind them and push down with added leverage. Archie demonstrates. He drapes the red-orange flesh over a curved piece of wood the size of a cutting board. He holds the skin with one hand and thrusts with the other—short, rhythmic chops. The blubber peels away in pieces as thick as Texas steaks. In his hands, butchering looks like a snap.

It's not. When it's my turn to cut, I realize just how heavy the sealskins are—they must weigh fifty pounds each, and they keep sliding off the board. It doesn't help that they're as slick as Crisco on black ice. With every jab of the ulu, oil spatters my face and hands. Needless to say, my cuts are clumsy, and leave far too much blubber on the skin. And while the portions Archie removes are smooth and flat, mine have a jagged topography, jerking up and down.

We hand each blubber steak to Janet, who stands at a nearby table with another ulu. Her job involves removing any bruised or bloody bits, which make the oil rancid. She chops the rest into one-inch strips, and drops them into plastic drums. Over the summer, as the temperatures in Shishmaref soar into the 40s, the strips will melt into oil, which Janet and Archie's family uses throughout the year. Four seals fill a 55-gallon drum. When I ask whether I can help chop, Janet's mouth falls open; I've clearly committed a faux pas, although I'm unsure what. Archie's fiancée Jessica finally explains: "This is women's work!" Then they all laugh at me. For whatever reason, women can both deblubber skins and slice blubber into strips, but men never do the latter.

People in Amaruq's time didn't use plastic drums to render blubber, of course. They used sealskin sacks called pokes. These were essentially hollowed-out seals whose entire body cavity became the bag, flippers and all.

Native Alaskans used sealskin bags, called pokes (above the swings), to store oil, meat, and other goods. (Credit: Dorothea Leighton M. D. Collection, UAF-1984-31-19, Archives, University of Alaska Fairbanks.)

In Shishmaref, I talk to a young man named Thomas who made a poke after finding a seal marooned on a beach. Given its limited locomotion on land, it seemed likely to starve, so Thomas took mercy and clubbed it. Upon seeing it was female, he decided to make a poke like his Iñupiaq elders used to. (Males make poor pokes because they bite each other and leave holes in the skin.)

First, he slit the seal's cheeks to widen the mouth hole and scooped out the skull. Then he rolled the skin down the body inch by inch, turning it inside-out and removing each organ in sequence. The flippers were especially tricky to hollow out because the anchored bones and ligaments there can tear the skin and ruin the poke. When I express surprise that a tiny mouth hole would stretch over a seal's ample belly, Thomas smiles. It took a hell of a lot of work, he says, but sealskin is impressively stretchy. After three hours, he had a saggy, inside-out skin covered in blubber.

To remove the blubber, he relied on an old trick. It's far easier to scrape blubber off when the skin is taut, so Thomas cinched the mouth hole shut and inflated the poke with air from a tire pump. In the old days, people blew air inside with their lips, like filling a balloon. (A polite man, Thomas skated over what orifice they used as the inlet valve. But an old-timer in town, Cliff, told me straight up: "You blow 'em up through the asshole." Then he cackled, lit his fifth cigarette in a row, and told me about 'Nam.) When Thomas finished scraping blubber, he let the air out and turned the skin right side out. Hey presto, a poke. Eventually, he turned it into a backpack to carry ammunition and binoculars on hunting trips.

Sadly, Thomas might be the last living Alaskan to have made a seal poke—it's a dying art. But for centuries, people used them to store seal oil, the fuel that drove Alaskan culture. To make them leakproof, they stitched the mouths and anuses shut or plugged them with wooden stoppers. People stored food in pokes as well. As long as every cranny was filled with oil to prevent bacterial growth, strips of meat could last for years inside them in the arctic cold, especially underground—a welcome cache of nutrition in times of emergency.

Since there's no smooth way to transition into this ("Speaking of orifices..."), let's just jump right into the poop-knife. Did I make that story up? No. No, I didn't.

The basic poop-knife tale, about an old Inuit man sharpening his feces into a blade, has been widely reprinted in archaeology textbooks, and many scientists assumed that it was based on a nugget of truth. In a culture that largely lacked metal and had little stone, perhaps a turd-knife was a viable tool. Other scientists cocked an eyebrow. I mean, come on—a poop-knife? Only one scientist, however, had the guts to put the story to the test.

That would be the irrepressible Metin Eren, who taught me how to knap stone and fling atlatl darts at Kent State. Eren first heard the poop-knife story on the radio in high school, and was so captivated that he wrote the station for a copy; he still has the cassette. As he got older, however, he grew skeptical, so he designed an experiment to investigate.

He started by switching to an all-meat diet similar to what arctic people ate, to make his turds the right consistency. (Day one was delicious, he recalled. By day three, he had a headache.) After a long week, he collected "the raw material," as he put it, froze it solid, and sharpened it with a file. He then tested the poop-knife on a pigskin. Would it slice through?

Not even close. No matter how long Eren carved or sawed, the fecal knife did little more than streak the hide "like a brown crayon." Poop-knives just don't cut it.

Overall, it seems that scientists were a wee credulous for taking the story seriously. Rather, it was probably just a funny, mischievous tale.* Perhaps it also imparted lessons about resourcefulness and persistence, and especially about not fearing the cold but turning it to your advantage—lessons that Amaruq would need to heed, and then some, to rescue her beloved Tuluk...

---

* A Danish explorer purportedly did save his own life with a poop tool once. After digging a pit to sleep in, he found himself trapped by snow in the morning. So he defecated, shaped the turd into a chisel while it was pliable, and chipped his way out when it froze. But as Eren points out, no one else witnessed this feat, so we have to take the explorer's word for it. Moreover, chiseling snow is easier than cutting flesh, so his tale proves nothing about the Inuit story.

Incidentally, the poop-knife study won Eren and his team an Ig Nobel Prize, a sort of bizarro Nobel awarded to research "that first make people laugh, and then make them think." Metin says that the founder of the Ig Nobels told him that the poop-knife report was one of the most nominated papers in the prize's history.

# Northern Alaska — AD 1000s

Amaruq's journey begins along the ice near the coast. Every quarter mile she hears a new snow bunting and its unique song. The ones near her village sound familiar, then increasingly less so. Eventually, she turns from the safety of the shore and heads out to the open sea ice.

She keeps the dogs zigzagging—*left! right!*—to avoid any gray patches of ice, which are rotten; solid ice looks powder blue. But the dogs, being inferior, often cut too sharply and the sled teeters; she has to dig her boots into the crust of snow to keep upright.

Before long, she feels her face getting wind-burned. She also needs to locate food for the dogs, who need far more fuel than she can carry— 10,000 calories a day. Her uncle told her the accident took place west of Qumanguaq, "The-shrugging-hill-with-no-neck." To reach Qumanguaq normally, she'd follow a curving coastal route that meanders from one reliable fishing hole to another. But given the emergency, she's gambling on a more direct route across a mosaic of islands and open ice. Where they'll find food along the way, she doesn't know.

Dogsleds have long been vital to life in Alaska. (Credits: Library and Archives Canada, PA-011706, and Michael Barera, Wikimedia Commons.)

Above all, she worries about getting lost. On a normal journey, she'd use a story-song to reach Qumanguaq. Arctic people name every possible feature along a route, even tiny creeks and ridges, and they do so with elaborate monikers—"the hillock whose streams resemble tattoos on the chin" or "two islands that look like seal teeth"—so there's no confusing one feature with another. Then they work those names into traveling songs, which tell stories that describe the journey from one landmark to another. The songs function as melodious maps.

These maps, however, don't work on open sea ice, one of the emptiest and most disorienting landscapes on Earth. Instead, Amaruq relies on subtler navigation clues. A warm breeze often blows from the southeast during summertime; so whenever she passes an island with a pond, she checks which corner is melted. Similarly, northwesterly gusts will carve snow into hard drifts shaped like tongues. By observing the orientation of these drifts, she keeps her heading true.

Unfortunately, there's far more rotten ice than she expected—a consequence, she knows, of mistreating seals. (That damn Meriwa.) As a result, it's slow going, and after some time—perhaps half a day; it's impossible to tell—the dogs are staggering with hunger. So when Amaruq spies a breathing hole in the ice, she halts the team, grabs her atlatl, darts, and leather rope, and creeps over to lie down and wait.

The delay frustrates her; every minute sitting still is a minute that Tuluk and her uncle are lying wet and exposed on the ice. But there's no way around it. The dogs need to eat, and she doesn't know of any food caches nearby.

Mercifully, the wait is short. An hour later, a ringed seal pops up and flops onto the ice. Amaruq remains perfectly still until it's settled and sunning itself. She's fifty yards away, and begins sliding forward on her belly. Every so often she scratches a barbed atlatl dart on the crusty ice, which intrigues the seal; it cocks its head curiously.

When she's twenty yards away, it rises up warily. She stops. Then it slumps down again, exposing its flank. Amaruq whispers a prayer of thanks. The seal is offering itself to her. She slowly loads a dart into the atlatl, formed from the penis bone of a sea lion. The shaft is tied to the leather rope, which she binds around her left fist. She takes a deep breath—then pops to her knees and flings.

The dart pierces the seal's back, a perfect shot. The seal yelps and tumbles into the hole with a splash. Amaruq is jerked forward onto her belly. But she holds the rope tight, and for once the rotten ice helps—her knees and elbows dig in and anchor her. The seal strains to dive, but can't get deep. It surfaces for air a minute later, then dives again.

They do this dance several times, the seal thrashing and barking in desperation. Finally, Amaruq feels it weakening—the tug on the line feels less sharp. When it surfaces again, she yanks it back onto the ice. It flops around wildly, heaving itself toward the water. But it's already lost too much blood. Within minutes, it exhausts itself. Amaruq reels it in, and dispatches it with a jab to the throat from another atlatl dart.

Her dogs are howling like banshees behind her. But she patiently performs the rites. She fills her mouth with ice and, ignoring the pain in her teeth, lets it melt. Then she dribbles the offering of water into the seal's mouth and prays.

Now to carve it. With a start, she realizes she forgot her ulu—in her haste to leave the village, she left it at the drying rack. So she uses the atlatl dart to split the gut, exposing the dark liver and intestines, plus two acid-cooked clams in the stomach. After downing a few morsels, she walks back, unties the dogs, and watches them bathe the ice in crimson.

Hours later, west of The-shrugging-hill-with-no-neck, Amaruq spots Tuluk and her uncle lying on the ice. There's no sign of the dogsleds, just a ragged hole. She stops her own sled and crosses the last hundred paces on her hands and knees to spread her weight out. The ice beneath her is dark, dangerously rotten.

Her uncle is dead. She briefly wonders whose child his soul will occupy—perhaps Meriwa's boy, if it lives. As a stocky man, his weight has pushed him several inches into the ice, further cracking it. He'll eventually fall through and be devoured by sea lice, a sad fate. He was a good uncle, funny and caring, and she'll miss him dearly—but she can't spare any emotion right now. Tuluk needs her.

Her nephew is lying ten yards away, shivering and delirious; her womb scar tingles as she scoots toward him. She can see the white birthmark in his hair peeking out from his hood, but his crooked half-smile has been replaced by tight white lips pulled into a rictus. His foot is twisted so grotesquely that she fears he might never walk again. There's also a wound where the sled runners punctured his boot; it's crusted in blood. She checks his eyes, and sees a white film from snow blindness. At the village, she'd treat this by plucking a louse from someone, leashing it to a strand of hair, and letting it crawl across his eyeball; the film sticks to lice legs. But she has bigger worries now. She ties the leather atlatl rope around his torso and starts towing him to stronger ice. At the first tug he groans, groggy with pain.

It takes twenty minutes to drag him to safety, and the strain leaves her arms and legs trembling. But she can't rest—they need shelter. As a girl, Amaruq once heard a story from some traders, the men from the interior who brought her people furs, antlers, and flint in exchange for baleen and sealskin. Around the fire, they told her village of a faraway people who made domed shelters from ice sometimes. Amaruq's people had never heard of these "igloos."* They seemed ludicrous, fantastical. But the notion always stuck with her, and inspires her now. She recalls a snowdrift a mile back, five feet high and compacted by the wind.

Slowly, painfully, she loads Tuluk onto the sled to backtrack to it. When they arrive, she dismounts and begins chiseling blocks of ice from the drift with her atlatl darts. Both points break eventually, but within three hours, she's scooped out a low cave. Then she stacks the blocks into a wall to seal it. She leaves a hole for an entrance tunnel and drags Tuluk inside with the last burning bit of strength in her arms.

She lays him atop a fur and gets a small oil fire going in the

---

* Historically, igloos were common in north-central Canada and parts of Greenland, but never in Alaska.

soapstone lamp-bowl she brought. Then she removes his boots and mittens. His foot looks bad — an angry swollen purple. Meanwhile, his toes and fingers look maggot-white, a sign of frostbite. She heads outside and scoops up handfuls of warm dog feces. Apologizing for the stench, she packs them around Tuluk's hands and feet. Better smelly hands than fingerless ones.*

Even when covered in another fur, Tuluk won't stop shivering. His poor state forces Amaruq to consider her options.

They could make a dash for the village — which might be smart, given that the summer sun will only continue to weaken the ice. But she doubts he'd survive the trip in his hypothermic state.

More brutally, she could abandon him — her people do that sometimes.† But only during famines, when the entire community's survival is threatened. Even then, she'd never abandon Tuluk.

Her best option, then, despite the weakening ice, is to find hot food to warm him and let him recover. The question is where to find the food. Her atlatl darts are broken, making seal hunting impossible. She hears the dogs yapping outside, and considers butchering one. But no; she'll need them all to pull the added weight of Tuluk back.

That leaves one choice — a baroque, multipart plan that could well fail. But if she hurries, she can have a hot meal ready in hours.

---

\* An Inuit cultural council in Canada has noted that Arctic people really did pack dog feces around people's extremities sometimes to prevent frostbite. Squirm all you want, but I bet you'd do the same rather than lose fingers or toes.

† Infanticide, invalidicide, and senilicide — leaving the young, sick, and old to die of exposure — have long-standing roots in cultures around the world. (Indeed, one of the founding sagas of European culture starts with the baby Oedipus being exposed in the wilderness and left to die.) The Thule resorted to such measures only in extremis. They dispatched people by stabbing them with harpoons, tossing them into frigid water, or walling them into ice. Female infanticide was especially common; some Arctic villages had boy/girl ratios of two to one. But the ratio usually evened out by adulthood, given the high rate of accidents among male hunters.

Oddly enough, my next bit of Arctic research takes me to Sin City—Las Vegas—to investigate the mystery of Alaska's "ugly pots."

This might be anathema to admit in an archaeology book, but I always found clay pots boring. I mean, yes, I know they're vital for the field. Most material goods from the past—clothing, bedding, art, wooden tools—disintegrate all too quickly, leaving no trace behind. Pottery lasts. Pottery also conveniently changes from place to place and era to era, making it diagnostic for different cultures. It even retains residues from food sometimes, allowing us to reconstruct what people ate. All important stuff.

But egad. The infinite taxonomy of jars with slightly different rims or handles or whatever never fails to make my eyes glaze over. If I were an archaeologist who'd been toiling in the dirt for months, my knees aching and lungs black from inhaling dust, and I had no necklaces or goddess figurines to show for it, just a pile of frickin' potsherds, I'd be homicidal. As with tax accounting, I'm glad someone studies that stuff, but I'm even gladder it's not me.

All my grumpiness vanished, however, when I learned about western Alaskan pots. These pots excited me, because they employed some pretty outré ingredients, things you won't find on the shelves of your local pottery studio—seal oil, and a whole lot of blood.

I first heard about Alaskan pots from Karen Harry, a ceramics archaeologist at the University of Nevada, Las Vegas. She's a boisterous Texan with a voice that climbs several octaves when she gets excited. She explains to me that cooking pots—vessels for heating porridges, stewed grains, and the like—generally share certain features across the world. They're thin-walled, to allow heat from a fire to warm the contents efficiently. They're hefty, to hold enough gallons of stew to feed a family. They're fired, to seal the walls and prevent water from seeping out. And they're round-bottomed, to extend their lifespan. That's because pots expand when heated, and sharp joints (think of the base of flowerpots) are high-stress areas that crack during expansion. Rounded cooking pots, which have no joints, last far longer.

Alaskan cooking pots lacked all those features: They were pint-sized, unfired, thick-walled, and cylindrical. Harry shows me some shards in her lab; they look like chunks of scorched dirt, what ceramicists call "crudware." In fact, the first time a colleague showed Harry the shards, she laughed and told him they couldn't possibly be the remains of cooking pots. And yet. Despite Harry's elegant, well-grounded objections, people in Alaska have been cooking with these ugly, squat, unfired pots since 1000 BC. Harry had to swallow her pride and figure out what was going on.

One mystery gave way easily—the small size. As noted, Arctic natives often ate meat raw to preserve the vitamin C. Still, there's something cozy about a hot meal, that warmth radiating from your belly. Hot food and beverages can also save someone's life when they're hypothermic. The problem is, big fires weren't an option in the upper Arctic. The area lacks trees, and the driftwood that arrives when the summer ice breaks up was too precious to burn, so roasting meat on spits was out. Instead, the natives started small fires using oil and tiny arctic twig-brush—just enough to heat a few cups of water and parboil some bites of meat. Harry calls it arctic fondue. The pots' small size, then, simply reflected Alaskan eating habits and their lack of fuel.

Other mysteries proved more stubborn. In general, clay pots need firing at high temperatures to harden the walls and prevent them from crumbling. But before being fired, the clay needs to dry. (If you fire still-damp pots, tiny pockets of water in the clay will expand into steam and crack the walls.) Unfortunately, the damp, chilly climate of coastal Alaska makes drying clay all but impossible. Alaskan potters therefore needed to make vessels that could hold water without being fired. But how?

Old anthropological accounts mentioned that they solved this problem with seal blood and seal oil. But the reports were vague on details. So Harry visited Alaska and started experimenting. She soon realized that Native Alaskans were masters of practical chemistry.

Initially, Harry tried blending oil or blood into the clay. It did not go well. Oil turned the clay into goop—a slimy mess that wouldn't hold its shape. Blood, meanwhile, clotted the clay and turned it into crumbly dough. Mixing these fluids into clay clearly wasn't the answer.

Then Harry had an inspiration. What if she *coated* the pots with blood and oil? Would that seal them well enough to boil some water for fondue?

To find out, I joined Harry at UNLV one spring morning to re-create her experiments. We meet in a concrete courtyard behind her lab that's outfitted with folding tables and some alarmingly wobbly office chairs. Harry pulls out a bucket of dark-gray Alaskan clay and begins mixing it with her hands; the muck soon reaches her elbows. Then she splats out a few handfuls each for me and a few graduate students on hand. The Alaskan clay feels crumbly, far coarser than silky commercial potters' clay. We mix in some water and temper and begin shaping it. We're making four sets of pots for our experiment: some that we'll coat in oil, some in blood, some in blood and oil, and some in nothing.

Native Alaskan pots were smeared with seal oil and seal blood (*left*), then used to cook fish or other foods, often with heated stones (*right*). (Copyrights: Sam Kean, Karen G. Harry.)

As we work, Harry explains another mystery. Native Alaskans made pots during the short Arctic summers, the driest time of their year. (Although "dry" is a relative term: "That means maybe it doesn't drizzle all day," she says.) Unfortunately, that flash of summer is also an overwhelmingly busy season in upper Alaska. People are gathering the only berries and greens they'll see for the next eleven months, as well as processing the seal and fish they'll need to survive the winter. There are a million chores to do, and potters lacked the time to lovingly craft each vessel. They simply cranked them out and moved to other tasks. That haste explains the shape and thickness of the pots. However superior for cooking, round pots with thin walls take hours to sculpt and require high-quality clay; they also need firing to hold their shape. Stumpy cylinders take twenty minutes and don't need good clay or firing. There's an old engineering saw about the inevitable trade-offs you face when manufacturing things: good, fast, cheap—pick two. Based on the unique challenges they faced, Alaskans opted for fast and cheap.

In the courtyard, Harry even suggests that the pots we're making *should* be a little crappy. They won't be authentic for our experiment otherwise. I perhaps take Harry's words a bit too much to heart, though, because my pot is truly a piece of garbage—a lumpy, uneven Quasimodo, like a sandcastle after the tide came in. It reinforces a gnawing fear of mine, one that came up over and over while writing this book—that if I'd been born pretty much any time before about 1900, I would have starved.

After we form the pots, it's time to apply the blood and oil. To my surprise, Harry stores her seal oil in a yellow bottle of chocolate Nesquik, just like you'd get at 7-Eleven. Inside, I see a viscous, rosé-colored liquid that smells vaguely herbal. One student eventually gets nauseous from the odor and has to leave, but I'm too curious not to dab some on my pinky and taste it. The strong fishy flavor crinkles my nose, and lingers on my tongue for hours.

I nevertheless pour some into a green plastic bowl and begin coating my pot using a paper towel as a blotter. It makes Quasi glisten. Only after finishing do I notice everyone else wearing gloves; I hadn't seen Harry set them out. I wonder how long my hands will smell like seal guts.

When half the pots are nice and shiny, we switch to blood. Harry doesn't have any seal blood handy, but one of her students used to work at a veterinary clinic that does transfusions for dogs that get struck by cars. That blood expires eventually, and the clinic has donated several pints to Harry. (The student later admits that the clinic, instead of discarding expired blood, usually donates it to CSI labs, who use it to train detectives on interpreting splatter patterns at crime scenes. We've deprived them of that month's supply. One student predicts, "Now there will be an unsolved murder because of us.") The blood comes in hospital drip bags. We snip them open with scissors and squeeze the fluid into a blue bowl. I'm normally not squeamish, but seeing that much free-flowing blood unsettles me.

Gloved this time, I begin dabbing blood onto Quasi with another paper towel. But unlike oil, which glided over the surface and oozed freely into every crevice, blood doesn't spread well. It's too sticky, forcing me to actively work it into every fissure. I finally resort to finger-painting.

By the time I'm done, the leftover blood in the bowl has congealed into a tacky glob, like a sunny-side-up yolk that's been sitting out on a diner plate overnight. But the pots themselves look striking—a deep glistening red, like animal hearts we've extracted for some mysterious rite.

At this point, we put the pots aside to let the coats set overnight. If I were Thule or Iñupiaq, I would have scrambled off to harvest some berries or process seals. As it is, I simply peel off my gloves—and find they've leaked. My hands are now coated in more blood and oil than some of the pots.

My cuticles still pink, I arrive the next morning on a surprisingly gusty day. Harry builds a fire in the courtyard's firepit, and hot embers are soon swirling in the wind, drifting around campus. We have four trials planned; in each case, we'll fill a pair of pint-sized pots with a few cups of water and try to bring them to a boil.

We start with the pots without any oil or blood. They stink. The untreated clay walls have no structural integrity, and holes open up within a minute. The water gurgles out as quickly as if you'd pulled the drain from a bathtub. A total failure.

The oil-only pots do better: the water begins simmering within about five minutes. Then things go sideways. Clay from the walls began sloughing into the water, turning it the color of mocha. When I dip a stick in the pot, it comes up coated in muck. Harry explains that oil conducts heat well and thereby transmits the fire's energy through the thick walls and into the water. But oil alone cannot seal clay. Water eventually invades the walls and degrades them.

Next up are the blood-only pots. One fails quickly due to poor construction; all the water gushes out. Thankfully, the other holds up well: blood forms a good seal, and the clay never degrades. But without the oil, the heat from the fire has trouble penetrating the thick walls, and the water inside never comes to a boil. The result, then, is the opposite of the oil-only treatment: clean(ish) water, but not enough heat to cook.

The final trial mimics Alaskan pots, with both blood and oil. We also toss a third pot into this round — Quasi. I'm not optimistic. The upper third cracked overnight, and it looks like something excavated from an earthquake. I'm not sure it will even hold water, much less stay intact long enough to bring it to a boil.

On this trial, one pot again fails immediately (not mine!) and the water rushes out. But the other pot starts boiling in minutes, and the water inside remains clean. Overall, this setup combines the best of both blood and oil: efficient conduction and a good seal. From her initial disbelief, Harry has come to appreciate how clever ancient Alaskan pots were. What they lacked in aesthetics, they made up for in engineering and chemistry.

The dual coating even redeems Quasi a little. Sure, the cracked upper rim tumbles down within a minute of entering the fire; but the base holds up well, and the half inch of water in the bottom is soon streaming with bubbles. That's when I know Arctic pottery is truly genius: it overcomes even my incompetence. Maybe I wouldn't have starved after all.*

As we wrap up in the courtyard, Harry recalls her own fumbling attempts to re-create Alaskan pots over the years. "I always picture the ancestors looking down from the afterlife and laughing at me. They're

---

* Okay, I would have.

saying, 'No, you idiot!'" She shrugs and sighs. "Then again, I say back to them, 'I'd like to see you drive the interstate at rush hour.'"

We all laugh. But we can afford to. When our ugly pots fall apart, we lose nothing but pride and labor, and even gain knowledge in the process. The stakes were far harsher for Amaruq on the polar ice sheets. There, a failed pot could cost someone their life...

From the igloo, Amaruq and the dogs make the long dash toward land. Her people often cache food after particularly good hunts, burying it onshore for later emergencies. Her uncles showed her one such cache several years ago, and she thinks she can find it again. She rehearses the map-song on the way, going through all the landmarks in her mind.

A light fog has obscured the shoreline when she arrives, but after some searching, she finds the rock she's seeking; it looks like a bloated seal flipper. Just behind it, she topples a cairn of stones and starts removing the heavy wet earth below with her scapula shovel, to reveal a trapdoor of whale bones. In a hollow beneath, on a bed of more stones, sits an intact seal poke. She hauls it to the surface.

The straps cinching the poke shut have hardened into rawhide, and it's tough work loosening them. The congealed oil inside—lard, really—smells a bit rancid; she prays the meat hasn't spoiled. On top sits a piece of blubber. She works it loose, takes a nibble—and just about swoons. It's delicious, chewy and rich with oil. Her appetite roars awake and she shoves the whole chunk in her mouth. Two more follow, and she licks the grease off her fingers. She tosses several pieces to her dogs, then cinches the poke shut and loads it on.

At the shoreline, she pauses to scoop up twigs and several handfuls of clay, then snaps the reins. A sharp jerk, and they're off.

Two hours later, she hops off at the igloo. Tuluk looks paler than ever. She has to hurry. She shapes the clay into a pot and slathers it with the seal oil for the lamp. Now for the blood. She needs it fresh, and with no seals around, she works a shard of baleen loose from the sled runner.

Then she rolls up her parka sleeve. When she jabs the shard home, she gasps, and a pulse of snow-blind-white pain flashes through her mind.* But she now has all the blood she needs.

After coating the pot, she adds oil and twigs to the lamp fire, and sets the crimson pot nearby to dry a bit. Twenty minutes later, she melts some snow for arctic fondue, and shakes Tuluk to wake him. He barely stirs. Begging his forgiveness, she winds up and slaps him. His eyes pop open, but she knows he's not all there.

She starts by dribbling warm oil into his mouth with her fingers; this will coat and soothe his throat, which has surely been scraped raw by the cold. Tuluk responds hungrily, his lips puckering like a suckling infant's.

After feeding him a cup of oil, she plops some meat from the poke into the boiling water. To her people, a perfectly cooked piece of meat would have a warm, soft surface and a crunchy frozen center. But Amaruq isn't aiming for culinary brilliance right now. She needs the meat hot, however overcooked.

Fearing he's too weak to chew, she does as mothers have done since time immemorial and mashes the meat with her own teeth first. It's scalding hot. Then she leans over and, as with the seal before, dribbles the mash into his mouth.

She chews up piece after piece, singeing her own mouth repeatedly. In all, she gets a pound of meat into his belly. By meal's end, a pink color has bloomed on his cheeks—a wonderful sign. He soon drifts off.

The sight of him sleeping all but drugs her; she's suddenly so drowsy she can barely sit upright. But she dutifully cleans him up first, scraping the dog turds off his extremities and replacing his boots and mittens. There's a funny lingering smell, perhaps from the excrement, but his fingers and toes look much better. She also feeds the dogs outside—they

---

* Honed baleen can be lethally sharp. Whenever polar bears harassed arctic villages long ago, mauling and killing people, the villagers eliminated the threat by coiling sharpened baleen inside a shank of seal meat and leaving it on the ice. Like many animals, polar bears don't chew their food as much as swallow it whole, and after the booby trap ended up in the bear's stomach, the acids there dissolved the meat and released the baleen shard—impaling the bears from within and killing them.

deserve it. Finally, at last, she crawls back into the igloo. And for the first time in far too long, she crashes into a deep sleep.

Uncountable hours pass in the same way. Journey, food, return, rest. Journey, food, return, rest. Amaruq visits a half dozen other caches she's heard songs about. Most are empty, but one holds another poke of seal meat. Tuluk seems stronger with every meal. The first time he flashes his crooked half-smile, she breaks down in tears.

She tries not to move him, even forming a second pot for a bedpan. And she's gratified to see how well her igloo is holding up. The warmth from the lamp even melted some ice on the ceiling and walls, which has since refrozen and created a tight seal against the wind. But despite her care, she senses that Tuluk is slipping. After one excursion, as she peels back the blanket, she can no longer ignore the smell. Reluctantly, she pulls off his boot. The odor staggers her. She's been telling herself that the smears of brown there were leftover dog poop, but she knows the truth. It's gangrene.

She'll need to amputate, and she curses herself again for not remembering her ulu. But she doesn't have time to wallow. She's got to get him home.

Outside, she blinks in the endless sun. Have they been out here two weeks? More? Packing up, she tosses aside the shovel, atlatl darts, and stone lamp, and squeezes the extra oil out of the poke—anything to cut weight. Then she bundles Tuluk in a blanket at her feet and takes off.

They don't stop this time, except for the shortest of breaks. She's even willing to run a dog or two to death—which will harm the community, but she pushes the thought from her mind. She'll confess in public, do her penance, whatever she needs to, so long as Tuluk pulls through.

They make decent time, despite the dogs' exhaustion. But a half day from home, the air grows ominously thick and white. She removes her snow goggles and wipes her eyes, hoping it's just snow blindness. It's not. A mile later, as if some skygod has dropped a shroud, she enters the worst possible weather—a dense fog.

## Northern Alaska — AD 1000s

Some of the Thule fear fog even more than blizzards; in blizzards you can at least orient yourself by the prevailing wind. Fogs are impenetrable, directionless, a white void. The safest strategy is to wait them out.

Amaruq tries this. A day passes, maybe two, but nothing changes. She eventually feeds Tuluk the last meat from the poke. But as she unwraps him, the smell gags her. She makes herself check, and the black rot is creeping toward his knee. She cannot wait. She bundles him back up and pushes on into the fog.

She finally sees some patches of mud flashing beneath the sled runners. They've reached shore, and can't be more than a few hours from home. But which direction?

She halts the dogs and glances left and right, uncertain. She scans her mental inventory of songs about fog, and comes up empty. Every last one advises the same thing—stay put. But like the old man in the poopknife tale, she refuses to listen.

She decides to skirt the coast, hoping to find a recognizable landmark. She picks left and travels for an hour, then doubles back to her starting point and tries an hour the other direction. The fog deadens every sense except sound, which seems magnified: she hears birds squawking everywhere. She shuts out the distraction and studies each rock and patch of dirt. But she can't see anywhere inland, where the real landmarks are. She finally stops near a cluster of alien boulders, and checks on Tuluk. He seems delirious again, confused; he's sinking. She paces around, scouring her memory. It's useless.

Amid this, she hears the call of a snow bunting and looks up. It's sitting on some nearby boulders; they often nest in such spots. But this one is apparently an interloper. A moment later, another snow bunting darts in, shrieking furiously. After a scuffle, the intruder flits off. The proud defender hops around, and bursts out in song. Amaruq is struck by how different he sounds than the snow bunting at home.

All at once, she knows what to do. She hops back onto the sled and tells the dogs to go—and this time, she listens carefully. She ignores most of the birdsong around her—the honks, squawks, and twitters of other species—but stops for every snow bunting. Sometimes she has to wait excruciatingly long; the birds are shy, and duck down between the rocks when her dogs approach. But eventually each one emerges and sings its aria. As soon as she senses it's unfamiliar, she drives on.

After hours of stop and go, she finally hears something. A vaguely familiar song. She hurries on, and the next tune is more familiar still. She whispers at Tuluk to hold on. The dogs are stumbling now—she's pushed them too hard—but she urges them forward faster.

Then she hears it—the song of the snow bunting from her village. Weary exhilaration washes over her. She hollers hoarsely at the dogs to turn left.

As they stagger into town, several dogs slump over, tongues dangling. Two women rush up, plus several men who've returned from hunting. Amaruq tries to explain all that's happened, but the fog seems to have invaded her brain. All she can manage is Tuluk's name. Her clanmates spring forward to unbundle him.

It's then that Amaruq notices her cousin. Meriwa's standing a few feet away, holding a baby. It must be her son—she of all people would not tend someone else's child. But the boy's been eerily transformed. He looks bubbly and active, with pink, plump cheeks. He's waving his arms and even smiling—a crooked half-smile that Amaruq has seen only once before. He also has some hair now, with the wisp of a white birthmark.

She feels the scar on her womb burning. Meriwa catches her wild-eyed stare, and swallows.

—Just in the past few days, he came to life.

Amaruq turns toward the sled, where two men are holding the limp body of Tuluk. The one on the right speaks, his voice a ghostly echo of the words already screaming through her mind.

—He's dead.

The snow bunting streaks by. Amaruq gazes at the boy they must name Tuluk now—bouncing in the arms of her silly, frivolous sister's namesake—and collapses onto the muddy ground.

# China — AD 1200s

The cultural continuity of China over the past 3,500 years is virtually unprecedented in human history, matched only by that of ancient Egypt. One reason for this persistence is an incredible record of technological innovation. Combined with good farmland and sophisticated irrigation systems—the Chinese loved canals every bit as much as the Romans did roads—this technical expertise allowed China to support a gigantic population. In the year 1000, China already had 100 million people, more than 40 percent of the quarter billion people alive at the time. (Overall, half of humankind lived in Asia then, with 20 percent in Africa, 20 percent in Europe, and 10 percent in the Americas. Oceania didn't even rise to 1 percent.)

The Chinese also excelled at trade, especially with the Middle East, whose wealth and cultural achievements between the 800s and 1200s rivaled China's. Islam had united the region a few centuries prior, bringing peace and prosperity, and scholars there were diligently studying ancient philosophy and literature, and making bold strides in astronomy, chemistry, mathematics, medicine, and physics.

During this golden age of Islam, Arab merchants traded goods east and west. The Crusades created markets in Europe for their millet, rice, lemons, apricots, shallots, and sugar. (Given the high cost of sugar in Europe then, people used it only sparingly, sprinkling it on meat and veggies like a spice.) Mostly, though, Europe was a backwater market. The real money involved trading with China. Arab merchants could reach East Asia along multiple routes, including the famous Silk Road. They also transported goods by sea. And upon arriving in China, many merchants decided to stay put. Even by the early 1000s, some Chinese cities had thousands upon thousands of Arabs living in them—a situation that mostly benefited both sides, but did introduce tension at times. When foreign merchants grew wealthy, and local merchants felt marginalized, riots often broke out in the Arab districts of Chinese cities. One anti-immigrant riot in Guangzhou killed roughly 100,000 foreigners.

As a result of this comingling of cultures, Islamic thought and philosophy gained footholds in China during this time, influencing Chinese life. And the reverse was of course true for Arab people living abroad: being immersed in Chinese culture influenced how they thought and acted. This was especially common with science-minded types like doctors, apothecaries, and those who indulged in the dark art of alchemy...

General Jiaolong, the eunuch, wakes before dawn to a scream—followed by a gush of piss in the alley below his window. A phantom pain stabs his groin. He listens to the doctors out there murmuring to the boy, telling him he's lucky—that however painful it is, passing urine means he'll survive. Jiaolong wonders if his own doctors told him the same thing thirty years ago. He can't recall and can't ask them; he's had each one executed in the decades since.

He's lying on a reed mat, his head resting on a ceramic pillow-block. His soldiers appropriated this hostel last night after they marched down from the front lines. The rooms are shabby, with uneven timber floors; he never would have stayed here except for the name, the Salamander Inn. Given today's task, this struck him as an auspicious omen. So he booted the occupants and took over, sleeping quite well until the scream shattered his rest.

He rises to dress. He steamed his red robe last night over the brazier, and can smell the sandalwood incense as he pulls the garment on. First thing, he affixes his *bao* purse—his treasure purse—around his waist. The frayed rope belt irritates him; he'll need to buy another at the market. Then he hangs the sheath for his dagger on the belt, and opens the shutters on the window. The boy and doctors in the alley have left, so he grabs the remains of last night's dinner, two eggs, and crumbles the shells onto the stones—a feng shui ritual for dispelling negative energy. This task complete, he turns his eyes upward. As he has every night for the past week, he scans for shooting stars. Thirty years ago, on the day Jiaolong became a eunuch, a meteor descended from the heavens—he saw it with his own bleary eyes, a streak of celestial fire. The rumor of the new meteor cannot be a coincidence. He pats his bao purse and nods at the silk print of the salamander on the wall. It's time to find Jabir.

He opens the room's door and ducks beneath the frame; like most eunuchs, General Jiaolong is quite tall. Halfway down the hall, he checks on the one resident of the inn he let stay last night, an old courtesan with bound feet. He sympathizes with such women: like eunuchs, they've suffered since they were children, through no fault of their own. She's sleeping, so he leaves some money and goes. Descending the steps, he feels his breasts and belly jiggle—eunuch fat, people call it, although never to his face. In the tavern downstairs, a half dozen soldiers are sprawled on the tables, snoring. He barks as best he can with his high voice, telling them to get up. He sees a new recruit smirk and fingers his dagger before thinking better of it. When they attack the rebels in Gung this afternoon, he'll put that fellow on the front line.

A bustling Chinese marketplace in medieval times. (Credit: Wikimedia Commons.)

They march for Jabir's home on the edge of town, passing through the market on the way. It never hurts for commoners to see the emperor's soldiers. He's also curious to know how much has changed in the decade since he was stationed here. He remembers the market as a bustling labyrinth of stalls, crammed with shopkeepers and farmers in pointed hats. During festivals, there were acrobats, musicians, sword-swallowers, and puppeteers, too. You could find jade trinkets from Formosa, pearl necklaces from Sri Lanka, vials of amber from the Baltic, ivory idols and rhinoceros horns from Africa, kingfisher feathers from somewhere beyond the sea. Today, with the civil war raging, the market's mostly bare—just tired old men selling tired old fish and ugly statues of Buddha. He can't even find a new rope belt.

Jiaolong winces upon entering the Arab district. It's still largely burnt from the riot; they pass several gutted, blackened homes. Outside Jabir's, the apothecary sign hangs crooked and the shutters are broken—very unlike him. Jiaolong tells his men to stand guard and enters without knocking. The rugs inside are dirty, the jars of drugs thick with dust. The smell makes him gag: the cloying tang of opium mingled with the rot of infection. Jiaolong fishes some clove oil from his bao purse and dabs it under his nose. Then he calls out, more nervous than he'd like.

—Jabir?

He's about to call again, more growlingly, when a groggy voice answers.

—I need to dress. There's tea in the pot.

Jiaolong puts the pot on the fire. As it warms, he studies the art on the wall and frowns. Jabir once had several beautiful portraits hanging, as well as paintings of birds and monkeys. Now it's all geometric designs. When the tea is warm, he pours some into a chipped bowl and sips. It's delicious. He much prefers tea like this, the old way, as a soup with ginger, scallion, mint, and dates—not thin, modern, powdered tea. It's something that first brought him and Jabir together.

Finally, a series of uneven clomps announces Jabir's approach. Jiaolong is surprised to see him wearing not the silk Chinese robes he once favored, but a white Muslim garment. His eyes look sleepier, with deltas of crows' feet. It pangs him to see the beard: as a eunuch, Jiaolong cannot grow one, and truth be told, he's been longing to see his dear friend's rosy cheeks. But the most startling change is the claw of Jabir's left foot. It's twisted painfully sideways, as gnarled as any foot-bound courtesan's; he lurches more than he walks. The foot's also covered in sores. When Jabir stops, two flies begin snacking on it.

Jiaolong realizes he's staring. His eyes dart up to find Jabir smiling. He winks and points to Jiaolong's waist.

—You don't see me gawking at *your* shortcomings. Give your old friend a hug.

Jiaolong lets him, and relaxes into the embrace. It feels right in a way he never thought possible again. But he quickly pulls back.

—I have business with you.

Jabir laughs and pats his chest.

—You're so serious, Jiao-y! Pleasure first, then business.

Jiaolong bats his hand away, but Jabir won't be deterred.

—Let me show you something. From the moment I saw it, I thought, *Jiao would love this.*

However annoyed, Jiaolong indulges him; Jabir takes pride in his hospitality, and Jiaolong will need his help later anyway. Jabir limps over to a chest and removes a white rag from a drawer. Jiaolong approaches and takes it; it feels slightly waxy and shines a bit like silk. Otherwise, it looks like a normal wool rag. He hands it back.

—So what.

Jabir motions for him to grab a jar of red Sichuan pepper paste on a nearby shelf. Then Jabir smears some onto the rag, grinding it into the fabric. He holds it up.

—How would you clean this?

—I don't know. Can we just…

—How would you clean it?

—With soap and water. Why?

Jabir's eyes light up with amusement.

—Water? How conventional. I would use water's elemental opposite.

Jabir does this sometimes, teases him with riddles, showing off his intelligence. More often than he'd like to admit, Jiaolong feels one step behind around him. But before the general can tell him to get to the point, Jabir flings the rag into the fire. Jiaolong tenses, expecting it to erupt in flames and smoke. It doesn't. The Sichuan pepper sizzles and fries, but the white cloth just sits there, pure and unburning. After a minute, Jabir plucks the cloth out by the corner and lays it on his palm.

—Asbestos. Completely fire-resistant. In fact, fire cleanses it.

The name seizes Jiaolong's attention. According to lore, asbestos cloth was made of the skin of salamanders. It's another sign.[*] He reaches

---

[*] Legend has it that two intertwined salamanders were the inspiration for the yin-yang symbol in Chinese culture, which portrays two opposing life forces that drive the cosmos.

for the cloth, but Jabir snatches it back and tucks it into his waistband. Jiaolong frowns down at him.

—That's actually related to the business I'm here for. Is the pleasure over?

—Is it?

Before Jiaolong can answer, Jabir leans up and kisses him.

He tastes just as sweet as Jiaolong remembers, like anise. Surprisingly, Jiaolong even enjoys the beard bristles scratching his cheeks. But a phantom ache in his groin brings him back to himself.* He tears his lips away, and shoves the shorter man off him.

Jabir stumbles on his bad leg and lands hard on an ottoman. Jiaolong is furious and ashamed all at once. He marches up and looms over his never-was lover.

—Business time. We attack Gung later today.

—The destruction of a beautiful city. The pottery works, the calligraphy, does that mean nothing to you?

—They sealed their fate when they rebelled. And you're coming with me.

—How can I resist such hospitality? May I ask why?

—We caught a spy last week trying to sneak past our encirclement. He offered us information about certain hidden treasures in town. Did you know a meteor landed there last year?

That gets Jabir's attention. He raises a skeptical eyebrow and searches Jiaolong's face.

—You can't be serious.

—I am. I have the other ingredients here.

He reaches into his purse and sets two minerals on the table in front of Jabir—a lump of bright-red cinnabar, and a lump of smelly yellow brimstone. Jabir shrugs.

—I have given up alchemy.

—That's not what I've heard.

---

* Similar to phantom limbs, phantom penis sensations do exist, including phantom pain and erections.

—Then you heard wrong.

—We make it our business to know everything in the province.

—So you're willing to spy on me, but you wouldn't even visit after my leg was destroyed.

The words pierce Jiaolong. He heard about the riot, of course. It started right after the war broke out, when rumors began swirling that Arab merchants were selling supplies to the enemy a few hours north. Jabir apparently took refuge in a mosque, which someone lit on fire; he leapt from the tower, and crushed his foot so badly that it nearly required amputating. Jiaolong had wanted to visit, sincerely, but could never bring himself to.

Losing patience, Jiaolong spits on the floor and grabs his dear friend's collar, wrenching him to his feet.

—You're coming to Gung. That's an order. It's also in your interest.

—Why?

—What's the one thing you want most in the world?

—To die.

Jiaolong's shock manages to overthrow his anger.

—What? Why?

—I think you can smell why. My foot's dying. Or do you always rub clove oil under your nose when making house calls?

When Jiaolong doesn't answer, he goes on.

—I loved life here once. Not now. My fondest wish, inshallah, is to return home to Arabia and hear the muezzin ringing out again over the rooftops. But the pain's too fierce. I can't travel and can't afford to. So I want to die and wake up in Paradise. If I could end my life today, I would.

—But you won't?

—Allah says that suicide is *haram*. Forbidden to Muslims.

Jiaolong feels a rush of relief. But he also looks his old friend up and down. The beard, the white robe, the geometric art. He never would have suspected a hedonist like Jabir to turn devout. But people respond to crises in all sorts of ways.

—What we'll find in Gung can heal you.

—I put my trust in Allah, not alchemy.

—Enough with Allah. Besides, Allah gave you your gifts in the first place. Surely He'd allow you to use your genius to heal yourself. With the meteor, you can make an elixir purer than any before.

—Perhaps. But I assume you didn't visit for the first time in a decade just to help your dear old friend. What's in it for you?

Jiaolong's shame burns even hotter now. But he reaches into his purse for his bao jar. He places it on the table.

—I want to be restored.

He explains his idea. It involves a sacred Chinese animal, the salamander. Jiaolong has long been obsessed with salamanders because they have the power to regrow lost limbs—among other body parts. And he knows alchemists can make elixirs, life-giving potions. With a meteor—a celestial substance, straight from heaven—Jabir can make the purest elixir yet. And by blending a salamander into this elixir, Jiaolong can absorb its power of regeneration.

At least he hopes so. He asks Jabir if it will work. Jabir's shoulders slump.

—I told you, I've given up alchemy.

—That's not what I asked. I know you have some salamander—you showed me once. So yes or no, can you heal me?

—Healing starts with the soul. Only then can you heal the body.

—No more riddles. Can you fix me? And you can heal yourself, too, don't forget that.

In answer, Jabir traces a finger along the bao jar, deep in thought. He finally speaks.

—I think I can give us both what we desire.

Jiaolong nearly hugs him in joy. Then he remembers himself and resumes the mien of a general. He orders Jabir to gather supplies for the journey to Gung—including the salamander. Jabir shakes his head and limps off. Jiaolong can't understand why he's so upset.

But no matter. Jiaolong replaces the cinnabar and brimstone in his purse, then grabs the jar with his bao—his treasure. He's carried it with him every day for thirty years, but he hasn't looked inside for twenty. Making sure he's alone, he removes the cork and peers down.

The brine is dark, and for one panicked moment, he can't see them. But when he tilts the jar in the firelight, they bob to the top—his bao, his treasure: his pickled penis and testicles.

No civilization in history can surpass China's record of technological achievement. Among other things, the Chinese invented paper, printing, movable type, cast iron, the compass, rudders, saddles, trousers, suspension bridges, wheelbarrows, blast furnaces, matches, mechanical clocks, paper money, rockets, and gunpowder. Read that list again. The history of Chinese technology is practically the history of modernity itself.

Beyond specific inventions, medieval China also pioneered mass production. By the year 1100, Chinese blacksmiths were making a quarter billion pounds of cast iron every year; even common folks owned iron cooking pots and had access to iron chisels, saws, hoes, and axes. The Chinese built iron skyscrapers, too, including a 294-foot pagoda with a 10-foot phoenix on top. Porcelain production was even more impressive. The transformation of raw clay into delicate cups and bowls required six dozen steps, each of which relied on specialized labor. Some 375,000 people worked in the porcelain industry, and so-called dragon kilns—brick tunnels two dozen feet high and a hundred yards long—could fire 30,000 pieces at once. Overall, the assembly-line nature of production in China, especially with regard to weapons, has drawn comparisons to the manufacturing of automobiles in Detroit in terms of output and efficiency.

For all the brilliance and sophistication, however, some aspects of medieval Chinese life seem bizarre today, even cruel. Take foot-binding. Tiny feet were considered attractive for women, so around age five, girls would have their feet bound with silk or cotton bandages to deform the bones. The foot would more or less fold in half as it grew, curling the toes under until the women were knuckle-walking on their feet. A length of three inches was considered ideal. At best, the women had a painful, mincing gait; at worst, they couldn't walk at all and had to be carried to their masters' bedchambers on stretchers.

# CHINA — AD 1200s

Then there were the eunuchs—little boys who'd been castrated. The procedure varied over the centuries, but the victims were either held down by grown men or bound to chairs with a hole in the bottom, similar to an outhouse bench. A surgeon—called a "knifer"—would smear Sichuan pepper paste onto their genitals to numb them. Then he made a quick, clean stroke with a curved knife. The penis and testicles were set aside for pickling (more on this below). Finally, the surgeon jammed a goose quill or pewter plug into the urethra hole, to keep it from swelling shut. The plug stayed in place for three days, held by bandages. During this time the new eunuch was not allowed to pee, which resulted in a huge reservoir of urine building up. After three days of agony, the plug was yanked out, along with any scabs, so they could relieve themselves at last. Sometimes, though, nothing happened: the urethra had swelled shut anyway, sealing off the bladder, and the eunuch would die from the buildup of toxins there. Usually, though, that Niagara of urine burst free, forcing the urethra open and keeping the little boy alive.

The Empress of China (*top*), carried by court eunuchs. A depiction of castration, from the Eunuch Museum of Beijing (*bottom*). (Top credit: Wikimedia Commons. Bottom copyright: Brian Salter.)

Peasant families subjected their sons to this torture because being a eunuch was a surprisingly promising career path. Traditionally, Chinese emperors lived in walled-off palaces, and few people had access to them. Eunuchs were an exception. For one thing, eunuchs could not seduce the emperor's concubines—a grave concern then. For another thing, the imperial court was something of a snake pit, full of sycophants scheming to advance their families and heirs. But eunuchs, unmarried and lacking heirs, were immune to such temptations. Finally, many Chinese mandarins considered eunuchs to be little more than gelded animals—unsullied by greed, wit, or ambition, and too dull and docile to do anything but dedicate themselves to the emperor's service. So for two thousand years in dynastic China, eunuchs served the emperor as cooks, butlers, gardeners, and attendants, some of the few people who saw his majesty on a daily basis.

In practice, of course, eunuchs did not lack greed, wit, or ambition. And as the centuries passed, and the emperors grew more remote and more Olympian, the eunuchs took full advantage of their access. If lords or wealthy merchants needed to pass a message to the emperor, the eunuchs demanded bribes and favors in exchange for delivery. Eunuchs also spied on the royal household for outsiders and helped raise the royal heirs, manipulating them from a young age and controlling them later on the throne. Top eunuchs could have people whipped, imprisoned, or tortured. Naturally, all this scheming and torture didn't make eunuchs popular. They often appeared as villains in novels and plays, and people mocked them as "bobtailed dogs," among other insults. (Many eunuchs were slightly incontinent and dribbled urine all day, which smelled bad. There was actually a saying in imperial China, "as smelly as a eunuch.") But however unpopular, by controlling access to power, many eunuchs grew immensely powerful themselves. During the reigns of certain weak or feebleminded emperors, eunuchs essentially ran the kingdom.

For a starving peasant family, then, castrating your little boy and sending him off to the palace wasn't a bad gamble. Eunuchs often rose to prominence outside of politics as well. The eunuch Cai Lun is popularly credited with inventing paper. There were powerful eunuch military commanders, and the eunuch explorer Zheng He sailed fleets as far afield as the Middle East. (There, he perhaps noticed the presence of Muslim eunuchs, who served the courts of local emirs—a point of commonality between Islamic and Chinese cultures.)

At the peak of their power, some 100,000 eunuchs were scattered throughout the Chinese empire. Most were easy to recognize because they had distinct physical features. They lacked muscle and carried extra fat on their hips, breasts, and bellies. They were unusually tall as well—which might sound strange, since eunuchs did not go through puberty, an age associated with growth spurts. But growth is a complex process. During normal puberty, the testicles produce testosterone and other hormones that kick growth into high gear: leg and arm bones get longer, the rib cage expands, and so on. But testosterone has another effect as well. Around age eighteen, it fuses shut the so-called growth plates in the arm and leg bones. These plates are where we gain most of our height and length during adolescence, and when those plates fuse shut, we stop growing. Overall, testosterone kickstarts growth early in puberty, but stops growth abruptly when puberty finishes.

Something different happened in eunuchs. Because they lacked testicles and testosterone, they grew more slowly in young adolescence. But the lack of testosterone also prevented the growth plates in their legs and arms from fusing shut. As a result, eunuchs kept growing deep into their twenties. Many stood well over six feet tall as adults—giants in ancient times—and had barrel chests and spidery long fingers, too.

Height wasn't the only physical symptom of castration. Elevated testosterone increases a man's chances of going bald, so eunuchs rarely lost their hair. Many had baby faces, in part because they couldn't grow beards. Most strikingly, testosterone in mammals is associated with dying younger, since testosterone takes a real biochemical toll on the body. As a result, low-T eunuchs tended to live significantly longer than the average man. (This wasn't simply because eunuchs led lives of leisure, either. One study found that Korean eunuchs lived around seventy years, while emperors—who did even less work—died around forty-seven. Eunuchs were also 130 times more likely to reach age 100.) Castration preserved a prepubescent voice, too. During puberty, men develop thicker cartilage near their voice boxes, which gives them an Adam's apple. Eunuchs had no Adam's apples. Testosterone also deepens the male voice by thickening and lengthening the vocal cords, stretching them from around 13 millimeters (a half inch) to 20 millimeters (four-fifths of an inch). But without testosterone, eunuchs kept their short, high-pitched vocal cords. People meeting them for the first time were often startled to shake hands with a giant, yet hear a pipsqueak voice.*

According to Chinese lore, eunuchs would be made whole again in the afterlife and have their severed penis and testicles restored—but only if they were buried with them. Otherwise, the god of the Chinese underworld, Yanwang, would turn them into female asses for eternity. Fear of this fate led to the macabre practice of eunuchs pickling their sex organs in jars of fluid; they called these bits their *bao*, or treasure. The more paranoid types reportedly carried their bao with them at all times in a purse, since you never knew when death would strike. But not all eunuchs were content to wait for death...

To Jiaolong's irritation, Jabir calls out yet again to rest. Jiaolong wheels on his white horse.

—We stopped ten minutes ago.

—And, Allah forgive me, my leg is aching again. This cart jolts like an earthquake.

Despite his grumbles, Jiaolong orders the peasant driving the water buffalo to stop; the cart with Jabir creaks to a halt. They're passing through beautiful hill country, with terraced rice paddies glistening in the sun like steps leading up to heaven. They're also an hour behind

---

* Another famous group of eunuchs was the castrati of Italy, little boys who were castrated to preserve their angelic singing voices. The last castrato died in 1922, and a recording of his voice still exists. (See episode 40 of my podcast, "The Sinister Angel Singers of Rome," for more: https://samkean.com/podcast.) Incredibly, the last Chinese eunuch lived even longer—dying only in 1996. There's a bonus episode of the podcast about his amazing life.

schedule. But there's nothing Jiaolong can do; north has always been the most inauspicious direction for him. Jabir does look gray anyway, and Jiaolong needs him at full strength when they find the meteor in Gung.

Jiaolong wouldn't feel so irritated if Jabir didn't turn every stop into a performance, doing magic tricks for the peasant and his small son. At the last stop, he took a stone from his apothecary bag and made it "weep," squeezing it until fluid dripped out. This time they've stopped near a blacksmith shop. Jabir takes a crucible from his bag and fills it with grasses and herbs. Then he motions for a flaming brand from the fire. He holds the crucible over the flame and mutters something. The plants crackle and smoke, and when he turns the crucible over, molten lead pours out. The boy shrieks with delight, and the peasant chuckles and bows. Jiaolong notices his guards grinning, too, and snaps at them to keep watch.

After each trick, Jabir reveals how he performed it. In this case, the crucible has a false bottom of meltable wax, with lead concealed beneath. Jabir explains these tricks, he says, to prove that the world is rational and intelligible to those who think. *Only Allah can perform miracles.* The preaching annoys Jiaolong, who snaps at Jabir to shut up—he's supposed to be resting. In truth Jabir used to perform such tricks for him privately, late at night, and he resents what's passed. They wait in silence afterward.

Restless, Jiaolong dismounts his horse to stretch his legs, but the belt for his bao purse gets snagged on his saddle and snaps. He tries tying the frayed ends together, but it won't fit around his pendulous belly.

Jabir watches him struggle and calls out.

—I can hold the purse for you.

Jiaolong shakes his head. He never lets his bao out of his sight.

—That's probably wise, Jiao-y. This cart's rickety, and there are so many grains of rice on the ground here. If the jar fell and broke, we'd never be able to find your shriveled bits among them.

It takes Jiaolong three strides to reach the cart. He backhands Jabir so hard he nearly tumbles out.

—Insult me again, and there will be worse.

To his surprise, Jabir does not cower or apologize. He looks up smirking, defiantly wiping the blood from his lips.

—You would kill me? It would be a blessing.

Jiaolong turns in disgust. He snarls at the peasant to beat the buffalo and get the cart moving—and not to stop the rest of the way. Jiaolong remounts his horse, cradling his bao, and snaps the reins.

There's little chatter the remainder of the trip, just the jolting of the cart and Jabir's whimpers. A mile from the front, things turn grim. They encounter several burned houses and farms, even bodies lying about. One has a belt-purse, so Jiaolong commandeers it and transfers his bao, dagger, and minerals. Approaching dead bodies is bad luck, and Jiaolong has no eggshells on hand, so he pours some water from a canteen around the body instead, to neutralize the bad fortune.

He hears his army before he sees it, a low roar like the ocean. The smell isn't far behind, a breeze of garbage and unwashed bodies. Then they crest a hill and spy Gung. The city's a thousand years old and looks every day of it, with nothing for defense except a single pagoda tower and a crumbling stone wall. The potters and calligraphers who live there have nevertheless proved feisty, hurling down stones and dumping buckets of boiling pitch whenever Jiaolong's troops approach. To take the city, he'll need to knock the wall down.

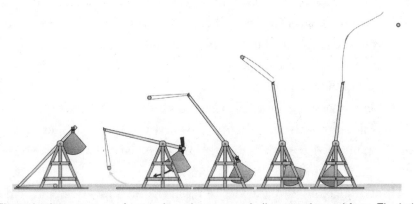

The trebuchet—a type of catapult—threw stone balls at castles and forts. The ball starts in a sling beneath the machine (*far left*), with the counterweight raised. As the counterweight drops (*near left*), the lever arm hoists the ball upward (*middle*) and starts to fling it. Moments later, the sling releases (*near right*) and the ball flies out (*far right*). (Copyright: Tom Roberts, courtesy of *Richmond Post-Dispatch*.)

To which end, his troops have built a dozen trebuchets—specialized catapults—a few hundred yards from the wall. Each one consists of two huge triangular frames supporting a forty-foot horizontal beam that pivots up and down like an off-center seesaw. At one end of the beam there's a massive lead counterweight that gets winched up off the ground; at the other end there's an oversized sling with a boulder inside. Initially, the sling rests in a trough on the ground between the frames. When the weight drops, the beam rises. This motion whips the sling up and around and hurls the boulder through the air.

Trebuchets can batter walls with devastating force—or so Jiaolong has heard. The basic trebuchet was an old Chinese invention, but engineers near Jabir's homeland improved it and developed this superior type, with the counterweight and sling, then imported it back into China. The technology is therefore new and untested here.

As Jiaolong's retinue approaches, he's greeted by his aide-de-camp, as well as his top engineer, Xi, a wiry man with a pinched rat nose and long hair. Xi assures him the trebuchets are finally ready. Jiaolong orders a test shot.

A dozen soldiers crank the five-ton lead counterweight into the air. In the distance, nervous potters and calligraphers gather on the walls of Gung to watch. Jiaolong notices Jabir watching keenly, too.

With the weight raised, the soldiers lock the beam into place and back off. Xi plays out a rope attached to a metal pin, which releases the locking mechanism. When everything's set, Xi grunts and yanks the rope. The pin snaps free, and the giant beam-arm lurches into motion.

But not for long. As the beam rises, the sling scrapes along the trough. But it gets tangled and twisted, and the boulder *smacks* the frame instead of lifting cleanly. This tears the sling open early and sends the boulder careering backward. Jiaolong and everyone around him throw themselves to the ground. The boulder streaks above their heads and crashes into a wooden latrine behind them, splintering the structure and splattering the poor soldier inside.

Jiaolong finds himself cowering ridiculously in the dirt. Worse, he can hear the potters and calligraphers hooting and ringing bells in

Gung, mocking their failure. Jiaolong scrambles to his feet and orders Xi arrested. He'll decide later whether to execute him.

Arresting Xi doesn't fix the trebuchets, however. Jiaolong pulls out his dagger for emphasis, and turns to face his troops. One of the few benefits of being a eunuch is his capacious chest, and he takes full, bellowing advantage now.

—Did anyone check the feng shui? Did it even occur to you? Where's my compass?

His aide-de-camp Deng, a squat, bearded man who can lift a whole barrel of rice wine over his head, scrambles off, and returns holding the compass in both hands. It's a small bowl of mercury with an iron fish floating on top. Due to some hidden life force in the universe, the fish always points south.* Around the rim of the bowl are directional markings, allowing Jiaolong to determine the orientation of any object.

A representation of an ancient Chinese compass. A magnetized spoon floated in a pool of mercury or another liquid. (Credit: Yug, Wikimedia Commons.)

He stalks over to the trebuchet, holds the compass up, and takes a sighting along the beam, his face so close to the bowl he can smell the metal. The beam is facing due southeast, a good omen. Then he bends

---

* By convention, Chinese compasses pointed south, not north.

down and takes a sighting of the trough between the frames. According to the compass, it's several degrees off-kilter. Not only does this violate the internal harmony of the trebuchet, it's causing the sling to drag unevenly and twist.

At least he thinks so. They'll have to do another test shot. He orders his men to straighten the trough and load another boulder.

—Jiao-y?

Jiaolong turns and grimaces to see the blood still crusted on Jabir's face. Jabir beckons him forward for a tête-à-tête. Jiaolong huffs but hands the compass to Deng and approaches Jabir.

—If you're offering prayers to Allah, I swear...

—A prayer never hurts. But you're also wrapping the sling around the waist of the boulder, side to side. Back home, we wrapped it top to bottom.

Jiaolong studies his face; he seems sincere. Jiaolong returns to Deng and points.

—Tell them to wrap the sling top to bottom.

As Deng relays the order, Jiaolong strides back twenty yards and watches the counterweight get winched into the air. Then he scoots back five more yards. At last, he gives the order to fire.

He braces for another smack. It doesn't happen. Instead, the beam shoots up cleanly and the sling whip-cracks around. The boulder is flung so high so fast that he loses track of it in the air. His soldiers gasp and aah, pointing excitedly downrange. But for several seconds Jiaolong can't see what they're hollering about.

Then he hears it—a *crunch*. In Gung, the roof of the pagoda tower explodes and crumbles, concussing several potters and calligraphers below.

Cheers erupt all around Jiaolong, who feels as triumphant as if he'd flung the boulder himself. He turns toward his men—but as fate would have it, catches Jabir's eye instead. He nods a short thank you. Jabir smiles sadly and turns away.

Jiaolong orders Deng to check the alignment of the other trebuchets with the compass—as well as, quietly, tell the soldiers to start wrapping the slings the other way. He wants every last machine firing on Gung within the hour.

The trebuchet ("treh-byu-SHAY") was the most fearsome weapon of medieval times, a giant catapult that could lay waste to any fortress by battering its walls with boulders. It's funny to think, then, that in terms of the physics, trebuchets were little more than children's seesaws: you raised a weight on one end, let it drop, and used that force to fling a projectile from the other end.

Of course, trebuchets differed from seesaws in a few ways. Instead of the pivot being in the middle of the beam, the trebuchet pivot was shifted toward the weighted arm. This gave the opposite, throwing arm more leverage and speed. The sling attached to the end of the throwing arm added still more juice: as with an atlatl, it effectively lengthened the throwing arm for more power. Modern re-creations of trebuchets have hurled boulders 125 miles per hour.

Basic trebuchets and catapults date back to ancient China; the catapult was actually a piece in the Chinese version of chess. These early trebuchets relied on human muscle. The short end of the beam had dozens of ropes dangling from it, and soldiers would grab hold and yank down hard. Advanced trebuchets, featuring slings and counterweights, first emerged in the Mediterranean in the 1100s, then trickled back into Asia via Muslim engineers hired by Chinese armies. (Contemporary Chinese sources referred to them as "Muslim trebuchets.") Counterweight trebuchets were far more powerful than earlier versions, and their invention kicked off an arms race across Eurasia. Fortresses and castles responded to better trebuchets by building thicker, studier walls. This encouraged military engineers to build even bigger trebuchets. Which encouraged castles to build even stouter walls, and so on. History's biggest trebuchets, from the 1300s, stood over 60 feet tall; they used counterweights of 33,000 pounds, and could toss a 300-pound boulder 300 yards. Engineers would have built even larger ones, except they couldn't find big enough timbers for the throwing arms. But no matter how thick walls got, the relentless pounding of trebuchets would eventually crack them.

The biggest trebuchets took months to build, and drew on the work of a dozen different professions: one historian listed "carpenters, carvers... engineers, framers, hewers, landscapers, masons, millwrights, riggers, seamstresses, smiths... stonecutters, and weavers." And while armies usually flung boulders, sometimes generals got creative. They might aim a little higher, at the town behind the walls, and fling beehives, scorpions, snakes, flaming pitch, rotten vermin, or excrement. One historian noted that, for a small trebuchet, "a good size projectile is, morbidly, a human head." And if it's the head of someone the townsfolk knew—a captured spy, maybe—all the better.

Sadly, zero medieval trebuchets survive today.* Drawings do exist, but artists back then didn't exactly hew to blueprint accuracy. (Historians have commented on the artwork's "Eschereque contradictions," including foreground-background confusion and impossible scaling.) The best way to study these wunderweapons, then, is through modern re-creations, built by trebuchet fanatics like Daniel Bertrand of Utah.

Bertrand's a tall, earnest fellow with a shaved head, mustache, and toothy grin. He says "man" a lot as emphasis. ("Good idea, man.") He first learned of trebuchets through his alma mater, Utah State University. The engineering school there holds a contest every fall where students build trebuchets (usually with modern materials) and chuck pumpkins. A history major, Bertrand had never even used a power drill before, but after watching the pumpkins fly, he became obsessed with trebuchets and decided to build an authentic medieval one.

Looking back on his first trebuchet, Bertrand laughs and admits he had no idea what he was doing. "It was sketchy, man. Someone could have gotten hurt." (Indeed, poorly built trebuchets did topple over

---

* One medieval trebuchet did turn up in 1890s Prussia, when parishioners unearthed it during the demolition of an old church. After spending several minutes marveling, they hacked it up for firewood.

sometimes in the Middle Ages, or fling boulders backward and kill the men who fired them—to the merriment of the besieged folks watching from their fortress walls.) Bertrand has since built several more, including one that stands thirty-three-feet tall. Despite some ups and downs—cracked beams, smashed frames, funding woes—his passion for them hasn't waned: he refers to them as if they were human ("He's firing well today"), and he's even thinking about moving to a different city to live closer to the thirty-three-footer. Historically, engineers named their trebuchets, as if they were battleships: War Wolf, Victorious, The Bad Neighbor. Bertrand initially named his The Black Widow, after a spider he found in the lumber used to build it, before switching to The Sentinel.

I got to see The Sentinel in action near Logan, Utah. It sits in a grass field amid cattle ranches; snow-capped peaks loom in the distance. I arrive on a warm Saturday in late October, one of those sunny, deep-breathing afternoons that make you wonder why you don't live in the mountains year-round. A crowd of twenty has gathered to see The Sentinel fire. It's nearly Halloween, so two little girls are dressed as a witch and a redheaded fairy; there's also a thirtysomething man in a puffy Tudor hat, doublet, and hose. A fuzzy white labradoodle darts around clawing holes in the dirt.

Bertrand is wearing a fluorescent yellow T-shirt that's filthy with tar. He's joined by his trebuchet-firing crew—young dudes in dirty jeans with chew-can rings worn into the back pocket; most study history or classics at Utah State.

I tip my head back and study The Sentinel. Its throwing arm, currently at twelve o'clock, weighs 360 pounds and is not hewn or sanded; knots are visible in the wood. (In medieval times, engineers sometimes left the bark on.) Opposite the throwing arm, at six o'clock, hangs the counterweight, a teardrop-shaped wooden bin ten feet tall. It weighs 500 pounds empty, and currently holds 1,400 pounds of sandbags. The throwing arm and counterweight are supported by two A-frames whose barn-red paint has faded badly. Miles of tarred rope bind everything together. When under stress, the ropes ooze tar, which attracts bees and wasps.

## China—AD 1200s

The Sentinel's target stands two and a half football fields away: a twenty-foot by sixteen-foot wall. It's made of pallets, plywood, and old cabinet doors that Bertrand nailed together and propped up with four-by-fours. I think of it as a faux fortress. The 250 yards is a typical distance for bashing castles, since the archers defending them struggled to hit targets much beyond that distance.

Firing a trebuchet involves dozens of steps, but the basic idea is simple. First, the counterweight needs to be raised to store energy. Three of the dudes and I begin ratcheting it up by turning two large wooden Xs attached to a thick rope. It's beastly work. We crank and crank and crank, stroke after stroke after stroke, straining like Atlas against the 1,900-pound counterweight. And with each turn of the X, it rises maybe one inch. Before long, my arms are burning and I'm sweating. My abs ache the next day.

Lifting the counterweight simultaneously lowers the throwing arm, and after ten long minutes of cranking, the counterweight has been raised to ten o'clock and the tip of the throwing arm is nearly kissing the ground. We lock the arm into place with a metal pin attached to a cord, then load the ammunition. Bertrand often flings bowling balls—flaming ones on New Year's Eve—and somewhat disconcertingly, there's a shattered bowling ball lying next to The Sentinel, cracked open like a geode. But today we're launching spherical garden stones. Each weighs twenty pounds, and they're streaked with grass stains from prior launches. For fun, Bertrand has drilled small holes into one of them, so it whistles as it flies.

Attached to the far end of the throwing arm is a rope-and-leather sling. We stretch it along a wooden trough on the ground beneath the frame and wrap it around a garden boulder. When the treb fires, the throwing arm will drag the sling and boulder down the trough—then whip them up and around, flinging the stone at the faux fortress.

It takes a good fifteen minutes to prepare each shot. When everything's finally ready, Bertrand hands me the trigger cord and tells me to pull.

Every person who fires the treb that day—yours truly emphatically included—gushes the same thing afterward: "Boy, that was satisfying!" The firing pin holds thousands of pounds of tension in place, and it takes a hard, full-body tug to release it. But when it goes, my God. It's like snapping your fingers and watching a dragon spring to life—hulking and graceful all at once. The counterweight plummets. The throwing arm flies up. The sling whips over the top with a whoosh. Then the garden boulder absolutely *snaps* into the air, as if Hercules hit it with a pitching wedge. It soars up and hangs there for four seconds, six seconds, eight— a black meteor impossibly high overhead, arcing against the blue sky.

Two-hundred fifty yards away, there's a splash of dirt, and we all *gaaah* in unison. I almost don't care that it missed the wall. Knowing I did *that* with one little tug was magic enough.

Bertrand, however, wants to smash the wall. We reraise the weight and reload the sling. He lets the witch and fairy pull the trigger cord this time; they yank so hard they topple onto their keisters. But downrange, it's another miss, another splash of dirt.

The ball is sailing high and left. To adjust the aim, Bertrand starts tweaking different components, and it soon becomes clear that aiming a trebuchet is more art than science. There are so many variables to consider: the mass of the counterweight, the mass of the boulder, the angle of release, the tension on the sling, the length of the sling, the presence or absence of wind, even whether a rough patch on a given boulder might cause it to curve left or right midair. One dude jokes, "If only I remembered any trigonometry, this would be easy." We chuckle, but after reflection, I think he's wrong. This is way harder than trig.

Ultimately, Bertrand removes a fifty-pound sandbag from the counterweight bin, tightens the sling, and realigns the wooden trough. We load for a third shot. My arms are noodles now from turning the ratchet wheels. Doublet-and-hose dude steps up to pull the trigger. The garden ball leaps up—and floats mere inches over the wall. Drat.

The fourth shot misses to the right. The redheaded fairy cackles, "Your hard work didn't pay off!" The rest of us groan. The novelty of flight has worn off. We want a hit.

Finally, around 1:00 p.m., we load up for a fifth shot—and nail that bastard. Fist pumps galore. But it struck the fortress wall awfully low, about knee height. To really rattle their teeth in there, Bertrand adjusts a pin that controls the sling's release point, to nudge the ball a bit higher. We hold our breath as we send the sixth shot aloft.

Due to the distance, we see what happens before we hear anything. When the ball hits, a black blot suddenly appears on the wall, as if someone hurled an inkpot. Then a *bang*, like a kernel of burst popcorn. We erupt in cheers, and run downrange to inspect the carnage.

There's a giant hole in the plywood now, and we stick our arms through like a magic trick. But the real delight comes around back. Two thick four-by-fours—house timbers—have been utterly splintered, and there's wood shrapnel sprayed all over the grass, like in news footage of tornados. Trebuchets truly are forces of nature. We can sense our enemies trembling.

Instead of trudging back to the treb for the seventh shot, a few of us linger downrange, about thirty yards to the side of the wall. I fall to chatting with a young married couple, Andrew and Kaeli. Andrew built his own metal trebuchet at Utah State a few years ago; he shows me video of it hurling a pumpkin at a derelict piano.

"How's life in northern Utah?" I ask.

"I love being in the mountains," Andrew says. "Great scenery, so many cool hikes. Plus, we can blow up microwaves in the gorge."

"You, uh—what?"

"Blow up microwaves, using Tannerite." (I google this later. It's an explosive.)

"Oh. Why microwaves? Just because they're strong and confined?"

Andrew nods. "Microwaves, ovens, stuff like that."

Kaeli adds: "We got the shrapnel really high with our last microwave."

At this point the trebuchet crew waves their arms, alerting us that they're about to launch. We settle in to watch.

The sling has whip-cracked before, but the noise seems more ominous now, with the shot coming toward us. We can see and hear the ball better, too, screaming down like a hawk. It's another direct hit, and my knees actually buckle at how loud the smack is. From the uprange, firing side, trebuchets are all majesty. From the fired-upon side, you appreciate the violence.

Overall, we smash the wall a half-dozen times. But there are plenty of misses, too, and at 3:00 p.m., Bertrand rallies us. "I want to end on a hit, man." One more good *smack*, and we'll go home.

Guess how many times we hit the wall after that. Sometimes we miss right, adjust for that, then miss farther right. Sometimes we miss right one throw, adjust nothing, and miss left the next one. Bertrand continues to tinker, and as the misses pile up, I revise my earlier opinion: aiming a trebuchet is neither art nor science, it's voodoo. Bertrand half-blames our bad fortune on him forgetting his lucky hat. Someone else suggests summoning a priest and holy water. But no matter how much juju we employ, we keep missing—just high, just low, just right. Again, this is how you get addicted to gambling. We finally give up.

To be fair, had this been a proper city or fortress wall, the just-lefts and just-rights would have been hits, and the just-highs would have clobbered the town beyond. Indeed, with a dozen trebuchets firing at once, a town like Gung stood little chance of resisting…

The trebuchets batter Gung for hours, concentrating on one section of wall and reducing it to gravel. Toward the end, to terrify the townsfolk, Jiaolong orders his men to lob other things—rocks dipped in pitch and lit on fire, baskets of snakes, clay shells packed with excrement from the shattered latrine. At last, the hole in the wall is wide enough. He orders his infantry to charge.

Jiaolong approaches the city's main gate—no clambering over crumbled walls for a general—and is gratified to hear that the laughter from before has given way to howls. He stops a hundred yards away and begins pacing; he's hyperaware of the bao purse clanging against his belly. Jabir sits in the cart behind him with his head bowed and prays.

At last, at dusk, the gate groans and swings open. The boulder-like Deng comes trotting out to announce that they've found the meteor, right where the captured spy said. Jiaolong mounts his horse, and orders a few troops to grab Jabir's cart and follow.

The town has been defeated, but Jiaolong can hear lingering skirmishes. As they pass an alley, he sees figures darting by—soldiers with swords pursuing artists armed with bamboo poles. Being amid a battle, even a lopsided one, always makes Jiaolong tense. Along the way, they pass the body of the guard who smirked over his voice at the inn this morning; he's impaled on a stake. Jiaolong forgot to send him into the melee first as punishment, but fortune has cut him down anyway. They also pass the body of a fellow eunuch, tall and flabby. His bao lies trampled by his side. A phantom pain flashes through Jiaolong's groin, and he tells Deng to make sure that the man is buried properly.

After they cross an arched bridge, Deng stops at a shop with a beautifully lettered apothecary sign. Jiaolong dismounts and helps Jabir down from the cart. Then he lights a lantern and enters, ducking under the doorframe. There are stone floors and a damp ceiling; it smells of mildew. Workbenches line the walls, piled with retorts, braziers, alembics, and other equipment. Above them sit shelves of porcelain jars with strange labels—mandrake, spiderwebs, pangolin scales, dried placenta, dragon's blood. Finally, Jiaolong swings the lantern toward the corner and spots it—a twenty-pound lump of charred black meteor. It has indentations like thumbprints on the surface.

—Jabir, look.

Jabir turns and sighs. Jiaolong cannot understand his friend's glum mood, but it doesn't dampen his own. He orders Deng to stand watch, and orders Jabir to begin the alchemy.

Alchemy encompassed much, much more than just turning lead into gold. The practice stretches back to ancient Greece, Egypt, and Persia, and key alchemical texts were often attributed (however wrongly) to luminaries like Pythagoras, Moses, and King Solomon. The fundamental aim of alchemists was to create the so-called philosopher's stone. This "stone"—variously described as liquid or powder, red, yellow, or rainbow, densely heavily or ethereally light—had the magical ability to change one substance into another. For grubbier types, this meant turning base metals into gold. Many alchemists, however, pursued the stone for high-minded philosophical reasons—to transform and purify their souls. For them, alchemy was more akin to a mystical religion than a get-rich-quick scheme.

The Chinese took an intermediate path. They pursued alchemy for medicinal reasons, whipping up elixirs to extend their lifespans and heal injuries and illnesses. Meteoritic iron would have been very tempting to include in such tonics, given its celestial origin and the association of meteors with supernatural omens. Other common ingredients included brimstone (sulfur) and cinnabar (a mercury ore). Chinese alchemists in the 800s even invented gunpowder accidentally while brewing elixirs. And if you're thinking that chugging mercury and gunpowder sounds a bit risky, you're right. Sinologists have estimated that at least a half dozen Chinese emperors died on the throne as a result of drinking elixirs. There was a fine line between immortality and death.

Typical alchemical equipment. (Credit: Wellcome Trust.)

## China—AD 1200s

Many historians today consider alchemy not a backward, Gargamel pseudoscience but something like a protoscience—a realm where people explored new ideas about nature, ran experiments, and formulated hypotheses that, however misguided they seem today, did advance knowledge over time. (Several giants of early science, including Isaac Newton, were committed and enthusiastic alchemists.) All that said, some alchemists played the game solely to make a buck. There were also alchemists who *claimed* the ability to change base metals into gold, then plied sleights of hand, chemical and otherwise, to swindle people. There were even a few alchemists who were neither wholly straight nor wholly crooked, and whose true aims remained obscure even to their dearest friends...

The noise of battle continues to echo in Gung, but to the two men in the lab, it seems distant, unimportant. Jiaolong has to refill the lantern with oil twice that night. He watches Jabir limp around the room, crumbling herbs into pots and dissolving colored powders in bubbling liquids. He's making Jiaolong's elixir first, and adds a dried-out salamander to one retort, a black beast with an orange face and tail and orange stripes. Whenever Jiaolong asks what he's doing, Jabir answers cryptically.

So instead they just... talk. It's awkward at first, as they feel each other out, but they fall into the old rhythm more easily than Jiaolong ever could have hoped. He tells Jabir of his adventures in the recent war—cities conquered, enemies outwitted, the burden of sending men to sure death. Jabir's stories are quieter. He talks of his lingering aches, his Muslim friends moving away, his loneliness. He relates this all with a smile—it's hard to suppress Jabir's spirit—but it pangs Jiaolong anyway. They seem to be having two conversations at once—one about their experiences, and one about what they might have experienced had they not drifted apart. The one chat is pleasant, the other painful, but Jiaolong wouldn't have missed either for the world.

All the while, Jabir continues stirring, distilling, extracting. At last, with the preparatory steps complete, he asks for the brimstone from the bao purse. He breaks the lemon-yellow mineral into crumbs with a mortar and pestle, then lights them with a punk. Jiaolong is awed to see the brimstone burn with a pale blue flame, yet melt into a crimson liquid, like blood. Jabir is just about to pour the liquid into a porcelain cup when Jiaolong interrupts.

—Can we use this one?

Sheepishly, he holds up a cup he took from the inn this morning, with a salamander on the side. Jabir smiles and touches his arm.

—Of course.

He pours the brimstone-blood in, and sets it near a flame to keep it warm.

Next Jiaolong hands over the cinnabar, a scarlet mineral covered in gray scales. Now for the meteor. Fascinated, Jiaolong watches Jabir rasp its surface with a file, producing a gram of dark-gray powder. This he adds to a glazed pot, along with a crumbling of cinnabar. He hangs the pot over some hot coals and places a lid on. A half hour later, Jabir lifts the lid to find liquid mercury—mingled with dissolved meteor—clinging to the lid's underside. He scrapes this into the salamander cup with a spoon.

Jabir then limps along the workbench, grabbing the tinctures he prepared before and splashing a bit of each into the cup—one yellow, another green, a third thick and crystal-clear. Finally, he approaches Jiaolong.

—I need two more ingredients.

To Jiaolong's surprise, Jabir takes his hand and pulls him toward the workbench. With a needle, he pricks Jiaolong's finger and squeezes several beads of blood into the cup. Jabir stirs this with the spoon. When he's finished, he turns his eyes up to Jiaolong's.

—Now to apply the elixir.

Jabir loosens the bao belt and begins to part Jiaolong's red robe.

—What are you doing?

—Something I've always wanted to, Jabir whispers.

His hand continues inside the robe, until he's touching Jiaolong's skin. Jiaolong tries to shove him back like this morning, but this time Jabir is ready. He grabs Jiaolong's arms in his powerful grip.

—I need to apply the elixir. Trust me.

He reaches inside the robe again. Jiaolong flushes red but says nothing, just turns his head aside. Jabir's fingers are hot, and he can feel them tracing down his stomach, then past his waist. For the first time in ages, he feels not a pulse of phantom pain but phantom pleasure—a ghostly swelling in his groin.

His robe now open, his shameful void exposed, Jiaolong cringes, fearing what Jabir will say. But the alchemist simply dips two fingers into the cup and dabs the elixir over his scars. The liquid is oily and warm and tingles his skin. He's left with a throbbing crotch and a dripping between his legs. It's the most breathless moment of his life.

Jabir sets the cup aside, washes his hand, then produces a white cloth from his waistband. He lays it open on his palm and gazes at Jiaolong.

—I applied half the elixir. Now I need your bao.

As if hypnotized, Jiaolong removes his bao jar from the purse and wriggles the cork free. From the day they entered the jar thirty years ago, he doesn't think he's ever touched them. He's surprised how rubbery they feel, and how cold. One by one he places all three treasures in the cloth, which Jabir rolls up tight.

As Jabir limps across the room to fetch a clean pot, Jiaolong asks what he's doing.

—They need purification.

—You're going to burn them?

—It's the only way.

Scarcely able to breathe, Jiaolong watches Jabir bundle the cloth inside a thick piece of linen; he mutters something, then places everything in a cast-iron pot and adds several coals to the fire beneath. Fearing the smell—his own flesh cooking—Jiaolong dabs clove oil beneath his nose. But the acrid smoke that emerges from the pot still stings his eyes.

When the smoke stops, Jabir scoops up the ashy remains of his bao. Then he dumps them into the elixir cup. He returns across the room, holding it up to Jiaolong.

—Do you hear the birds? It's dawn outside. A propitious time to drink. Alchemists say that like seeks out like, and the elixir in the cup will seek out the elixir between your legs and carry the bao with it. When they connect, your organs will grow again.

Jiaolong peers down into the cup. The smell of sulfur and metal overwhelm the clove oil. But with Jabir's hand guiding his, he begins drinking.

The powdery layer of ash on top chokes him, like too much pepper on soup. Then the liquid comes, thick and slimy as it courses down his throat. He can feel the dense mercury plopping into his stomach like slush off a sloped roof. He barely finishes.

Afterward, Jabir opens the shutters and leads Jiaolong to some chairs in the corner. There's no more noise of battle outside. They're both talked out, so they sit silently and hold hands and watch the sun rise.

All the world has collapsed to the warmth of Jabir's fingers. It's wonderful. But as the minutes pass, Jiaolong can feel other sensations creeping over him. His head grows light, as if he's floating slightly. His lips begin tingling, not unlike the pepper paste that numbed his bao the day he was snipped. But why a tingling in his lips, and not down there?

After a few more minutes, his stomach starts burning—fiery pins jabbing his bowels. When he finally speaks, his speech sounds slurred.

—What's burning? The mercury?

—The salamander. It's poisonous.

Jiaolong's watery eyes flash open. Jabir looks strangely, deadly serious.

—It's a sham recipe. Chemical sleight of hand.

An instant later, Jiaolong's military training takes over—all Chinese generals fear being poisoned. He jams two fingers down his throat, trying to make himself retch. But he's coughing too much to expel anything. He feels Jabir's hand caressing his back.

—You'll only cause yourself more pain. The poison is a potent muscle relaxer. You'll never get enough up.

—You said you'd heal me.

—I said I would give us both what we desire, love.

The word enrages Jiaolong. He draws his dagger. Or tries to—he feels clumsy, fumbling with the clasp on the sheath.

He finally gets the dagger free and turns. For some reason, Jabir hasn't backed away. Jiaolong flails at him. The resulting stab is superficial, barely an inch deep. But when Jiaolong tries to withdraw the blade and stab again, he once again feels the warmth of Jabir's hands around his. He glances up—and is confused to see Jabir smiling. He looks peaceful, even saintlike. And instead of withdrawing the dagger, he grips Jiaolong's hands tighter—and helps him drive it in to the hilt.

An orgasm of pain flushes across his face. He chokes out a whisper.

—Thank you.

Jiaolong lets go of the dagger, but the front of Jabir's robe is already flooded red. The blade still in place, Jabir drags himself across the room to the cast-iron pot and fishes around inside. His hand emerges with an unburned white cloth. Then he limps back, even more labored this trip, and places the wrapped bundle in Jiaolong's hand.

—We shall both be whole in Paradise, love.

He tries to say more, but crumples down to the ground instead.

In Jiaolong's clouded mind, the whole scene seems surreal. His head is swimming, his vision narrowing to tunnels. Every breath takes conscious effort.

He almost forgets the cloth in his hands. But when he runs a finger over it, he recalls the waxy texture of asbestos. He unwraps it slowly, and smiles to find his bao—perfectly preserved and warm to the touch, ready for reunification in Paradise.

# Mexico — AD 1500s

༄༅

Reciprocal trade demands drew Arab goods and technology into China in the 1200s, and sent Chinese goods and technology coursing back to the Middle East. African, Indian, and European ideas and merchandise were swirling in the mix as well. Globalization was not just a feature of the late twentieth century.

All the while, the people of the Americas were developing their own cultures and civilizations in isolation—until they weren't. In the late 1400s, Spain especially began sending hordes of sailors and soldiers across the Atlantic Ocean, first to the Caribbean and then to the American mainland. There, disastrously, they encountered the Aztec.

Properly speaking, the people we call the Aztec consisted of multiple related but distinct ethnic groups, each with unique cultures and traditions. One of these tribes, the Mexica ("meh-SHEE-kah"), wandered down into Central Mexico in the 1200s from the deserts farther north. (The name of their possibly mythical homeland, Atzlán, gave rise to the term *Aztec*.) The Mexica eventually settled on an island in the middle of a lake, supposedly after seeing a sign from their gods: an eagle gripping a snake, which landed on a cactus. An image of this bird still adorns the Mexican flag today. As a long-term home, the island seemed unsuitable: swampy and choked with reeds. But the Mexica built there one of the wonders of world history, the achingly beautiful city of Tenochtitlán. Imagine Venice dropped into the Alps, with a dash of Egypt—a bustling metropolis crisscrossed with canals, surrounded by snow-capped peaks, and studded with gleaming white pyramids.

The Aztec city of Tenochtitlán (*top*) had a thriving marketplace (*bottom*). Notice the canals. (Credits: Diego Rivera, Wikimedia Commons, and Joe Ravi, Wikimedia Commons.)

The political relationships among the kingdoms of ancient Mexico were as complicated as anything in Renaissance Italy, rife with coups, assassinations, and backstabbing rebellions. But after a civil war in the 1420s, Tenochtitlán and two other cities, Texcoco and Tlacopan, rose to power and formed the so-called Triple Alliance. These allies quickly gobbled up 80,000 square miles of territory, conquering tribe after tribe. Over time, Tenochtitlán and the Mexica began to dominate the alliance, becoming first among equals.

The Aztec mostly left conquered cities to their own affairs, allowing them to retain local rulers, gods, and traditions. But the empire's four hundred vassal cities did have to pay steep tributes twice a year in the form of goods, everything from feather cloaks to precious jade to live spoonbill birds. More painfully, some cities had to supply human beings for ritual sacrifice. The Aztec were highly refined in many ways—they had rich traditions of poetry and art, and cultivated flowers of all kinds—but they also sacrificed human beings to their gods at alarming rates. This naturally induced a lot of hatred among the subjugated, hatred that ultimately doomed the Aztec.

When the Spanish arrived in Mexico in the early 1500s, word about their atrocities on Caribbean islands probably preceded them, and the people they encountered near the coast fought them off and nearly wiped them out. (It's likely a myth that native Mexicans viewed the Spaniards as gods and welcomed them—the opposite, if anything.) Eventually, however, a few Machiavellian-minded tribes flip-flopped and allied themselves with the invaders in order to achieve the more pressing goal of ousting the Aztec from power. Indeed, a few hundred ragtag Spaniards never would have defeated the Aztec, or even come close, without their Native allies.

The following chapter focuses on those allies. It takes place in an imagined Aztec city whose people aped the dominant Tenochtitlán culture in food, in dress, in religion, and especially in their zeal for butchering and sacrificing other human beings...

Huehmac wakes to the rasping of corn on a *metate,* stone grinding stone. He rolls over on his mat, his head aching and joints sore. He's been imprisoned in this hut for five days now.

There's little furniture here beyond a wooden chest. Coyote skins hang on the walls beneath the low reed roof. When a conch trumpet sounds from the temple, marking dawn, he sits up groggily. He slept in a loincloth, and now dons a plain cloak of agave fibers—a far cry from the cotton robes trimmed with green feathers that his aristocratic family once wore. But a decade ago, the Aztec conquered his people and banned the importation of cotton and many other goods. Now even princes like him dress like farmers.

There's a loose chink in a corner of the adobe-brick wall, and Huehmac works it loose to spy on the world outside. Nearby, he sees his guards—an eagle knight in a feather suit and beaked hood, and a jaguar knight in a full-body pelt and hood with fangs. They're heavily tattooed and stand as still as statues. The only time they move is to toss food inside onto the dirt floor—the same cruel "meal" every time, tiny red chilis called *tepíns,* "fleas." They're blazingly spicy, and he resisted eating them for three days before gobbling them down with predictable results: choking tears, a burning mouth. He's been feverish and queasy anyway since his capture, and promptly threw the chilis up. He wasn't allowed to wash himself afterward, either, a pointed insult. His people are scrupulously clean.

Beyond the guards, the Sacred Precinct is coming to life: merchants carrying baskets of goods, trappers dragging cages of turkeys, old women sizzling tortillas on clay griddles, children running to school. He despises every one of them. The Aztec are interlopers, having migrated to this land a mere dozen generations ago. (His own family stretches back hundreds.) And while they did conquer the region, they did so through lies and trickery, and now exact steep tributes from their

An Aztec eagle warrior (*left*). A modern depiction of a jaguar warrior headpiece (*right*). (Credits: Gary Todd, Wikimedia Commons, and Carlos Valenzuela, Wikimedia Commons.)

subjects twice a year—heaps of jaguar pelts, bolts of cloth, gourds full of gold dust, honey and cocoa, live snakes, and much more. He grows bitter just thinking about it. Above all, he hates the Aztec for being bloodthirsty, for demanding a constant supply of captives to sacrifice. Huehmac's stiff-necked people have suffered more than most, having provided thousands of men and women and even children to be slain. One of Huehmac's sisters was flayed alive, her skin turned into a costume for their amusement. The Aztec are vermin—cockroaches—and Huehmac wants them exterminated from the earth.

Hence his people's alliance with the Spanish. No one knows where those smelly,* fish-belly-looking foreigners came from, and after hearing tales of their crimes elsewhere, Huehmac's people marched out to battle them. They were in fact on the verge of annihilating them when

---

* The Aztec always burned incense around the Spaniards, which the Spanish took as a great honor. In reality, the Aztec couldn't stand how ripe the invaders smelled, a result of the Spanish habit of sleeping with their clothes on, as well as a general indifference to bathing.

Huehmac's father, a counselor to their king, had a vision. Why not pit the Spanish against the Aztec instead? Use their horses and cannons and dogs to destroy the real enemy? So, after months of battle, the Spaniards became their allies.

Huehmac embedded himself with the Spanish as an adviser, and soon picked up enough of their language to speak with them directly. After six months, he became the right-hand man of Don Ferdinand, lieutenant to the conquistador himself. Burning with zeal, Huehmac even began volunteering to sneak into Aztec towns to spy on the fortifications before attacks. He completed six successful missions—until five days ago, when an old man taking a midnight piss spotted him and raised the alarm. He's been confined to this hut ever since. He just hopes Don Ferdinand and the rest of his Spanish friends haven't forgotten him.

Huehmac soon grows weary, and replaces the chink in the wall to lie down again. He sleeps for an hour, until a commotion outside wakes him. The guards pull back the wooden palisade blocking the door, and in walks a man Huehmac has never met, but knows by reputation—the "fire priest" in charge of human sacrifice. He's wearing a black cassock embroidered with skulls. He has grotesquely long fingernails and black hair down to his waist that's matted with dried blood. His ears are scarred and swollen from the self-inflicted cuts of daily blood-letting rites. Some fire priests, Huehmac has heard, even slice their own penises.

The priest introduces himself as Topiltzin, and tosses a cotton bundle at Huehmac's feet. Huehmac opens it reluctantly—what fresh torture is this? Inside, he finds tortillas stuffed with turkey and squash. They're even garnished with crispy insects—grasshoppers and beetles and other delicacies. Huehmac's mouth waters, but he eyes the priest warily. Topiltzin laughs.

—We don't *poison* people. Eat.

Huehmac's clan never eats before midmorning, but he tears into the food now. Topiltzin waits patiently while he finishes. Then the priest removes a black disk from his robe, a polished obsidian mirror. He hands it to Huehmac and tells him to look. Huehmac gazes at his sorry reflection. Upon capturing him, the Aztec removed his earrings and nose ring and lip labret; instead of sparkling with jewels, his face sags.

—Why am I looking in this?

In response, Topiltzin claps his hands, and the jaguar and eagle knights enter. The jaguar is carrying a human skull, at least partly. There's a long chert blade sticking out where its nose once was, and another blade for a tongue. It has giant white clamshell eyes with polished pyrite pupils. Topiltzin turns it in the light.

—It's a ritual of ours. The mirror shows your present. Here is your future. I believe this is a brother of yours. They say you look just like he did. You will again soon.

Huehmac almost makes it. He darts forward, arms outstretched, furious to seize the priest and tear his disgusting hair from his scalp. But he's a step slow. The two knights intercept him and slam him down, knocking his breath out. Huehmac can only hiss from the ground.

—The Spanish are coming. I'll see your skull before you see mine.

—The Spanish are filthy monkeys who care only for gold. They have a few hundred soldiers. We Aztec have a hundred thousand.

—You're forgetting the millions who hate you. The Tepanec, the Tlahuica, the Xochimilca, the Purépecha…

—Enough. You think the Spanish are your saviors? What of the pestilence they bring?

Huehmac's face hardens; he worries about the pestilence, too. The Spanish call it *viruela*—smallpox. People get queasy and feverish, and then ugly, pus-filled sores erupt all over their bodies, even the face. There was an outbreak among the Spaniards last week—Don Ferdinand fell ill—and Huehmac feels fortunate to have dodged it. But he can't let the priest get the last word.

—The plague is no match for our will.

—Why align yourself with thugs who will only betray you in the end?

—We can wipe them out whenever we please. And once we've defeated you, we will.

The priest shakes his head in disgust.

—There's no use arguing with a fanatic. I shall see you at noon. Make sure to rest.

—Why?

—You're playing in a ballgame then. And when you lose, I plan to sacrifice you.

Sadly, the native cultures of Mesoamerica, including the Aztec, were largely wiped out in century after the Spanish conquest. Still, you can catch glimpses of the old ways here and there, especially in two realms— food and the famous Aztec ballgame.

The Mesoamerican diet revolved around corn (maize) in all forms— tortillas, tostadas, tamales, pupusas, hominy, atole, and more. According to some native religions, humanity itself sprang from corn. But Mesoamericans didn't cook the kernels the way most people do nowadays, by boiling them. They used a process called *nixtamalization*.

Nixtamalization involved heating dried kernels in water with dissolved minerals, usually lime. When the kernels were perfectly al dente, cooks removed them from the heat and let the mixture rest overnight. The next day, the kernels were ground on stone metates into a dough called *masa*. Overall, this process did two things: The lime made the kernels more nutritious by breaking down certain indigestible molecules and freeing up nutrients like niacin. Nixtamalization also released pectin from the kernels, which made the masa dough sticky and elastic for foods like tortillas.

One of the few places that still nixtamalizes corn is Maizajo, a tortilleria north of Mexico City. I dropped by to speak with founder Santiago Muñoz, a chef with tiny eyes, an off-kilter grin, and a bun of brown hair. When I ask what first sparked his interest in nixtamalized corn,

Muñoz chuckles and says he wishes he had a romantic origin story to tell me — his *abuela* bouncing him on her knee, patting out tortillas while she sang lullabies, or what have you. In truth, he simply visited a farmer's market one day as an adult, and the sheer variety of kernels blew him away — red ones, blue ones, yellow ones, some as delicate as baby teeth, others as thick as gorilla molars. He started researching heirloom corn, then learned about nixtamalization and began tinkering with recipes.

However delectable it smells, Maizajo's kitchen looks like an auto-mechanic shop, with concrete floors and peeling paint. It also has industrial-sized stoves and aluminum pots so big you could play hide-and-seek in them. One chef stirs a vat of simmering corn with a spatula the size of an oar. After soaking, the chefs mash the softened kernels in an electric grinder, a disappointingly modern touch — until Muñoz removes a panel and shows me that the grinding plates inside are made not of metal but black volcanic rock. (Metal ruins the corn's consistency.) The plates weigh twenty-five pounds each and have grooves carved into them; when the grooves wear down, Muñoz sends them to an old man in a village nearby to chisel in new ones. The end result of the grinding, the masa dough, has a silky, elastic feel to it, like edible potter's clay.

Sensing my stomach growling, Muñoz takes pity and cooks me a tortilla. He chooses blue-corn masa that's a beautiful turquoise color. He squashes a ball of it flat with a handpress, then fries it on a flat grill, letting it sizzle softly. When he flips it with his fingers, the tortilla puffs up like a balloon, as if it's about to float to the ceiling. A moment later, Muñoz whisks it off and hands it to me to taste.

I immediately burn my fingers. (As a chef, Muñoz's hands are immune to heat in a way that mine are not.) But after some huffing and puffing and hot-potatoing, I bite in. Even plain, it's delectable, like silky cornbread; I could eat a whole meal of just this. And if you think you can get something like it at Whole Foods, think again. Fresh nixtamalized tortillas simply don't keep beyond a day, and even most "authentic" restaurants make their tortillas from dried flour that Muñoz politely calls crap. I agree: they taste nothing like this. And unlike all the other corn tortillas I've ever eaten—which inevitably split in half and spill my taco guts everywhere—nixtamalized tortillas are strong and flexible. You can bend, crease, crumple them, and fold them into origami cranes, and they never crack. They're the nutritious, delicious wonderfood that fueled the Aztec empire.

Muñoz sells fresh tortillas to restaurants around Mexico City. One former client is Los Danzantes, a gourmet bistro that now nixtamalizes corn in-house. After visiting Maizajo, I have lunch at Los Danzantes, although I'm not there for the tortillas. I'm there to eat insects.

Many people nowadays swoon like Victorians at the mere suggestion of eating bugs. But virtually every culture in history did so, because they're nourishing, environmentally friendly, and packed with far more protein than meat—nearly two-thirds by weight, as opposed to one-quarter for beef. If prepared right, they're also darn tasty.

Mexicans eat roughly a quarter of the world's 2,100 edible insects, several of which Los Danzantes serves up during an entomophagy festival each summer. Chef Alejandro Piñón usually blends them into haute cuisine dishes, but I ask for mine plain—more like what ancient Mesoamericans ate. Piñón brings them to my table on a thick black platter, divided into piles by species. It's a surprisingly colorful array—pink, white, yellow, black.

I start with chinicuiles, roasted moth larvae found on agave plants. (You've probably seen these "worms" in cheap mezcal.) I have to admit, despite being pro-insect, I hesitate to try one. They look like the mealworms my siblings and I had to feed to our pet chameleons

Like most cultures around the world, the Aztec ate insects. (Copyright: Sam Kean.)

as kids, with pink, segmented bodies and no obvious differences between head and anus. I expect them to start squirming any moment. But they surprise me. Mercifully, there are no squishy guts inside. They're essentially hollow shells, and they taste like fried Rice Krispies. Not bad.

I move on to cocopaches, an insect found on mesquite trees that looks suspiciously like a stinkbug. Piñón has charred them, and they taste like blackened toast. They'd make a nice texture contrast in a salad, a protein-packed crouton—albeit with legs that sometimes get stuck between your teeth.

Next in the rotation are hormigas chicatanas—roasted ants with big fat booties. I love them. They crunch beneath my molars like the salty, half-popped kernels at the bottom of a popcorn bag—a tantalizing bit of resistance, then a delicious *snap*.

The fourth course is escamoles, edible ant pupae. They're often called Mexican caviar, so I have high hopes. Alas, they lack much flavor. To be fair, they're normally slathered with butter, but these plain ones taste like goopy white tapioca pearls—soggy little spheres that dissolve to nothing in my mouth.

Up next is the most commonly eaten insect in Mexico—chapulines, grasshoppers. Even today, many Mexicans munch on them with cervezas, like bar pretzels. But that undersells chapulines; they're not just filler. They have a rich chipotle flavor, zesty and savory. After my second spoonful, I'm convinced they've been seasoned with some proprietary blend of herbs and spices, but Piñón insists no—just salt and lemon. I'm floored.

I save the most dubious-looking insect for last, caterpillars called meocuiles. To be blunt, they look like obese maggots with stumpy legs—bright white and two inches long. But wow. When I finally try one, the soft crunch and burst of flavor make my spine sag. They instantly remind me of something I've eaten before, but I can't recall what. I plug my ears and shut out the chatter at the restaurant while I munch another one, probing the Proustian byways of my memory. Then it hits me: meocuiles taste like McDonald's French fries, back when they cooked them in beef tallow. Specifically, they taste like the large, flat fries that were golden brown and slightly leathery. I came to Los Danzantes to be transported to Aztec times, and largely was—the bugs are a revelation. But for a few seconds, I find myself transported back to my boyhood instead, a spacey little mop-haired kid devouring Micky-D's fries by the fistful.

One aspect of Aztec life that wowed the conquistadors was the Mesoamerican ballgame, *ullamaliztli*.* Different regions played different versions of the game, but in general, teams of four or five players batted heavy rubber balls back and forth with their hips, and scored points by either hitting an unreturnable shot or knocking the ball past their opponents' end line. Teams could also bump the ball through an elevated stone ring—a feat so difficult that it ended the game automatically.

After the Spanish banned ullamaliztli as subversive in the 1500s, the game lived on in remote provinces of Mexico for centuries, and has only recently been revived in Mexico City. Its biggest champion there is Arturo Sánchez, a burly, talkative robotics engineer turned ullamaliztli archaeologist; he also wrestles lucha libre on his days off. Sánchez started playing the ballgame at age twenty-six, which he says is too late to get truly good at it—less because of any need to develop special skills and more because you have to toughen your body up for the abuse it takes during games. Players fling themselves around with abandon, and regularly break ribs and dislocate elbows. Even normal gameplay leaves them bruised and limping.

I met Sánchez and several other players—including a dentist, a

---

* The ballgame went by different names in different times and places—*pitz, pok-ta-pok, ulama*. To keep things simple, I'm using the Aztec term *ullamaliztli* here. Archaeologists don't know exactly how the scoring worked, although they do know it was complicated. (Scoring twice in a row, for example, might cause your opponents to lose points.) Ullamaliztli was so popular that one town, as part of its annual tribute, had to provide 16,000 rubber balls to the Aztec every year.

psychologist, a physical therapist, and the PT's eleven-year-old daughter—at a plaza in Mexico City one Wednesday night. After some holas, Sánchez shows me the cantaloupe-sized balls. He gets them from a man in Guatemala who slashes rubber trees in the jungle and collects the sap that oozes out. They're dark brown, quite soft, and bounce as well as basketballs. (The Spanish, watching them boing around, thought they were either alive or contained a devil.) But however soft, the balls are ominously heavy. Sánchez's weighs eleven pounds, and they range up to fifteen. In essence, we'll be batting around a small bowling ball with our hips.

Next, we gear up. This involves binding our groins into girdles of leather and cotton. Players do this partly to protect the crotchal area from blows, but mostly to compact and harden the hips so they can endure a beating. I watch a player named Giovanni dress. Most players are right- or left-hipped, he says, so that side gets extra padding. After he finishes wrapping himself, he tells me to punch him in the right hip; it's as dense as a speedbag. Then Giovanni kits me up. I keep having to suck my stomach in while he cinches the wraps tighter and tighter. My groin is soon crushed to a neutron-star density. Equally bad, the girdle restricts my movement to a penguin waddle; I struggle to bend over and tie my shoes. The other players, meanwhile, remove their shoes to play barefoot on the concrete plaza.[*]

While everyone else divides into teams to start a match, Sánchez and Giovanni teach me how to hit the ball. The basic motion involves stepping forward, flinging your arms back, and thrusting your hip out to punch the ball. As Sánchez says, "Just do a Shakira." Then Giovanni starts lobbing balls at me. I'm pretty gun-shy at first. When you see the equivalent of a bowling ball arching overhead, it's hard to persuade yourself to smash your pelvis into it. But I soon get the hang of things.

---

[*] For television "documentaries," producers sometimes ask modern ullamaliztli teams to don war paint and elaborate headdresses while playing. That's all baloney, Sánchez says. The Aztec did enjoy dressing in costumes, even during battle, but they played ullamaliztli unadorned.

And to my surprise, the ball really *pops* when I connect right, darting back like a line drive. It's immensely satisfying.

Then comes the torture. Gameplay usually starts with an aerial serve, and the players bounce the ball back and forth with their hips for a minute. Inevitably, though, the ball loses momentum and starts dribbling along the ground. Unfortunately, you cannot scoop it up or lift it with your feet — you still have to return it with your hips. And the only way to do that is to drop to the ground, jam your wrist into the pavement, wrench your body sideways, and swing your hip like a club. If you're good, you drop and strike in one fluid motion, and the ball zips back, or even pops up and starts bouncing again. If you're me, you flop down, smash your butt, and just kinda get in the ball's way as it caroms off you. I blame my struggles on the girdle; I feel as flexible as a mannequin. Once, flustered, I mistime my drop and land smack on top of the ball. Sánchez flinches. A player he knows recently cracked her tailbone that way. I make a mental note to keep the ball in the air as much as possible during games.

Speaking of which, however unprepared, I'm soon thrust into the match the others are playing. I'm told to choose which team I want to join — classic playground torment. I glance back and forth, cringing with the knowledge that whichever team I pick will likely lose. Cravenly, I select the side with the eleven-year-old girl, hoping she'll shoulder some of the blame.

In the version of the game we're playing, you score points by knocking the ball past your opponents' end line. I quickly realize that this style of ullamaliztli is really two sports in one, depending on whether the ball's bouncing or not. When it is, it's a madcap combination of soccer and tennis. As with soccer, you're pressing down the court as a team toward your opponents' goal. But as with tennis, you're batting the ball back and forth. Also like with tennis, it's a game of angles. You're constantly trying to wrong-foot your opponents, either by knocking the ball into awkward spots or making them scurry side to side. Finally, you punch the ball in for a score. It's fast, fun, and dynamic.

Unfortunately, when the ball loses momentum and starts rolling along the ground, the game deteriorates into *Wide World of Sports* as presented by Dante. Again, you can only return the ball with your hip, which means throwing yourself to the ground over and over. Imagine doing a hundred burpees in a row, except every time you drop, you have to hip-punch a bowling ball the other direction. My thighs begin quaking from all the up-and-down, and my already poor form deteriorates into flailing. I feel like somebody trying to play hockey without knowing how to ice-skate.

My legs are soon so shot that I can't even enjoy the aerial gameplay. The other team is also putting weird English on the ball now, spinning it like pool sharks. This throws off the timing of my Shakira thrusts. One time, the ball slams into my shin. Another time it blasts my ribcage. When it smacks my quad once, I crumple like a sniper took me out; it stings like hell.

Shamefully, I spend the last half of the game hiding behind the eleven-year-old and letting her return shots meant for me. In my defense, she's far better than me—fearless and agile. I've never before felt like the proverbial Little League right-fielder—silently begging the other team not to hit the ball my way, making irresponsible promises to God. But that's what the Aztec ballgame has reduced me to.

After an hour, I have to quit. Sánchez says I did well for my first time, but in some ways the torture is just starting. I limp home to my hotel, and spend the next two days shuffle-walking around the city, my quads clenched in one continuous cramp. I moan every time I see stairs. My arm aches, too, and after jamming it into the ground a hundred times, my palm is bruised, which I didn't even know was possible.

Cruelly, on Saturday—the very day it no longer feels like my legs need amputating—Sánchez has scheduled another game of ullamaliztli for us. This time, we play on an authentic Aztec court with stone hoops. It's a forty-five-minute drive northeast of Mexico City, set in a field amid cactuses and white horses rolling in the dirt.

# Mexico — AD 1500s

A Mesoamerican ball court. In this version, players tried to knock the ball through the rings atop the sloped walls. (Copyright: Sam Kean.)

The court is shaped like a squat capital I. On either side, a stone slope rises up to a stone ring roughly ten feet off the ground. Unlike basketball hoops, the rings are rotated 90 degrees to the vertical. Historically, the rings varied in height from bike tire–sized to dumptruck tire–sized, and often had holes barely bigger than the ball, like a crooked carney game. The courts themselves also varied in size. Some accommodated two-on-two matches, while others required as many as fifty-two players per side.

Our court today is on the small side, just twenty yards long. Happily, it's not made of goddamn concrete, although it's not in great shape. Sánchez tells me that, in ancient times, groundskeepers would pound the dirt until the courts were as soft as felt. This court is overrun with weeds and clumps of sod; it's been sadly neglected since Covid.

It's nevertheless exciting to step onto a real Aztec ball court. Gameplay on this style of court resembles less a soccer-tennis hybrid than a volleyball-basketball mash-up. As in volleyball, each team stays on its half of the court and bounces the ball back and forth; if the other team fails to return, you score a point. But like in basketball, you can also (try to) bank the ball through the ring, which ends a match automatically.

Despite my excitement, I once again prove a disgrace to ullamaliztli, partly because I do a poor job of swaddling myself in the protective leather girdle. Remembering how constrained I felt last time, I decide to prioritize mobility this match, and leave my wrappings far too loose; I'm essentially wearing lederhosen. As a result, every volley feels like a sledgehammer to my hips. I can't get any pop, either: Whenever I aim for the stone hoop, the ball skips no more than a yard up the slope before it trickles back down to the weeds. It's humiliating.

The other players are of course much better than me. And while no one actually sinks a shot — a rare and celebrated achievement even in Aztec times — several rattle in and out.

After the match, I peel off my hosen to find a huge angry spiderweb of broken blood vessels on my hip. It looks like a mule kicked me, and doesn't feel much better. Sánchez mentions that Aztec ballplayers often suffered from huge volcanic welts of blood that erupted beneath their skin; doctors released the pressure with obsidian scalpels. My hip doesn't warrant that, but the car-ride home is agony. The seat belt rubs right across it, and every time we turn, the strap bites in.

If I had to conclude one thing from playing ullamaliztli, it's that the Aztec were tough bastards. Especially considering that games could last an entire day, or even stretch across multiple days, until one team either scored enough points or collapsed in exhaustion. It was excellent training for the hardships of war.

To the Aztec, the game also had spiritual dimensions. The rubber ball symbolized the sun, which meant it could never stop during play — because if the sun stopped, the world would end. Tellingly, this was also the justification the Aztec used for sacrificing human beings. They believed that, without the constant shedding of human blood, the sun would quit coursing through the sky, and the apocalypse would follow. In the Aztec mind, then, ullamaliztli and human sacrifice were deeply linked — which explains why those who lost games were sometimes rounded up and ritually murdered...

At noon, with the sun at its peak, the jaguar and eagle knights march Huehmac from his hut in one corner of the Sacred Precinct to the ball court in the opposite one. On the way, they pass several temples, as well as a fountain, a zoo with caged wildcats, a military academy for local children, and a marketplace that leaves Huehmac both gawking and scowling. Hundreds of merchants sit on mats in orderly rows, peddling feather blouses, baskets of flowers, mounds of cocoa beans, cactus leaves, snakeskins, dried deer flanks, copper necklaces, and more. Every item has been extorted or stolen from a different corner of the Aztec Empire.

The ball court has bleachers made of black volcanic basalt. Several hundred people are already packed into them, and when they see Huehmac's retinue approach, they begin hooting and jeering. The court itself is twenty yards long, with vertical stone rings mounted high on either side. Despite the hostile reception, as he steps onto the pitch, Huehmac is impressed at how spongy the pounded dirt feels beneath his feet.

From a throne atop the bleachers, the fire priest Topiltzin rises and announces that Huehmac will be playing the jaguar knight—Maxicatzin, they call him. Two attendants strip off the knight's spotted pelt to reveal a tawny torso blanketed with tattoos—frogs, monkeys, ferocious bears. The attendants start cinching his hips into leather girdles. Huehmac looks around for whoever will be dressing him. When the crowd laughs at his confusion, he realizes the truth—he won't be getting any padding.

Maxicatzin serves first. He smacks a sharp, low shot, and Huehmac scrambles to return it. Although a fine ballplayer as a child, Huehmac has neglected the game lately in favor of fighting the Aztec. On top of that, he's stiff and weak from his imprisonment in the hut. But as the first rally proceeds, it pains him to admit that, even at full health, he's never been the equal of Maxicatzin. The knight's quick and aggressive, with hips that whip around like an alligator's tail. Pride keeps Huehmac hustling; he manages to return a few dozen volleys. But his legs eventually stumble, and the ball scoots past him for a point. He stands and pants in the searing sun as the crowd cheers.

He falls two more points behind over the next ten minutes. Feeling cocky, Maxicatzin punches one volley up toward the stone hoop. It just rims out, and the crowd groans. He loses a point for the attempt, but easily scores another during the next rally. The game is getting out of hand.

Huehmac considers aiming for the hoop himself. But if he misses without having scored first, the rules say he loses automatically. He'd prefer to finish the match and go down with dignity. He whispers a prayer to his gods to look after his family, then slowly, fitfully, resigns himself to death.

He's just about to serve again when he hears laughter. The Aztec abhor drinking outside of sanctioned festival days, and routinely put people to death for consuming too much *pulque*, their fermented cactus sap. Only the elderly are exempt from this ban. *Let them soothe their pains and unwind a little*, the thinking goes. But because the Aztec drink so little growing up, they have low tolerances even as adults. Huehmac has heard several slurred insults already, and one old man finally takes things too far. He rises on wobbly legs, drops his loincloth, and wags his sagging buttocks at Huehmac.

It's the vilest insult in Mesoamerican culture, an unspeakable slur. The stadium falls silent in shock—then quakes with laughter. Even Topiltzin smirks from his throne. The only one not laughing is Maxicatzin, who tries to hush the crowd. But his decency falls on deaf ears. Huehmac has rarely been so furious in his life; he's shaking with indignation. The heat no longer matters. His aches and thirst evaporate. He's going for the stone hoop.

But he needs to set the shot up. All game long, he's been playing conventionally and serving to Maxicatzin's left side, a harder return for a right-hipped player. In fact, he can see Maxicatzin leaning left now in anticipation. So Huehmac crosses him up, serving to his right. It wrongfoots Maxicatzin. He still reaches the ball, but for the first time all game, he doesn't punch it toward a corner. He returns a fat, slow volley down the middle of the court instead.

Praying he hasn't lost his touch, Huehmac darts forward and slaps the ball with topspin, curving it towards Maxicatzin's left. The ball caroms like a stone off water, zipping right where he'd intended. There's little the knight can do except send a high lob back in Huehmac's direction.

It floats like a second sun. Huehmac will never get a better chance. He leaps up and whips his body around.

The ball crushes his hip when he connects. It's agony. It's also the most beautiful shot of his life. It skips up the slope, banks off the wall, and whistles through the ring.

The stadium detonates. Maxicatzin runs two empty steps and crumples to his knees on the court. Huehmac almost feels sorry for him. But the snarling of the crowd—and several more drunken buttocks—hardens him again. He limps over to the ball, hoists it over his head, and turns toward Topiltzin.

The priest holds his arms up to quiet the mob, his long fingernails outstretched.

—Congratulations to our challenger. He will have the honor of being sacrificed second.

Huehmac is thunderstruck; the ball slips from his grasp.

—You promised that the winner would be spared!

A smile twists Topiltzin's face.

—I merely said the loser would die. But if the gods decree two sacrifices today, so be it.

The Aztec were obsessed with war, an attitude inculcated from birth. Midwives sounded war whoops after a successful delivery, and umbilical cords were sometimes buried on battlefields to draw future warriors there. The main role of women in Aztec society was to breed warriors, to the point that women who died in labor received full military honors no different from fallen soldiers.

During battle, Aztec soldiers armed themselves with everything from bows and arrows to obsidian-studded clubs, but the most revered weapon—because their gods used it—was the atlatl, the same ancient throwing-stick that people in the Andes and many other cultures used.

The word *atlatl* actually comes from the Aztec language of Nahuatl ("NAH-wah-tuhl"), and Aztec warriors could hurl their bone- or obsidian-tipped darts with enough force to pierce Spanish chain mail. The Spanish feared the atlatl more than any other Aztec weapon.

Beyond the usual motivations for waging war, like gaining land and seizing wealth, the Aztec also used wars to secure prisoners to sacrifice. To be sure, people in every corner of the ancient world practiced human sacrifice. We've already touched on bog bodies in Europe, and we can add Turkey, Canada, Australia, China, Polynesia, Arabia, and West Africa to the list. But no one made a spectacle of sacrifice like the Aztec did.

The most dramatic sacrifices involved removing someone's heart with a flint or obsidian blade. The victims were often dressed in costumes to resemble gods and marched to a temple atop a huge stone pyramid. There, four priests bent them backward over a large stone and held down their limbs. A fire priest then stepped forward to vivisect them. Reportedly, some priests could remove a heart so quickly that it was still beating when they held it up. Some texts imply that priests could even feed a few bites of the organ to the victim before he expired. So much blood gushed out that vampire bats often flocked to the temple grounds afterward.

The Aztec performed this rite for religious reasons. According to various myths, a god of theirs spilled his blood long ago and sacrificed himself to save the cosmos from destruction. Afterward, he transformed into the sun. In return for his sacrifice, the sun god demanded that humankind repay him by spilling its own blood every day. (Indeed, the Aztec did not think of the sacrifices as killings, but as making good on a sacred debt.) The blood also nourished the god on his trip through the heavens, and without daily blood sacrifices, the Aztec believed that the sun would stop moving and the world would end. On a more pragmatic level, the rites bound the Aztec people together in worship, and sent a not-so-subtle message to conquered groups about who was in charge. In fact, the Aztec often refrained from annihilating their enemies during battles in order to ensure a steady supply of future victims, effectively keeping what one historian called a "human stockyard" of sacrificees.

Depending on what other gods needed appeasing, the Aztec also sacrificed people by beheading them, drowning them, tossing them on pyres, shooting them with arrows, or subjecting them to gladiator duels where a fully armed eagle or jaguar knight would attack some poor shmuck wielding only a cudgel whose "blades" consisted of bird feathers; often they'd tie him to a heavy stone for good measure. Post-sacrifice, the limp bodies of victims might be tossed down the steps of the local pyramid. In other cases, the Aztec flayed the bodies and peeled off the skin to make costumes. Sometimes they ate the limbs with chilis.

An Aztec warrior battles a captive bound to a stone (*top*). The Aztec also sacrificed captives (*bottom*) and extracted their hearts to appease their sun god. (Credits: Wikimedia Commons.)

Many victims' skulls were turned into art. Sometimes this involved tiling the surface with turquoise and other gems. More macabrely, they might jam stone blades into the face for a "tongue" and "nose," then add ghastly googly-eyes of shell and pyrite. Other times, the Aztec turned skulls into public monuments. Towers called *tzompantli* consisted of hundreds or even thousands of skulls mounted on wooden rods and arranged in rows—tier after tier after tier of death. One tzompantli was the size of a basketball court, with skulls piled thirteen feet high.

The Aztec often made art from the skulls of sacrificial victims (*left*), or skewered their skulls on rods and displayed them in public (*right*). (Copyrights: Sam Kean.)

Firm numbers are elusive, but some archaeologists estimate that the Aztec sacrificed thousands of people per year.* Naturally, the sheer scale of death provoked a lot of hatred among people like Huehmac, people whose tribes supplied the majority of victims. This hatred in turn

---

* Given that conquistadors often lied in official reports, some earnest and well-meaning archaeologists have insisted that the Aztec never indulged in human sacrifice, and that stories to this effect were mere propaganda to make the Aztec look barbaric and thus justify Spanish atrocities. But both the historical and archaeological record—as documented by the Mexican people themselves—contradict that claim. However despicable, the conquistadors were not lying about the vast number of human sacrifices, nor about what the Aztec did with the bodies afterward. For all their sophistication, the Aztec were as bloodthirsty as the Romans at times.

drove them to align with the Spanish for revenge. The Spanish, however, had their own agenda...

A phalanx of priests, armed with obsidian machetes, marches Huehmac and the disgraced Maxicatzin to the top of the gleaming-white stone pyramid. The wide staircase is flanked with fanged stone snakeheads every fifteen steps, painted blue and red. The steps themselves are steep and narrow, forcing Huehmac to turn his feet sideways and never directly face the temple on top. Every time he lifts his battered leg, he winces.

Huehmac and Maxicatzin have been dressed in the garb of gods — eagle headdresses, green feather cloaks, sandals with gold trim, tinkling anklets of silver. They've filled Huehmac's sagging lip- and earholes with turquoise and amber plugs. Throngs of people stand on both sides of the stairs, saluting the gods as they ascend. Among them, junior priests wave banners of feathers strung between poles, and courtesans in white tunics burn bundles of incense. The women have tinted their skin yellow with mashed bug paste, and dyed their teeth red with crushed beetles. Each one's black hair is piled into two small horns. Whatever their failings, Huehmac thinks, the Aztec can put on a pageant.

Atop the pyramid looms a temple with a gaping serpent's mouth for an entrance. In front sits a brazier with a fire, as well as a black boulder a yard tall and a yard wide. A carpet of thousands of yellow marigolds and purple dahlias lies around it. But in the gaps between blossoms, Huehmac can see the paving stones stained red.

Several priests in black cassocks step forward. Two are holding orange jugs of pulque, the fermented cactus sap, to ensure their victims are drunk and docile. Maxicatzin gulps his obediently. Huehmac refuses — he won't submit. But he pays for his stubbornness: four priests wrench him backward and force his mouth open; the milky white booze gushes down his throat. It's got a sour citrus tang, but any notion of tasting it is quickly overwhelmed by the sheer volume of liquid. It fills Huehmac's mouth, then floods his nose. He snorts and chokes, but they empty the whole jug.

The victims ready, Topiltzin emerges from the temple's mouth like a shade from the underworld. He slips three jade beads beneath Maxicatzin's tongue—payment for passage to the afterlife—then gestures at the massive black rock. Maxicatzin lies down on top, his arms and legs splayed. Four burly priests pin him down.

Another priest, wrinkled and half-blind, shuffles forward holding a snakeskin pillow with a ceremonial knife on top. Topiltzin accepts it. The handle is a jaguar inlaid with gems; its teeth grip the blade, which is made of a special gold obsidian. His head swimming from the alcohol, Huehmac surveys the crowds. There are hundreds of people on the steps, thousands more milling below. All watching in silence.

Topiltzin holds the knife up to the sun.

—Death brings forth life, he chants. By our sacrifice, we keep the universe alive.

He lowers the knife to Maxicatzin's chest, the tip indenting his skin. The knight squirms and sweats.

—This precious water nourishes the fires of the sun.

Quick as a snake, he splits Maxicatzin from sternum to groin. The knight gasps, but Topiltzin's tone neither rises nor falls.

—Now we liberate the sacred cactus fruit of man.

Topiltzin's forearms plunge into Maxicatzin's body. The knight screams at the sky, but the priests hold him firm. Blood overflows his stomach, pooling on the paving stones and soiling the flowers, but after a few deft cuts, Topiltzin emerges with the heart, still pulsating softly.

Huehmac nearly throws up his pulque. But he's glad for the drunkenness—it makes the moment seem less real.

Heart in hand, Topiltzin turns toward a stone statue of a jaguar. There's a basin carved into its back, circled with jade. As if gently lowering an infant, he places the dripping organ inside, then wipes his hands on his hair. The half-blind priest hands him a brand from the brazier, along with dried cactus kindling. Topiltzin starts a fire over the heart and lowers his head in prayer.

The blast of a conch trumpet startles Huehmac. As if woken from a spell, the assembled crowd cheers. Priests on the steps begin slicing their ears with agave thorns and flicking the blood at the temple above. The courtesans begin dancing, jangling the copper bells on their dresses. The universe will survive another day! Meanwhile, three junior priests grab Maxicatzin's limp body and hustle it off to harvest his skull for the tzompantli rack.

By the time the heart has burned to ash, Huehmac can barely stand upright. As the last embers drift away on the wind, Topiltzin raises his eyes and extends his arms. The dancers stop; the masses fall silent. Then the four burly priests grab Huehmac, wrestle him down onto the sticky sacrificial stone, and put his limbs into headlocks.

Huehmac barely resists; he's too exhausted. Topiltzin raises the knife and repeats his prayers. Dimly, Huehmac realizes that they've put no jade beads into his mouth. No passage to the afterlife for him—one last insult. He almost laughs.

The fire priest lowers the knife and presses it into Huehmac's skin. But just as he's about to slice, a gigantic *boom* shatters the heavens.

Huehmac feels the pressure on his limbs lift, and the top of the pyramid erupts into chaos—people screaming, knocking into each other, running amok. He sits up woozily. When another boom sounds, someone shouts that it's a thunderbolt from the sun god.

Huehmac knows better. He's heard that noise before. It's the sound of a Spanish cannon.

Ask someone how the Spanish conquered Mexico, and they'll probably guess something along the lines of guns, germs, and steel. Not quite. At the time, guns were pathetic weapons. They were heavy, clumsy, slow to load, wildly inaccurate, and misfired all the time. They didn't even have sights, because why bother? Even into the Napoleonic Era—three centuries later—it still took armies several hundred musket shots on average to inflict a single casualty. Throw in the impossibility of securing ammunition abroad, and guns seem like a shaky advantage at best for the Spanish.

Steel was a different story, because of armor and cannons. In Europe, cannons replaced trebuchets in the 1400s as the weapon of choice for battering castle walls, even though cannons (as large guns) faced some of the same limitations that muskets did. But in the Americas, cannons added a new dimension to warfare—terror. Short of thunderstorms, Mesoamericans had never heard anything so ear-splittingly loud and violent. They imagined cannons as Promethean thunderbolts stolen from heaven, and often crumpled to the ground or fled headlong in fear. Few Spanish soldiers in the Americas had ever fought a real battle before—they were hardly elite troops—but the psychological edge of cannons provided a significant boost.

To learn more about such artillery, I visited the United States Military Academy at West Point, where a trio of professors—historian Cliff Rogers and chemists Dawn Riegner and Tessy Ritchie—have been re-creating medieval gunpowder recipes and firing a replica cannon. We meet on a stunning autumn morning at one of the campus artillery ranges, number 8. It's a wide, green lawn at the foot of a shaggy, weed-strewn hill; the hill's dotted with derelict tanks that students use for target practice. Trees with warm red leaves surround us on three sides.

Gunpowder first arrived in Europe, via China and Arabia, in the 1200s. Modern gunpowder contains 10 percent sulfur, 15 percent charcoal, and 75 percent saltpeter, but those ratios varied widely in the past. Some old recipes also called for head-scratching additives like vinegar, camphor, brandy, mercury, ground amber, and arsenic. Rogers, Riegner, and Ritchie are testing different recipes to determine which ones bang and which ones fizzle.

Surprisingly, preliminary lab tests at West Point have determined that a few of the older recipes pack more punch than modern ones. But that wasn't necessarily a good thing, since the added punch also boosted the odds of a cannon bursting. Rogers suspects that recipes varied over time because medieval gunners had to adjust each era's powder for the strength of the available steel. "But I could be wrong," he admits. "Maybe they just didn't know what the hell they were doing."

On a workbench at the range, amid a spread of donuts and cookies, the chemists Ritchie and Riegner uncap an orange bucket labeled FLAMMABLE SOLID and unload a dozen Tupperware containers of gunpowder that their students made in lab last week. Some batches contain brandy (Paul Masson Grande Amber) or vinegar (Heinz). While they're all black powders, they vary drastically in texture, ranging from chunky sugar to fine volcanic ash. I pry one container open and accidentally spill some on my hands; I'm glad I'm not flying soon.

When Rogers shows me the replica cannon, I'm severely underwhelmed. Instead of the twenty-foot bazooka I've been envisioning, it's just fifteen and a half inches long—a cap gun. It's also oddly shaped, less a cylinder than a goblet, with a contoured body and six-inch-wide mouth. It's bolted into a steel frame, which will soon be anchored to a wooden pallet; the pallet is in turn anchored to the lawn with green fence-post stakes. Given the cannon's puny size—I could wear it as a party hat—all this reinforcement seems like overkill.

Before we start, Rogers gathers a dozen cadets in fatigues and boots for a briefing. He explains that small cannons like this were used to attack troops and knock down wooden structures. For ammunition, artillery teams often used stone balls. Rogers then reaches into a paper bag and produces today's ammo—marble garden ornaments from Crate & Barrel, $30 each. They're softball-sized, cool to the touch, and shiny. They weigh just four pounds—a third of the weight of the rubber balls I played ullamaliztli with. This doesn't lessen my skepticism.

After Rogers finishes up, we get a safety briefing from an army demolition expert, Captain Hunter. He stresses that "eye pro" and "ear pro" are mandatory—earplugs and safety glasses. ("Pro" = protection.) Sadly, we can't recover the marble ball after each shot, because the shaggy hillside we're firing into is littered with "UXO"—unexploded ordnance that could detonate if disturbed. We're pretty much shooting into a minefield.

For the day's first shot, Rogers opts to use modern gunpowder, to serve as a basis of comparison for other recipes. While an AV tech fusses around with some GoPros to record everything, Hunter and another demolition expert load the cannon. First, they tip it upright and pour gunpowder into the mouth; it settles into a skinny chamber at the bottom. Then they hammer a wooden bung into place to seal the chamber off. Rogers explains why this is necessary. Cannonballs are propelled by rapidly expanding gases. In a regular cannon with a long barrel, the gases have several feet to push the ball. In contrast, in our snub-nosed cannon, the gases don't have much of a runway; they have to propel the ball instantly. Plugging the chamber allows this to happen, because the gas stays confined longer and can build up to a critical pressure. At that point, the gas pops the bung out and blasts the ball forward. Rogers compares the process to Bruce Lee's famous one-inch punch—a quick, short thump that nevertheless packs a wallop.

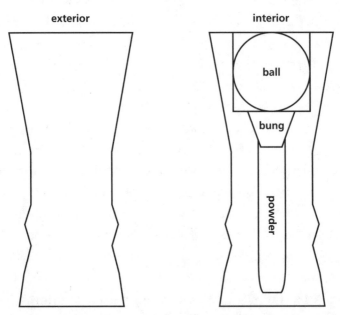

A diagram of the cannon used to test medieval gunpowder at West Point. (Copyright: Sam Kean.)

With the gunpowder and bung in place, a black marble cannonball is dropped into the cannon's mouth and wedged in with wooden shims. At this point the cannon is laid horizontal on the pallet, and the metal frame around it is weighed down with fifty-pound bags of landscaping gravel. Finally, Hunter pours a gram of priming powder—extremely fine, extremely flammable gunpowder—into a quarter-inch-wide touchhole atop the cannon. When lit, the priming powder catches fire quickly, then ignites the other gunpowder inside.

Lighting cannons in medieval times was a dicey job. You tiptoed up with a glowing-hot iron, kissed the touchhole with it, and ran like hell.

(In contemporary drawings, the unlucky soldiers who drew this duty are visibly cringing, their bodies leaning backward at steep angles.) In a bow to safety, the West Point crew is using a remote igniter today: Hunter snakes electrical wires into the touchhole and connects them to a blue plastic trigger-box twenty yards distant. The box looks surprisingly cheap, like something from a Lego set.

We scatter to await the firing. I'm standing thirty yards left of the muzzle, and with my ear pro in, the scene is eerily muted. When I scratch my stubble, it sounds like someone rubbing sandpaper on a microphone.

When everyone's settled, Hunter turns left, right, and center, and yells: "Fire in the hole! Fire in the hole! Fire in the hole!" Then he counts down and punches the Lego button.

I've been warned that medieval cannons don't blow instantly; the gas pressure needs time to build up. First, smoke curls up from the touchhole. Then a pause. One-Mississippi. Two-Mississippi.

When things finally happen, they happen fast. A jet of fire shoots up from the touchhole. A moment later, the mouth of the cannon burps white smoke. Then—BANG! I feel the blast more than I hear it; it shakes my ribs like a defibrillator. I have no idea where the ball escaped to. It zips along too fast to track.

But all that's secondary to the real drama. Because however tiny, the dwarf cannon absolutely obliterates the whole setup. When the shot goes, the cannon and steel frame leap up as if they stepped on a landmine, cartwheeling backward onto the grass. The blast also shreds the bags of landscaping gravel—rocks spew into the air like surf crashing into a cliff. I half expect to see a smoldering crater.

After a stunned pause, we creep forward to check on things.

It's bad news. The cannon is dinged and scraped to hell, and the surrounding steel frame looks even worse: multiple welds have failed, and it's as twisted and gnarled as if some giant beast chewed it up and spit it out. The screen of the nearest GoPro is shattered, too, and every bag of gravel has been torn open at the gills, scattering a blast radius of gray rock. We spend the next ten minutes shaking our heads and marveling. As an experiment, it seems like a bust. As theater, it's magnificent.

The worst news involves the steel frame: it looks rickety now, hazardously so. (Without a frame to anchor the cannon down, it will tumble backward dangerously in recoil.) Rogers, Riegner, and Ritchie fear they might get only one more shot in today.

After some debate, Rogers decides to use the earliest gunpowder recipe, from 1336, for this potential finale. Captain Hunter orders the cadets to start lugging sandbags over from an adjacent artillery range to replace the shredded bags of stone. The cadets also uproot the green stakes holding the pallet down, then thread those through the twisted frame and pound them into the ground for further support. Whether this will hold, we have no idea.

When the cannon is loaded again—gunpowder poured in, bung and ball in place—Hunter yells "Fire in the hole!" thrice more and punches the blue button. Smoke puffs out of the touchhole. We hold our breaths and wait...and keep waiting. It's a fizzle.

Per army regulations, we have to pause fifteen minutes after the misfire before approaching the cannon. The cadets shoo bees away from the donuts and chat about the upcoming Army–Air Force football game. Finally, the demo guys reload the cannon for another shot.

The next shot isn't a fizzle, but the ball doesn't exactly rocket off. Like one of my Aztec ballgame volleys, it merely dribbles out of the muzzle and rolls into the nearby weeds. Rogers pronounces it a "crappy shot." He looks nervous, fretful he won't get any usable data today. At least the frame held.

Rogers needn't have worried. For the fourth shot, we use a gunpowder recipe called *bellifortis*, "strong in war." True to its name, it produces the single biggest explosion I ever hope to witness—a teeth-rattling, sinus-clearing, ear-pro-barely-mattering detonation that I'm convinced could have flattened a building. BOOM. Best of all, the ball plunks one of the tanks on the hillside—there's a huge metallic *thwack*, and we all cheer.

The fifth, sixth, and seventh shots prove no less deafening—the hillsides echo, and the sulfury smell of brimstone fills the air. Scientifically, too, the shots yield some interesting data about different gunpowder mixes: brandy slows shots down for some reason, while vinegar speeds them along. The top shots reach speeds of 240 yards per second, nearly 500 miles per hour.

However puny, the cannon destroyed every scrap of restraining equipment during testing. (Copyright: Sam Kean.)

Eventually, the frame looks so mangled that we can't continue firing safely. Patches of grass surrounding the pallet are scorched black. There's rock and sand strewn everywhere, and the green fence stakes are twisted like pipe cleaners, destined for the scrap heap. Overall, it looks like someone bombed the summer garden display of a hardware store. But my fears about wimpy cannons have been laid to rest. Medieval artillerymen did know what the hell they were doing.

Given their thunderous sound, even tiny cannons terrified the Aztec. Still, citing cannons as the main reason for Spanish military victories would be misleading. Beyond cannons, the Spanish also derived huge psychological advantages from their horses and ferocious dogs, which the Mesoamericans feared as almost mythological beasts.* And again, the Spanish never would have conquered a single square mile of Mexico without the help of thousands of people like Huehmac, who hated the Aztec and willingly joined the Spanish cause to fight them.

Perhaps most important of all, the Spanish spread diseases like smallpox, which native Mexicans had no resistance to. To recast the tired "guns, germs, and steel" cliché, a more accurate summation of why the Spanish defeated the Aztec would be animals, steel, and germs, germs, germs...

The words pulse through Huehmac's mind as he picks his way down the pyramid stairs. *The Spanish have not forgotten us!*

Despite his excitement, it's a slow descent. He's still sore from the ballgame, not to mention drunk, and he fears casting himself down headlong like a sacrificial victim. He stops halfway down to tear off the costume, tossing the feather shroud and headdress aside—back to an honest loincloth. And before descending further, he drinks in the bedlam below.

The Spanish cannons are still booming, the flying balls pummeling the schools and temples in the Sacred Precinct. The marauders on horseback are causing even more terror. Huehmac remembers his own dread upon first seeing the Spanish cavalry last year—swift, monstrous man-beasts clad in steel. He's less skittish now, but marvels anyway as the soldiers tear through the marketplace, swords flashing, busting up stalls and hacking down anything that moves. One rides along the tzompantli rack smashing skulls, sending teeth flying like corn kernels. Huehmac winces. Some of those skulls no doubt belonged to people he knew.

---

* One Spanish horseman was so skilled that the Aztec initially thought he was a man-horse hybrid, like a centaur. The death of a few horses during one battle plunged the Spanish into crisis, since their enemies now knew that the creatures were mortal. They quickly buried the corpses to prevent the Mesoamericans from inspecting them and learning their anatomical vulnerabilities.

As he resumes descending the steps, Huehmac hears growling and snapping. A pack of Spanish dogs is attacking several priests near the bottom, drawn by their blood-soaked robes. His own people raise tiny, hairless dogs as pets and food, and often sacrifice them to escort the souls of the dead to the underworld. But those little quivering mutts are nothing like these Spanish curs, vicious as jaguars.

However vicious, though, they're also obedient. Huehmac spent hours playing with them in the Spanish camps, and as he reaches the base of the pyramid, he whistles sharply. Several come bounding up to him, their snouts smeared red. They bark in excitement, and he knows how they feel. All around them, there's vandalism, plunder, death. It's maximum retribution, maximum humiliation for the Aztec.

Huehmac then spots a familiar face—Don Ferdinand, the lieutenant he translates for. He's glancing around imperially from his horse, as if weighing what to destroy next. Huehmac cries out to him in Spanish and limp-jogs over, the dogs following.

When Huehmac is a few yards away, Ferdinand turns. Back at the camps, whenever Ferdinand talked about crushing the Aztec, he smiled a hard smile, his nostrils flaring over his trim beard. He's wearing that same expression now, and raises an iron-gloved hand in greeting.

—Don Ferdinand, if we hurry...

Before he can finish, the iron hand sweeps down and catches Huehmac in the teeth. He's sprawled on the dirt before he understands what happened—understands that the hand was raised not in greeting but menace. By then, Ferdinand has unsheathed his sword.

The dogs howl. Huehmac whimpers to Ferdinand that it's *him*, his faithful ally. But, flustered, he sputters this in his native tongue—the sound of which only swells the disgust on Ferdinand's face. He raises the sword.

In taking his eyes off the battle, however, Ferdinand has made a grave mistake. A second later, an atlatl dart smashes into his breast, piercing the rusted chain mail. He gasps, and as he crumples, his horse panics and bolts. Ferdinand is dragged along behind like an effigy, bound to the horse by a spur tangled in a stirrup. The dogs chase merrily after.

Huehmac struggles to his feet and staggers toward a quiet corner of the Sacred Precinct. There's a hut there, and he's already inside before he realizes it's his prison. He hides in the corner, and removes the chink in the wall to watch the battle.

It's a tense few hours, but he's been mercifully forgotten in the fray. He sees the fighting ebb and flow, sometimes near, sometimes distant. Exhausted, he even nods off once, and has a wild dream of revenging himself on the fire priest Topiltzin—filleting him and tearing his black heart out. Otherwise, Huehmac rubs his sore jaw and worries over the encounter with Don Ferdinand. He'd spent untold hours with the man, both translating and simply talking—like men, like friends. He's pained to think that Ferdinand didn't recognize him, that he was just another Indian. He's even more pained to think that Ferdinand did recognize him, and attacked anyway.

The sun is sinking toward the horizon by the time the battle dies away. Huehmac emerges to survey the city around him.

Ruin—it's the only word. Fires are burning everywhere, and every building of substantial size is a heap of rubble, including the bleachers of the ball court. Goods lie smeared and trampled in the marketplace—dented baskets and muddied pelts, scorched mats and dead turkeys. He can't even count the number of dead Aztec. It's destruction beyond his wildest hopes. But he feels little excitement. He wonders which way the Spanish have gone.

Not knowing what else to do, he grits his teeth against the ache of his hip and climbs the pyramid to look around, step by painful step. At the top, he sees torn, trampled flowers scattered everywhere. He also finds the fire priest sprawled across the sacrificial stone, his hand still gripping the ceremonial knife plunged into his gut. Suicide—or nearly so; he's still wheezing. But the man who spilled so much blood is now stained with his own. Several long fingernails are broken, presumably from clawing the rock beneath in pain.

Huehmac turns and shades his eyes against the sun, squinting in several directions. There's no sign of the Spanish. He wonders if they're amassing for an assault on the Aztec capital, like he's always dreamed of.

But if so, he doubts he could catch up to them right now. He's feeling more run-down than ever, more queasy, and his hip throbs from the climb. He touches the welts there and grimaces. By now there's a huge knot of blood pooling under the skin. He's got to slice it and relieve the pressure.

He approaches Topiltzin warily. He tries to remove the knife from his gut without disturbing him, but the priest groans and shifts. Huehmac steels himself and tries again—then jumps in fright when Topiltzin opens his eyes. The priest even tries to speak, but all that emerges is a mouthful of blood.

They stare at each other for a long moment, the defiant new order gazing down into the eyes of the old. Huehmac realizes he could exact whatever revenge he wished right now. But he's oddly repulsed by the idea. He simply bats Topiltzin's groping hands aside, ignoring his moans, and retrieves the blade.

He retreats several paces and peels down his loincloth—exposing his buttocks to the high holy priest, an insult returned. He then makes a few sharp slices over the welts. This hurts less than he feared—obsidian is sharper than even Spanish steel. And as the blood oozes out, he feels distinct relief. He's still weak, but he can rest tonight, get some food in him, and figure out what to do tomorrow. However pained by his treatment at the hands of the Spanish, the Aztec have suffered grievously, which is the most important thing.

But as he's finishing up, he hears the last sound he ever expected—Topiltzin laughing. He turns to see the priest grinning through blood-stained teeth. He speaks in a hiss, and has to stop and suck in breath halfway through.

—You really expected... them to save you.

He reaches out a shaky finger, pointing. Huehmac looks down at himself, seeing little but streaks of blood.

—Your... backside.

Confused, Huehmac raises his arm and twists around, craning his neck. That's when he sees it. A constellation of smallpox sores on his skin.

By the time he can tear his eyes away, Topiltzin has fallen still. The Aztec priest and everything he stood for is dead.

*The fall of the Aztec Empire was swift and stunning. "From morning to sunset," said one contemporary source, "priests did nothing else but carry the dead bodies and throw them into the ditches." Thanks to war, the disruption of food supplies, and especially diseases (smallpox, measles, typhus), as much as 90 percent of the population died, a drop from 20 million people to 2 million. A comparable drop in the United States today would result in 300 million deaths.*

*The conquest of Mesoamerica helped inaugurate the world we know today. Had the Spanish been repulsed, Europe likely would have remained a secondary power. Instead, emboldened by their success, the Spanish began colonizing the Americas in earnest and siphoning off resources, boosting their wealth and prestige. Fear of being left behind prompted other European powers to follow suit with their own colonial programs, in the Americas and elsewhere. And while it's far beyond the scope of this book, all the other trends of modern life followed from this — including mass migration to colonies around the world, voluntary and forced, as well as the establishment of the economic, intellectual, and political regimes that we all inhabit today.*

# Conclusion

∽⌒∽

At the beginning of this book, I bellyached about how flat and even dull modern archaeology seems sometimes—people spending whole days digging broken pots out of the ground, or sifting literal tons of soil to find a few measly seeds. In doing so, I was being a bit dramatic, obviously. I still believe archaeology is the most stirring field in science, and I wouldn't have been able to fill these pages with so many rich details about the lives of our ancestors without the patient work of thousands of men and women over the past few centuries. We know most of what we know about our collective past because of them.

Hopefully, though, this book proves that there's more to archaeology than moving dirt. If the field truly wants to fulfill its promise of illuminating the lives of everyday people in the past, then it needs to focus on what people in the past actually *experienced*—the moments that filled their days, and filled their lives. Experimental archaeology can do that, is already doing that. Whether it's the leathery feel of mummy flesh or the squishy taste of blubber, the crash of trebuchet balls or the queasy odors of half-tanned hides, such moments bring the past alive in ways that no armchair theorizing can. Archaeology needs the concrete and homey, too.

I also hope this book can highlight, despite the many boons and benefits of modernity, what humankind has lost in modern times, or at least lost sight of. Not to get too highfalutin, but we increasingly live in a world of the ghostly: flickering images, information abstracted to bits. Ultimately, however, human culture and human life depend upon solid objects—on food, garments, tools, walls, things you can run your hand along, things you can feel the grain of. To be sure, there's nothing inherently wrong with modern passing fancies. And human beings will always have our abstract side, our ideologies and religions and mad passions. But civilization requires finding a balance between the abstract and the material, and it's not crazy to wonder whether things have swung too far in one direction lately.

Experimental archaeology—getting out and actually doing and making things—can provide a welcome corrective, in part by making the material a little more spiritual. That tool or textile or meal isn't just something you bought, something you'll dispose of without a care. It's something crafted, something hewn, something raised from seed. You put some part of yourself into it and thereby transformed it. Experimental archaeology certainly isn't the easiest field. As one practitioner told me, "This is hard shit. It's not comfortable all the time." But the past wasn't comfortable for our ancestors, either. And yet, they stuck things out. Indeed, we wouldn't be here if they hadn't. The least we can do is try to understand them on their own terms.

Finally and above all, I hope this book can reveal what unites us today with people from long ago, and help us understand that they *were* just people, no different than us. Archaeology has always focused first and foremost on artifacts, for understandable reasons. Artifacts are physical proof of what people in ancient times achieved, achievements that the past few centuries of colonization and other upheavals cannot erase. But artifacts sitting on a shelf, behind glass, can take us only so far into the minds of other people. To reach them on that deeper level, we need to inhabit their world—to endure their frustrations and revel in their joys, to touch and smell and taste everything they took for granted, so we no longer will. We need to feel the fabric of their lives. And experimental archaeology—living archaeology—can do that in a way that no other field can.

# A Thank-You and a Bonus

I hope you enjoyed *Dinner with King Tut*. If you want to learn more, I've compiled some bonus material—macabre anecdotes, amusing asides, and more—at samkean.com/books/dinner-with-king-tut/extras/notes/. There are bonus photographs as well, at samkean.com/books/dinner-with-king-tut/extras/photos/. If so moved, you can also drop me a line at samkean.com/contact. I love hearing from readers…

# Acknowledgments

More than anything I've ever written, completing this book required the generosity of others. So many people helped me in so many ways that I'm not sure where to begin. I suppose with the usual suspects.

Thanks first to my family, and to friends spread far and wide across the world. I'm lucky to have you all in my life.

There's also my agent Rick Broadhead, who's been a steadfast partner throughout. At Little, Brown, I'd like to thank my editor Morgan Wu for her dedication and care in honing and crafting what I hope will be the first of many books we'll do together, as well as Ben George for acquiring the book and providing vital early feedback, and Maya Guthrie and Evan Hansen-Bundy for their editorial help. Michael Noon deftly guided the book through production, and Mike Fleming's sharp copyediting caught some embarrassing booboos and helped polish many passages to a shine.

Scores of people let me tag along and poke around their labs, field sites, kitchens, breweries, salons, tattoo parlors, artillery ranges, ball courts, and more, giving this book the rich detail it has. These fine folks include (in no particular order) Metin Eren, James Watson, Randy Haas, Douglas Baird, Gökhan Mustafaoğlu, Laura Wolfer, Caelan Dunwoody, Kate Wentworth, Mark Arnott-Job, Roger Larsen, Seamus Blackley, Jefferson Nestor, Mac Sasao, Dalani Tanahy, Chris Haynes, Hazel Riggall, Nathalie Roy, Sally Grainger, Janet Stephens, Elle Festin, Daniel Bertrand, Kaare Erickson, Karen Harry, Dennis Sinnok,

## Acknowledgments

Janet Kiyutelluk, Daniel Perez, Dawn Riegner, Tessy Ritchie, Cliff Rogers, Joe Curran, Arturo Sánchez, and Santiago Muñoz.

I also conducted vital in-depth interviews with dozens of others. Not everything made the final cut, but it all helped shape the text. These people include Christina Lee, Freya Harrison, Aaron Deter-Wolf, Salima Ikram, Bob Brier, Maya Sialuk Jacobsen, Daniel Riday, Tiffany Treadway, David Falk, Theresa Kamper, Lyn Wadley, Shahina Farid, James Van Lanen, Joe Mellen, Sloan Mahone, Jon Hather, Richard VanderHoek, Claudio Aporta, Bill Schindler, Daniela Moreno, Duygu Çamurcuoğlu, Marie Hopwood, Adel Kelany, Delwen Samuel, Li Liu, Matt Gibbs, Ron Leidich, Ali Haleyalur, Hoturoa Barclay-Kerr, Hollie Funaki, Robert Arnott, Eline Schotsmans, Elizabeth Pennefather-O'Brien, and Austin Mason.

If I've forgotten anyone, I remain thankful if embarrassed. And a second time, to all: Thank you.

# Works Cited

## Chapter 1: Africa

Biesele, Megan. "Ju/'Hoan Women's Tracking Knowledge and Its Contribution to Their Husbands' Hunting Success." *African Study Monographs,* Suppl. 26 (March 2001): 67–84.

Biesele, Megan, and J. D. Lewis-Williams. "Eland Hunting Rituals among Northern and Southern San Groups: Striking Similarities." *Africa: Journal of the International African Institute* 48, no. 2 (1978): 117–34.

Ben-Dor, Miki, Raphael Sirtoli, and Ran Barkai. "The evolution of the human trophic level during the Pleistocene." *Yearbook of Physical Anthropology* 175, no. S72 (August 2021): 27–56.

Lee, Richard Borshay. *The !Kung San: Men, Women, and Work is a Foraging Society.* Cambridge, UK: Cambridge University Press, 1979.

Liebenberg, Louis. *The Art of Tracking: The Origin of Science.* Cape Town and Johannesburg: David Philip, 1990.

———. *A Field Guide to the Animal Tracks of Southern Africa.* Cape Town and Johannesburg: David Philip, 1990.

Lombard, Marlize, and Katharine Kyriacou. "Hunter-Gatherer Women." *Oxford Research Encyclopedia of Anthropology.* September 28, 2020. Accessed April 27, 2024. https://oxfordre.com/anthropology/view/10.1093/acrefore/9780190854584.001.0001/acrefore-9780190854584-e-105.

Steyn, H. P. "Southern Kalahari San Subsistence Ecology: A Reconstruction." *South African Archaeological Bulletin,* 39, no. 140 (December 1984): 117–24.

Wadley, Lyn. "Were snares and traps used in the Middle Stone Age and does it matter?: A review and a case study from Sibudu, South Africa." *Journal of Human Evolution* 58 (2010): 179–92.

Wilkins, Jayne, Benjamin J. Schoville, Robyn Pickering; Luke Gliganic, Benjamin Collins, Kyle S. Brown, Jessica von der Meden, et al. "Innovative *Homo sapiens* behaviours 105,000 years ago in a wetter Kalahari." *Nature* 592 (2021): 248–52. doi.org/10.1038/s41586-021-03419-0.

Willoughby, Pamela R. "Early humans far from the South African coast collected unusual objects." *Nature* 592 (2021): 193–94.

## CHAPTER 2: ANDES

Arriaza, B. T., V. G. Standen, V. Cassman, and C. M. Santoro. "Chinchorro Culture: Pioneers of the Coast of the Atacama Desert." In *The Handbook of South American Archaeology*, edited by Helaine Silverman and William H. Isbell. New York: Springer. doi.org/10.1007/978-0-387-74907-5_3.

Grund, Brigid Sky. "Behavioral Ecology, Technology, and the Organization of Labor: How a Shift from Spear Thrower to Self-Bow Exacerbates Social Disparities." *American Anthropologist* 119, no. 1 (January 3, 2017): 104–19.

Haas, Randall, James Watson, Tammy Buonasera, John Southon, Jennifer C. Chen, Sarah Noe, Kevin Smith, Carlos Viviano Llave, Jelmer Eerkens, and Glendon Parker. "Female Hunters of the Early Americas." *Science Advances* 6, no. 45 (November 4, 2020).

Lee, Richard Borshay. *The !Kung San: Men, Women, and Work in a Foraging Society*. Cambridge, UK: Cambridge University Press, 1979.

Lombard, Marlize, and Katharine Kyriacou. "Hunter-Gatherer Women." *Oxford Research Encyclopedia of Anthropology*. September 28, 2020. Accessed April 27, 2024. https://oxfordre.com/anthropology/view/10.1093/acrefore/9780190854584.001.0001/acrefore-9780190854584-e-105.

Malpass, Michael A., *Ancient People of the Andes*. Ithaca, NY: Cornell University Press, 2016.

Mann, Charles C. *1491: New Revelations of the Americas Before Columbus.* New York: Knopf, 2005.

Osorio, Daniela, José M. Capriles, Paula C. Ugalde, Katherine A. Herrera, Marcela Sepúlveda, Eugenia M. Gayo, Claudio Latorre, Donald Jackson, Ricardo De Pol-Holz, and Calogero M. Santoro. "Hunter-Gatherer Mobility Strategies in the High Andes of Northern Chile during the Late Pleistocene–Early Holocene Transition (ca. 11,500–9500 CAL B.P.)." *Journal of Field Archaeology* 42, no. 3 (2017): 228–40.

Raff, Jennifer. "Finding the First Americans." *Aeon*, December 22, 2022. https://aeon.co/essays/the-first-americans-a-story-of-wonderful-uncertain-science.

Venkataraman, Vivek. "Women were successful big-game hunters, challenging beliefs about ancient gender roles." *The Conversation*, March 10, 2021. https://theconversation.com/women-were-successful-big-game-hunters-challenging-beliefs-about-ancient-gender-roles-153772.

Whittaker, John C., and Kathryn A. Kamp. "Primitive Weapons and Modern Sport: Atlatl Capabilities, Learning, Gender, and Age." *Plains Anthropologist* 51, no. 198 (May 2006): 213–21.

## CHAPTER 3: ÇATALHÖYÜK

Atalay, Sonya, and Christine A. Hastorf. "Food, Meals, and Daily Activities: Food Habitus at Neolithic Çatalhöyük." *American Antiquity* 71, no. 2 (April 2006): 283–319.

Balter, Michael. *The Goddess and the Bull: Çatalhöyük: An Archaeological Journey to the Dawn of Civilization.* Florence, MA: Free Press, 2010.

Çatalhöyük Research Project. "Archive Reports." https://www.catalhoyuk.com/research/archive_reports.

Dietrich, O., M. Heun, J. Notroff, K. Schmidt, and M. Zarnkow. "The role of cult and feasting in the emergence of Neolithic communities: New evidence from Göbekli Tepe, south-eastern Turkey." *Antiquity* 86, no. 333 (September 2012): 674–95. doi:10.1017/S0003598X00047840.

Hodder, Ian. *The Leopard's Tale: Revealing the Mysteries of Çatalhöyük.* London: Thames & Hudson, 2011.

Newitz, Annalee. "An Ancient Proto-City Reveals the Origin of Home." *Scientific American,* March 1, 2021. https://www.scientificamerican.com/article/an-ancient-proto-city-reveals-the-origin-of-home/.

———. *Four Lost Cities: A Secret History of the Urban Age.* New York: W. W. Norton, 2021.

Wright, Katherine. "Domestication and inequality? Households, corporate groups, and food processing tools at Neolithic Çatalhöyük." *Journal of Anthropological Archaeology* 33 (2014): 1–33.

## CHAPTER 4: EGYPT

Fisher, David E., and Marshall Jon Fisher. *Mysteries of Lost Empires.* London: Channel 4 Books, 2000.

Ikram, Salima. *Divine Creatures: Animal Mummies in Ancient Egypt.* Cairo: The American University in Cairo Press, November 14, 2015.

Kemp, Barry J. *Ancient Egypt: Anatomy of Civilization.* London and New York: Routledge, 2018.

Quigley, Christine. *Modern Mummies: The Preservation of the Human Body in the Twentieth Century.* Jefferson, NC: McFarland, 2015.

Samuel, Delwen. "Brewing and Baking." In *Ancient Egyptian Materials and Technology,* edited by Paul T. Nicholson and Ian Shaw. Cambridge, UK: Cambridge University Press, 2000.

Stocks, Denys A. *Experiments in Egyptian Archaeology: Stoneworking Technology in Ancient Egypt.* London and New York: Routledge, 2003.

## CHAPTER 5: POLYNESIA

Bond, Michael. *From Here to There: The Art and Science of Finding and Losing Our Way.* Cambridge, MA: Belknap Press, 2021.

Kāne, Herb Kawainui. "In Search of the Ancient Polynesian Voyaging Canoe." 1998. https://archive.hokulea.com/ike/kalai_waa/kane_search_voyaging_canoe.html.

Lewis, David. *We the Navigators: The Ancient Art of Land-Finding in the Pacific.* Canberra: Australian National University Press, 1972.

Nickum, Mark. "Ethnobotany and Construction of a Tongan Voyaging Canoe: The Kalia Mileniume." *Ethnobotany Research and Applications* 6 (2008).

Polynesian Voyaging Society. *Hawaiian Voyaging Traditions.* https://archive.hokulea.com/.

Richter-Gravier, Raphael. "Manu narratives of Polynesia: A comparative study of birds in 300 traditional Polynesian stories." Department of Social Anthropology and Ethnology. University of Otago; Université de la Polynésie française / Ecole doctorale du Pacifique, 2019.

## CHAPTER 6: ROME

Beard, Mary. *Pompeii.* London: Profile Books, 2008.

Bergmann, Bettina, Stefano De Caro, Joan R. Mertens, and Rudolf Meyer. "Roman Frescoes from Boscoreale: The Villa of Publius Fannius Synistor in Reality and Virtual Reality." *Metropolitan Museum of Art Bulletin* 67, no. 4 (2010).

Fisher, David E., and Marshall Jon Fisher. *Mysteries of Lost Empires.* London: Channel 4 Books, 2000.

Grainger, Sally. "What's in an Experiment? Roman Fish Sauce: An Experiment in Archaeology." *EXARC* no. 1 (2012). https://exarc.net/issue-2012-1/at/whats-experiment-roman-fish-sauce-experiment-archaeology.

Newitz, Annalee. *Four Lost Cities: A Secret History of the Urban Age.* New York: W. W. Norton, 2021.

Paardekooper, Roeland. "The process of fulling of wool: Experiments in the Netherlands." *EXARC* no. 2 (2005). https://exarc.net/sites/default/files/exarc-eurorea_2_2005-the_process_of_fulling_of_wool.pdf.

Stephens, Janet. "Ancient Roman Hairdressing: On (hair)pins and needles." *Journal of Roman Archaeology* 21 (2008): 110–32.

Stewart, Doug. "Resurrecting Pompeii." *Smithsonian Magazine*, February 2006. https://www.smithsonianmag.com/history/resurrecting-pompeii-109163501/.

Szu, Annamária, Ildikó Oka, and Orsolya Madarassy. "Reconstructing the Roman and Celtic Dress of Aquincum." *EXARC* no. 4 (2007). https://exarc.net/sites/default/files/exarc-eurorea_4_2007-reconstructing_the_roman_and_celtic_dress_of_aquincum.pdf.

Trimble, Jennifer. "The Zoninus Collar and the Archaeology of Roman Slavery." *American Journal of Archaeology* 120, no. 3 (July 2016): 447–72.

Wilner, Ortha L. "Roman Beauty Culture." *Classical Journal* 27, no. 1 (October 1931): 26–38.

## CHAPTER 7: NATIVE AMERICANS

Curran, Joseph B., and David E. Raymond. "War Clubs in Southern California: An Interdisciplinary Study of Blunt Force Weapons and Their Impact." *Journal of Archaeological Method and Theory* 28 (2021): 1200–1223.

Dixon, Roland B. "Northern Maidu." *Bulletin of the American Museum of Natural History* 17, no. 3 (May 1905): 119–346.

Deter-Wolf, Aaron, Danny Riday, and Maya Sialuk Jacobsen. "Tattoos using different pre-modern Tools." *EXARC*, January 24, 2022. https://exarc.net/history/ea-award-tattoos-using-different-pre-modern-tools.

Grund, Brigid Sky. "Behavioral Ecology, Technology, and the Organization of Labor: How a Shift from Spear Thrower to Self-Bow Exacerbates Social Disparities." *American Anthropologist* 119, no. 1 (March 2017): 104–19.

Jones, David E. *Poison Arrows: North American Indian Hunting and Warfare*. Austin, TX: University of Texas Press, 2007.

Lombard, Marlize. "The tip cross-sectional areas of poisoned bone arrowheads from southern Africa." *Journal of Archaeological Science: Reports* 33, no. 1 (October 2020): 102477.

Mann, Charles C. *1491: New Revelations of the Americas Before Columbus.* New York: Knopf, 2005.

Mason, Sarah L. R. "Acorns in Human Subsistence," PhD diss., Institute of Archaeology, University College, London, 1992. https://discovery.ucl.ac.uk/id/eprint/10121575/1/Mason_10121575_thesis.pdf.

St-Pierre, Christian Gates. "Needles and bodies: A microwear analysis of experimental bone tattooing instruments." *Journal of Archaeological Science: Reports* 20 (2018): 881–87.

## CHAPTER 8: VIKINGS

Anonye, Blessing O., Valentine Nweke, Jessica Furner-Pardoe, Rebecca Gabrilska, Afshan Rafiq, Faith Ukachukwu, Julie Bruce, Christina Lee, Meera Unnikrishnan, Kendra P. Rumbaugh, et al. "The safety profile of Bald's eyesalve for the treatment of bacterial infections." *Scientific Reports* 10, article number: 17513 (2020).

Archaeological Institute of America. "Bodies of the Bogs." *Archaeology,* 1997. https://archive.archaeology.org/online/features/bog/.

———. "Pathologies of the Bog Bodies." *Archaeology,* 1997. https://archive.archaeology.org/online/features/bog/physical.html.

Arnott, Robert. *Trepanation: History, Discovery, Theory,* edited by Chris Smith, Robert Arnott, and Stanley Finger, Centre for the History of Medicine. Lisse, The Netherlands: Swets & Zeitlinger, 2003.

Cornelius, Keridwen. "Did Processed Foods Make Us Human?" *Sapiens,* November 24, 2020. https://www.sapiens.org/archaeology/paleo-processing-foods/.

Dell'Amore, Christine. "Who Were the Ancient Bog Mummies? Surprising New Clues." *National Geographic,* July 18, 2014, https://www.nationalgeographic.com/history/article/140718-bog-bodies-denmark-archaeology-science-iron-age.

Hansen, Valerie. *The Year 1000: When Explorers Connected the World—and Globalization Began.* New York: Scribner, 2020.

Reade, Ben. "Bog butter: A gastronomic perspective." *Nordic Food Lab* (blog), October 2, 2013. https://nordicfoodlab.org/blog/2013/10/bog-butter-a-gastronomic-perspective/.

Sammut, Dave, and Chantelle Craig. "Bodies in the Bog: The Lindow Mysteries." Science History Institute. *Distillations,* July 23, 2019. https://www.sciencehistory.org/stories/magazine/bodies-in-the-bog-the-lindow-mysteries/.

Treadway, Tiffany, and Clement Twumasi. "An Experimental Study of Lesions Observed in Bog Body Funerary Performances." *EXARC* no. 3 (August 26, 2021). https://exarc.net/issue-2021-3/ea/experimental-study-lesions-observed-bog-body-funerary-performances.

## CHAPTER 9: INUPIAT

Aporta, Claudio. "The Trail as Home: Inuit and Their Pan-Arctic Network of Routes." *Human Ecology* 37 (2009): 131–46.

Aporta, Claudio, and Eric Higgs. "Satellite Culture: Global Positioning Systems, Inuit Wayfinding, and the Need for a New Account of Technology." *Current Anthropology,* 46, no. 5 (December 2005): 729–53.

Bond, Michael. *From Here to There: The Art and Science of Finding and Losing Our Way.* Cambridge, MA: Belknap Press, 2021.

Canadian Social Sciences and Humanities Research Council. "The Thule." Newfoundland and Labrador Heritage website. Accessed September 15, 2024. https://www.heritage.nf.ca/articles/indigenous/thule.php.

Daviss, Betty-Anne. "Heeding Warnings from the Canary, the Whale, and the Inuit: A Framework for Analyzing Competing Types of Knowledge about Childbirth." In *Childbirth and Authoritative Knowledge.* Berkeley, CA: University of California Press, 1999. doi.org/10.1525/9780520918733-019.

Engelhard, Michael. "Inuit Wayfinding: Sky above, sea below." *Above and Beyond—Canada's Arctic Journal,* 2018.

Eren, Metin I., Michelle R. Bebber, James D. Norris, Alyssa Perrone, Ashley Rutkoski, Michael Wilson, and Mary Ann Raghanti. "Experimental replication shows knives manufactured from frozen human feces do not work." *Journal of Archaeological Science: Reports* 27, article no. 102002 (October 2019).

Frink, Liam, and Celeste Giordano. "Women and Subsistence Food Technology: The Arctic Seal Poke Storage System." *Food and Foodways* 23, no. 4 (2015): 251–72.

Frink, Lisa, and Karen G. Harry. "The Beauty of 'Ugly' Eskimo Cooking Pots." *American Antiquity* 73, no. 1 (January 2008): 103–20.

Gadsby, Patricia, and Leon Steele. "The Inuit Paradox: How can people who gorge on fat and rarely see a vegetable be so healthy?" *Discover* online, January 19, 2004. https://www.discovermagazine.com/health/the-inuit-paradox.

Kjellstrom, Rolf. "Senilicide and Invalidicide Among the Eskimos." *Folk* 16–17, no. 1 (1974–75): 117–24.

Lombard, Marlize, and Katharine Kyriacou. "Hunter-Gatherer Women." *Oxford Research Encyclopedia of Anthropology*. September 28, 2020. Accessed April 27, 2024. https://oxfordre.com/anthropology/view/10.1093/acrefore/9780190854584.001.0001/acrefore-9780190854584-e-105.

Moran, Emilio F. "Human Adaptation to Arctic Zones." *Annual Review of Anthropology* 10 (1981): 1–25.

## Chapter 10: China

Al-Dīn, Jamāl, and ʿAbd Al-Raḥīm Al-Jawbarī. *Book of Charlatans*, translated by Humphrey Davies, foreword by S. A. Chakraborty. New York: New York University Press, 2022.

Bertrand, Daniel A. "Building the Medieval Trebuchet." Utah State University, Undergraduate Honors Capstone Project, 2019. Accessed December 11, 2024. https://digitalcommons.usu.edu/honors/421.

Dale, Melissa S. *Inside the World of the Eunuch: A Social History of the Emperor's Servants in Qing China*. Hong Kong: Hong Kong University Press, 2018.

Fisher, David E., and Marshall Jon Fisher. *Mysteries of Lost Empires*. London: Channel 4 Books, 2000.

Hansen, Valerie. *The Open Empire: A History of China to 1800*. New York: W. W. Norton, 2015.

———. *The Year 1000: When Explorers Connected the World—and Globalization Began.* New York: Scribner, 2020.

Levin, Ed. "The Highland Fling: Timber Framing." *Journal of the Timber Framers Guild* no. 50 (December 1998): 12–20.

Peschel, Enid Rhodes, and Richard E. Peschel. "Medical Insights into the Castrati in Opera." *American Scientist* 75, no. 6 (November–December 1987): 578–83.

Rosselli, John. "The Castrati as a Professional Group and a Social Phenomenon, 1550–1850." *Acta Musicologica* 60, no. 2 (May–August 1988): 143–79.

Temple, Robert, and Joseph Needham. *The Genius of China: 3,000 Years of Science, Discovery, and Invention.* Rochester, VT: Inner Traditions, 2007.

## CHAPTER 11: AZTEC

Aguilar-Moreno, Manuel. *Handbook to Life in the Aztec World.* Oxford, UK: Oxford University Press, 2007.

Aliseda, Andrea. "Unlocking Nixtamalization." *Epicurious,* June 23, 2021. https://www.epicurious.com/ingredients/what-is-nixtamal-article.

Hansen, Valerie. *The Year 1000: When Explorers Connected the World—and Globalization Began.* New York: Scribner, 2020.

Mann, Charles C. *1491: New Revelations of the Americas Before Columbus.* New York: Knopf, 2005.

Ritchie, Tessy S., Kathleen E. Riegner, Robert J. Seals, Clifford J. Rogers, and Dawn E. Riegner. "Evolution of Medieval Gunpowder: Thermodynamic and Combustion Analysis." *ACS Omega* 6, no. 35, (2021): 22848–56.

Soustell, Jacques. *Daily Life of the Aztecs on the Eve of the Spanish Conquest.* Palo Alto, CA: Stanford University Press, 1961.

Thomas, Hugh. *Conquest: Cortes, Montezuma, and the Fall of Old Mexico.* New York: Simon & Schuster, 1995.

Wade, Lizzie. "Feeding the gods: Hundreds of skulls reveal massive scale of human sacrifice in Aztec capital." *Science* 360, no. 6395 (June

22, 2018). https://www.sciencemag.org/news/2018/06/feeding-gods-hundreds-skulls-reveal-massive-scale-human-sacrifice-aztec-capital.

Zhang, Sarah. "A New Clue to the Mystery Disease That Once Killed Most of Mexico: The evidence comes from the 16th-century victims' teeth." *The Atlantic* online, January 15, 2018. https://www.theatlantic.com/science/archive/2018/01/salmonella-cocoliztli-mexico/550310/.

## INTERVIEWS

In addition to the sources above, I conducted in-person interviews with Metin Eren, Randy Haas, James Watson, Douglas Baird, Gökhan Mustafaoğlu, Laura Wolfer, Caelan Dunwoody, Kate Wentworth, Mark Arnott-Job, Roger Larsen, Seamus Blackley, Jefferson Nestor, Mac Sasao, Dalani Tanahy, Chris Haynes, Hazel Riggall, Nathalie Roy, Sally Grainger, Janet Stephens, Elle Festin, Joseph Ash, Daniel Bertrand, Christina Lee, Freya Harrison, Kaare Erickson, Karen Harry, Dennis Sinnok, Janet Kiyutelluk, Archie Kiyutelluk, Daniel Perez, Dawn Riegner, Tessy Ritchie, Cliff Rogers, Joe Curran, Arturo Sánchez, and Santiago Muñoz. Thank you so much for your time!

Others gave me just as much time and help, although we couldn't meet in person. I conducted phone or video interviews with: Salima Ikram, Bob Brier, Aaron Deter-Wolf, Maya Sialuk Jacobsen, Daniel Riday, Tiffany Treadway, David Falk, Theresa Kamper, Lyn Wadley, Shahina Farid, James Van Lanen, Joe Mellen, Sloan Mahone, Jon Hather, Richard VanderHoek, Bill Schindler, Daniela Moreno, Claudio Aporta, Duygu Çamurcuoğlu, Marie Hopwood, Adel Kelany, Delwen Samuel, Li Liu, Matt Gibbs, Ron Leidich, Ali Haleyalur, Hoturoa Barclay-Kerr, Hollie Funaki, Elizabeth Pennefather-O'Brien, and Austin Mason.

# INDEX

Note: Italic page numbers refer to illustrations.

acorns
  acorn flour, 249
  acorn wars, 258–59
  cultivation of, 243–44
  experimental archaeology and, 261–64, 262n
  fictional account of oak grove, 291
  as food, 261, 263–64
  processing of, 261–62
  storage of, *249*
  tannins in, 262–63
acorn weevils, 262, 262n
*Acrocanthosaurus*, 129n
adhesives, in hafting process, 65
Africa
  arrows found in, 272
  fictional account of hunting in, 21–29, 27n, 32–35, 37–41, 77
  *Homo sapiens* originating in, 9, 15, 43
  hunter donning ostrich skin, *38*
  megafauna of, 75n
  population of, 347
  spears of, 66
  stone tools from, *17*
agriculture
  civilization and, 134
  grains and, 134
  intensive agriculture, 117
  Maidu and, 243–44
  sedentary living and, 117, 134
  shift from foraging and, 263
  traditional agricultural societies, 243
Alaska. *See also* native Alaskan culture; Thule people
  dogsleds and, *331*
  Inuit/Iñupiaq people of, 315, 315n, 325–26, 328, 335n, 340
  land bridge to Russia, 43
alchemy, 373–75, *374*
alcoholic beverages, 133. *See also* beer
aliens, and pyramid construction, 153n
alien tribes, fictional account of interactions between strangers, 48, 54–58, 70–72, 74
alpacas, meat of, 51–52
Americas. *See* North America; South America
Anchor Brewing, 131–32n
ancientbiotics, 288–89, 289n
ancient people, recreating life experiences of, 5, 6, 11–14
animals, mummification of, 136, *136*, 139, 143–45

## Index

antimicrobial resistance, 288
archaeology. *See also* experimental
    archaeology
  as field of science, 3, 4, 6, 419, 420
Arctic. *See also* native Alaskan culture
  cultures of, 315, 315n, 326–27
Artemisinin, 289n
Ash, Joseph, 252–57, 257n
ash, as preventive for ticks, 12, 14–15
Asia. *See also* China
  land ownership and, 244
  North and South American
    colonization and, 43
  population of, 347
  trade of, 141
atlatls
  Aztec soldiers armed with, 401–02
  for big game, 74
  bones of vicuñas and, 68, 71
  bows and arrows replacing, 270–71
  darts of, 67
  depiction of female hunter from the
    Andes with, *52*
  experimental archaeology and, 67, 76
  expertise in use of, 68–69, 271, 366
  fictional account of, 58–60, 332–34
  firing of, 67
  hafting process and, 64–66, 71
  hunting strategies and, 75n, 271
  illustration of, *59*
  loading of spur-hook, 67
  widespread use of, 67
atolls, 192
Atzlán, 381
Australia, megafauna of, 75
Aztecs
  atlatl used by, 401–02
  Aztec eagle warrior, *386*
  ballgames (*ullamaliztli*) of, 3, 393–98,
    393n, 394n, *397*
  cities conquered by, 383
  fall of, 418
  fictional
    accounts of ballgames, 399–401
  fictional accounts of conquests of,
    385–89
  fictional accounts of human sacrifice,
    405–07
  fictional accounts of Spanish conquest
    of, 414–18
  human sacrifice of, 383, 398, 402–04,
    *403*, 404n
  insects as food and, *391*
  jaguar warrior headpiece, *386*
  skulls as art, 403–04, *404*
  skulls as public monuments
    (*tzompantli*), 404, *404*
  Spain and, 381, 383, 386n, 389,
    404–05, 407–08, 413–14
  warriors of, 401–03, *403*

Baird, Douglas, 95n, 111
Baker, Josephine, 226
*Bald's Leechbook,* remedies of, 285–88,
    288n
beds and bedding
  discovery of, 14, 15
  experimental archaeology and, 14–15
  fictional account of, 11–12
beef tallow, for Roman hairdressing,
    224, 227
beer
  of Egypt, 1, 131–33, 132n, 134
  grain-based beer, 133–34
Bering Strait, 43
Bertrand, Daniel, 367–72
biltong jerky, fictional account of, 32
biofilms, 288
bird species. *See also* seabirds
  'elepaio bird, 184

extinction of, 187n
long-tailed cuckoos, 186–87
wallata bird (Andean goose), 72
Blackley, Seamus, 129–31, 129n
bog lady, fictional account of, 291–93
bogs
    bog bodies preserved in, 295–96, 298, 402
    bog body studies, 296–99
    butter buried in, 295, 295n
    fictional account of, 290–94
    organic objects buried in, 295
    overkill deaths in, 296, 298
    pagan traditions and, 298
    Tullund Man of Denmark, *296*
Bolivia, potato domestication in, 49
Boncuklu, Turkey
    ancient landscape of, 95
    burial customs of, 97–98, 111
    communal work area of, 95n
    DNA evidence from buried bodies of, 94–95
    replica homes of, 95–97
    smoke within, 97
    soil surrounding, 96n
bone tools, *17*
bows and arrows
    atlatls compared to, 270–71
    hunting strategies and, 264, 267–68, 270, 271, *271*
    males favored in use of, 271–72, 271n
    poison used with, 272–73
    technology of, 270
Brier, Bob, 139–43, *140*
Buckland, Frank, 232, 232n
Buckland, William, 232, 232n
Buckland Club, Birmingham, England, 232–33, 237–39
burial practices
    experimental archaeology and, 111

fictional account of Çatalhöyük practices, 85–87, 99, 109–11, 113–16
sedentary living and, 85–88, 90–92
Bushmen, 9

Caesar, Julius, 148
Cai Lun, 358
California native tribes
    acorn cultivation, 243–44
    acorn flour and, 249, 262n, 263
    acorn soaking and, 263
    acorn storage and, *249*
    acorn wars of, 258–59, 264
    bows and arrows of, 264, 267–68, 270, 271, *271*
    fictional account of acorn gathering, 248, 249
    fictional account of attack, 249–50, 258–61
    fictional account of hunting, 248, 249
    fictional account of nuptial rites, 247–48
    fictional account of pregnancy and, 245–46, 248
    fictional account of revenge, 268–70, 273–75
    fictional account of tattoos, 245, 258, 274
    fictional account of tracking, 259–60, 261
    fictional account of tule basket making, 248–49, 268–69
    grubs eaten by, 262n
    individual oak trees claimed by, 244
    oak groves claimed by, 243–44, 249, 259
    shock weapons of, 264, 267–68
    tule baskets, *249*
Canada, 105n

# Index

canoes
  building process, 184–85, 193
  canoe plants, 176
  design of, 166–67, 183–84, 192–93
  fictional account of canoe voyage,
    169–74, 178–83, 189–91, 195–99
  model of ancient canoe (*drua*), *166*
canopic jars, 142
Cape quince tree, 14
Çatalhöyük, Turkey
  architecture of, 82–83, 92–93
  art of, 112, 112n
  Boncuklu settlers as ancestors of, 95
  bucrania skulls of, 93, 94, 99
  burial customs of, 97–98, 111–12, *114*
  egalitarian nature of, 82, 83, 88–89,
    98, 117
  fictional account of burial practices,
    85–87, 99, 109–11, 113–16
  fictional account of figurines of,
    99–100
  fictional account of food of, 93, 96n,
    102
  fictional account of funeral rites, 113–14
  fictional account of life in, 85–92,
    98–102, 113
  fictional account of making tunic from
    hides, 100–102
  fictional account of rebuilding of
    homes in, 98–99
  figurines of, 112
  fires of, 112–13
  founding of, 82
  horticulture of, 117
  interiors of homes and, 96
  leopards and, 99–100, 110, 112, 114
  modern reproduction of, *94*
  murals of, 112, *114*
  plaster used in, 93, 94
  population of, 82
  rebuilding of homes, 94
  replica houses of, *88*, 93
  ritual magic of, 112
  skeletons from, 97
  smoke within, 97
  soil surrounding, 96n
  splinter settlement across river and,
    83, 98
catapults, 4, 213, 363, 366
Charlemagne, 277
chert
  fictional accounts of, 12, 28, 37–38,
    45, 47, 71
  hafting and, 64–65
  knapping stone tools and, 16, 17, 18,
    19, 31, 35, 45, 47
China
  alchemy and, 373–75
  Arab merchants in, 347–48, 381
  canals of, 347
  castration in, 357, *357*, 359–60
  catapults of, 363, 366
  compasses of, *364*, 364n
  cultural continuity of, 347
  Empress of China carried by court
    eunuchs, *357*
  eunuchs' *bao* or treasure, 360
  eunuchs of, 349–50, 353n, 357–60,
    360n, 372–73
  fictional account of eunuchs, 349–55,
    360–66, 375–79
  foot-binding in, 356
  Islamic thought and philosophy in, 348
  mass production in, 356
  medieval marketplace of, 350, *350*
  meteors and, 158n, 374
  Silk Road and, 347
  South American population from, 43
  technological innovation of, 347, 356
  trade of, 347–48, 350, 381

trebuchets of, *362*, 363, 366–67
yin-yang symbol, 352n
chlorophyll, 49
Christian monasteries and priories
   fictional account of Viking siege of, 279–85, 289–94, 299–303, 310–14
   as target of Viking raids, 278
cities
   agriculture and, 134
   Aztecs' conquering of, 383
   founding of Çatalhöyük in central Turkey, 82
   sedentary living and, 81–82
civilization
   agriculture and, 134
   beer-making and, 133–34
   in Middle East, 277, 347
   New World clashing with Old World, 243
clay sauce
   for detoxifying potatoes, 50–51
   fictional account of, 47
climate change, and megafauna extinctions, 76, 77
coca leaves
   fictional account of, 55–56, 58, 72
   taste of, 62
Columbus, Mississippi, 150
Cook, James, 166
copper blades, and Egyptian mummification, 140
coral reefs, 187
Costello, Elvis, 250
Costner, Kevin, 325
Crusades, 347
Curran, Joe, 264–67, 265n
cuy, meat of, 51–52, 55, 58

Darwin, Charles, 15
David, Larry, 150

death. *See also* burial practices
   fictional account of beliefs concerning burial, 85–87
   overkill deaths in bogs, 296, 298
   Thule people's death by exposure, 335n
Deter-Wolf, Aaron, 250–51, 250n, 252
dinosaurs, extinction of, 75
dire wolves, in North America, 75
dormice, as Roman food, 239n
Douglass, Frederick, 151
Dunwoody, Caelan, 102–03, 106–08
dust devils, 45

Easter Island, 165, 187
Egypt
   agriculture of, 134
   alchemy and, 373
   bakeries of, 128–29
   beer of, 1, 131–33, 132n, 134
   bread of, 129–31, *130*, 132, 134
   embalming process of, 139–43
   emmer grains of, 128, 130, 131
   experimental archaeology and, 129–33, *130*, 132n, 139–43, *140*
   fictional account of brewer-baker of, 119–27, 135–39, 146–48, 159–63
   fictional account of mortuary temple, 124
   fictional account of mummification, 135–37, 139, 146–48, 163
   fictional account of Overseer, 120, 121–22, 126–27, 138, 159–62
   fictional account of pharaoh's burial chamber, 124–25
   fictional account of pyramid construction, 120–22
   fictional account of robbing pharaoh's tomb, 119, 121, 123, 124–27, 137, 138, 146–48, 159–63

Egypt (*Cont.*)
  fictional account of scribe of, 119–21, 123–27, 137, 138, 139, 148, 159–63
  food of, 128–29
  laborers of, 117, 131, 149–50, 153, 157
  lack of wheels used by, 150, 150n
  model bakery and brewery of, *128*
  mummies of, 117, *135*, 136, *136*, 139–43, 144, 145, 250
  pyramid construction and, 117–18, 128, 131, 148–51, 150n
  querns of, 128
  Roman Empire compared to, 201, 212
  royal tomb construction in, 149
  scribes of, 117
  social hierarchy in, 117
  straws of, 131n
  treasures of, 117–18, 134–35, 159
  Valley of the Kings, 158
'elepaio bird, 184
elephants
  archaeological remains of, 29–30
  butchering experiments and, 29–32, 76
  fictional accounts of hunting of, 29
  hunting of, 76–77
El Greco, 295
emmer grains, 87, 89, 102, 128, 130, 131
Eren, Metin, 15–21, 16n, 20n, 64–68, 65n, 330, 330n
Escher, M. C., 82
Eskimos, as antiquated term, 325n
Eumachia, 216
Europe. *See also* Vikings
  bogs of, 295, 298–99
  cathedrals of, 117
  fall of Roman empire and, 277
  food of, 129, 232
  hunting and, 271
  infanticide and, 335n
  land ownership and, 244

medical care in, 305
navigation of, 183n
population of, 347
technology of, 212
trade of, 141, 347
experimental archaeology
  acorns and, 261–64, 262n
  alpaca meat and, 51–52
  animal mummification and, 143–44
  atlatls and, 67, 76
  Aztec ballgames (*ullamaliztli*) and, 393–98, 394n, *397*
  beds and bedding and, 14–15
  bog bodies and, 296–98
  Boncuklu dig site and, 95–97
  burial practices and, 111
  cannons and, 408–14, *410*, *413*
  canoe building and, 193
  coconut processing and, 177–78
  Egyptian beer and, 132–33, 132n
  Egyptian bread and, 129–31, *130*, 132
  Egyptian mummification and, 139–44, *140*
  elephant butchery and, 30–32, 76
  elephant hunting and, 76–77
  indigenous cultural traditions and, 5
  native Alaskan clay pots and, 337–42, *338*
  ostrich eggs and, 35–36
  Polynesian navigation and, 191–94
  poop-knife and, 330
  potato-processing and, 50
  pyramid construction and, 150–58, 151n, *152*, 153n, *156*, 158n
  recreation of past and, 3–7, 419, 420
  remedies from *Bald's Leechbook* and, 285–88, 288n
  Roman concrete and, 214–16
  Roman food and, 233–39, 237–38n, 239n

## Index

Roman hairdressing and, 224–28, *225*, *226*
Roman road construction and, 212–14
shock weapons and, 264–67, *265*, 265n
tanning animal hides and, 102–09, 105n, 176, 257, 306
tapa/kapa cloth making and, 174–76, *175*
tattoos and, 252–57
traditional archaeology compared to, 4, 414
trebuchets and, 367–72
trepanation of skulls and, 108, *304*, 305–10
vicuña observations and, 52–53

Falk, David, 133
*Field of Dreams* (film), 325
Fiji, 167
flint, knapping stone tools and, 16, 17, 19, 20n, 31
flint blades, 306–07
foraging, 263
forfex (Roman scissors), *226*, 227

garum (Roman fish sauce), 234–35, *234*, 236
gathering
  fictional account in South America, 47–48, 48n
  fictional account of acorn gathering, 248, 249
  native Alaskan culture and, 338
  of wild grains, 133–34
giant sloths
  archaeological evidence of hunting of, 77
  disappearance of, 75
  fictional account of hunting of, 59, 71–72, 74, 78–80
  photograph of, *73*
  physical characteristics of, 77–78
  size of, 75
Ginsberg affair, 30–32
glycoalkaloids, 49, 50
golden plovers, 186–87
*Goodfellas* (film), 267
Grainger, Chris, 239
Grainger, Sally, 233–39, 239n
grains, gathering of wild grains, 133–34
Greece, 373
Greenland
  meteor weapons of, 158n
  tattoo methods of, 251
  Viking excursions to, 315
Grünewald, Matthias, 295
gunpowder
  modern gunpowder, 409
  recipes for, 408–09, 412–13
  trade in, 408

Haas, Randy, 61–64, 63n, 64n
hafting
  fictional account of, 47, 71
  steps in process of, 64–66
hairdressing
  fictional account of, 219, 222–23
  hairstyling dummies, 224, 224n, *226*
  Roman hairdressing, 223–28, *225*, *226*
Harrison, Freya, 285–86, 288–89, 288n, 289n
Harry, Karen, 336–42
*Hatari!* (film), 30
Hawaii
  canoes of, 184, 191
  discovery of, 187
  navigation and, 192
  Polynesia and, 165, 176

443

# Index

Haynes, Chris, 235–36, 239
hide-scrapers, 19, 20
*Homo sapiens*
   origins in Africa, 9, 15, 43
   spread to Middle East and Eurasia, 43
hops, 131, 131–32n
Hughes, Bart, 309–10n
human colonization, 43–44, 43n, 192
Hunter (Captain), 409–12
hunter-gatherers
   egalitarian lifestyle of, 117
   hafting techniques of, 64–65
   lifestyle of, 9, 11–14, 47–48, 48n, 81, 315, 316
   shift to sedentary living, 81–82
hunting
   archaeological evidence of hunting of giant sloths, 77
   with bows and arrows, 264, 267–68, 270, 271, *271*
   counterattack of wounded beast and, 69
   depiction of female hunter from the Andes, *52*
   fictional account of African hunting, 21–29, 27n, 32–35, 37–41, 77
   fictional account of hunting giant sloths, 59, 71–72, 74, 78–80
   fictional account of hunting turtles, 34
   fictional account of hunting vicuñas, 46, 48, 48n, 56, 59–61, 70, 72
   fictional account of South American hunting, 46, 47, 48, 48n, 56, 58–61, 69–70, 78–79
   fictional account of Thule people's hunting practices, 319, 320–21, 323–24, 332–33
   hunter's kit, 62, 278
   of mammoths, 76, 77
   megafauna extinctions and, 76
   persistence hunts, 27n
   of sea mammals, 315
   skills of, 76
   women's abilities in, 62–64, 63n, 69

igloos, 334n
Ig Nobel Prize, 330n
Ikram, Salima, 143–44
indigenous groups, and experimental archaeology, 5
infanticide, 335n
insects, as food, 391–93, *391*
interactions between strangers,
   fictional account of, 48, 54–58, 70–72, 74
Inuit/Iñupiaq people, 315, 315n, 325–26, 328, 335n, 340
invalidicide, 335n
Iron Age weapons, 297
Islam, golden age of, 347
Italy, castrati of, 360n

Jacobsen, Maya Sialuk, 251
Japan, 105n
Jesus Christ, 148

Kalahari, San people of, 9
kapa art, 176
Kent State University, 15, 16
Kisii (Kenyan tribe), 304–05n
knapping stone tools
   chert and, 16, 17, 18, 19, 31, 35, 45, 47
   counterfeiting tricks and, 65n
   experiments with, 15–21, 16n, 31
   fictional accounts of, 13–14, 47
   flint and, 16, 17, 19, 20n, 31
   hammerstones and, 16–17

injuries from, 20–21
obsidian and, 17, 19, 19–20n
Kombucha, 133
kudu
fictional account of hunting of, 21–29, 33
prints of, 24, *24*

Lake Titicaca, 49, 50
Larsen, Roger, 150–58, *152*, 153n, 154–55n, *156*, 158n
Las Vegas, Nevada, and Arctic research, 324–25, 336
Lebanon, temple built in, 149n
Lee, Bruce, 410
Lee, Christina, 286, 287, 288–89, 289n
leeches
doctors referred to as, 285, 285n
as medical treatment, 285n
Leidich, Ron, 192–93
Lennon, John, 309n
Liebenberg, Louis, 27n
llamas, 52
long-tailed cuckoos, 186–87
Los Danzantes, Mexico City, 391–93

Mafia, 239n
Maidu, acorns cultivated by, 243–44
Maizajo tortilleria, 389–91
mammals, evolution of, 75
mammoths
hunting of, 76, 77
on islands off Siberia, 148
Massachusetts Institute of Technology, 215
medical injuries, bizarre medical injuries, 20n
Mediterranean trade, 277
megafauna

hunting strategies for, 75n, 76
species extinction and, 74, 75–76, 77, 270
Melanesia, as region of Oceania, 165
Mellen, Joe, 309–10n
Mesoamerica. *See also* Aztecs
fictional accounts of, 385–89, 399–401, 405–07, 414–18
kingdoms of ancient Mexico, 382–83
Spanish conquest of, 389, 407, 414, 418
meteor weapons, 158, 158n
Mexica people, 381, 383
Micronesia, 165, 183n, 192
Middle East
civilization in, 277, 347
eunuchs of, 359
*Homo sapiens* spreading to, 43
trade of, 347, 381
*Moana* (film), 192
Moose Ridge Wilderness School, Maine, 102–09, 264n, 306
Moses, 148
mosquitos, plants as prevention for, 12, 14
mudbrick homes, 96
mummification
of animals, 136, *136*, 139, 143–45
cultures making of mummies, 139n
experimental archaeology and, 139–44, *140*
fictional account of, 135–37, 139, 146–48, 163
fraud in, 145
natron and, 135, 136, *140*, 141–44
tattoos of Egyptian mummies, 250
Muñoz, Santiago, Maizajo tortilleria of, 389–91
Mustafaoğlu, Gökhan, 96, 96n

# Index

Nahuatl language, 402
native Alaskan culture. *See also* Thule people
  baleen, 324, 334, 342–43, 343n
  clay pots of, 337–42, *338*
  cooking fires of, 337
  cuisine of, 326–27
  dog feces used by, 335, 335n
  dried whale and seal meat on wooden racks, *322*, 325, 326, 327
  gathering of berries and greens, 338
  poop-knife and, 15, 323, 329–30, 330n
  raw meat eaten by, 325–26, 325n, 337
  rendering blubber into oil, 327–28
  seal pokes (sealskin bags) of, 328–29, *328*
  summer chores of, 338–39
  tattoos and, 257n
  terminology for, 325n
  traveling songs of, 332, 342, 345
Native American warclubs, replicas of, *265*
natron, and mummification, 135, 136, *140*, 141–44
navigation
  fictional account of, 182–83, 189–91, 199
  Polynesia and, 183, 183n, 185–88, 191–92, 278
  seabirds and, 186–87, 194, 278
Neolithic Revolution, 81. *See also* Boncuklu, Turkey; Çatalhöyük, Turkey
Nestor, Jefferson, 177–78
neurosurgery, history of, 303
New Mexico
  fossilized footprints in, 43n, 77
  giant sloth footprints in, 77
Newton, Isaac, 375
New World civilizations, clash with Old World civilizations, 243

New Zealand, 187
nixtamalization of corn, 389–91
noddies (seabird), 186, *186*, 194
North America
  cultures of, 243
  human colonization of, 43, 43n
  megafauna of, 75
  population of, 347
  slavery in, 201
  Viking excursions to, 315
Northern Pacific rattlesnakes, 272, *272*
Nyachoti, 304–05n

obsidian blades
  Aztec sacrifices and, 402
  Egyptian mummification and, 140
  fictional accounts of, 47, 59, 70, 71
  knapping stone tools and, 17, 19, 19–20n
  in Peru, 54
  steel blades compared to, 19–20n
obsidian flakes
  as tattoo tools, 250, 251, 252
  trepanation with, 306, 308–09
ocean currents, 187–88
Oceania, 165, 347
ochre
  photograph of, *17*
  purposes of, 23, 54, 58, 62, 112n, 245
Olaf I, King of Norway, 183n
Old World civilizations, clash with New World civilizations, 243
organs, and Egyptian mummification process, 136, 141–42, 143
ornatrices, 224, 226–27
ostrich-egg canteen
  experimental archaeology and, 36
  fictional account of, 12–13
  photograph of, *35*
ostrich eggs

## Index

engraved ostrich eggs, *17*
experimental archaeology and, 35–36
fictional
   account of collecting of, 40–41
ostriches
   fictional account of hunting of, 34–35, 37–39
   hunting of, 39n
ostrich skins
   African hunter donning, *38*
   fictional account of, 37, 38–39
Ötzi (5,300-year-old "Iceman"), tattoos of, 250–51, 252, 256, 257

Palau, 192–94
*pallas* (Roman mantles), 223
Papua New Guinea, 305
parrots, clay eaten by, 50
*patina* (flaked sea bass), 235, 237–38
Perry, Heather, 309–10n
Persia, 373
Peru
   altiplano of, 53, 54, 56–57, 61, 63, 66, 67–68
   bofedales (spring-fed wetlands) of, 53–54, 72
   coastal cultures of, 54–57, 55n, 57n
   femur fragment from an ancient Andean hunter, *63*
   grave of female hunter in, 61–64, 63n, 64n
   *ichu* grass of, 53
   meats of, 51–53
   potato domestication in, 49
   potato-processing in, 50–51
   trade networks of, 54, 55–56, 58, 66
   Wilamaya Patjxa archaeological site, 61–64, 63n
   yareta shrubs (cushion plants) of, 53–54, 65, 70, 78

petrels (seabird), 186
Piñón, Alejandro, 391–92
Plains Indians, 270
poison arrows
   archaeological evidence of, 272
   eating game hunted with, 273
   fictional account of, 269–70, 273–74
poisons, venoms distinguished from, 273n
Polynesia
   bark cloth artwork, *175*
   canoe building process, 184–85, 193
   canoe design of, 166–67, 183–84, 192–93
   canoe plants and, 176
   clothes made of tapa/kapa and, 174–76, *175*
   coconuts and, 177–78
   colonization of, 192
   cultures of, 165, 167
   currents and, 187–88
   extinction of bird species and, 187n
   fictional account of canoe voyage, 169–74, 178–83, 189–91, 195–99
   fictional account of navigation, 182–83, 189–91, 199
   fictional account of sharks, 179–81
   fictional account of tattoos and, 170, 171, 174, 198
   fishing barks of, 183
   life-sized recreation of *drua*, *166*
   model of ancient canoe (*drua*), *166*
   navigation and, 183, 183n, 185–88, 191–92, 278
   poi and, 177, 177n
   as region of Oceania, 165, 176
   sailors of, 165–66, 167, 188, 243
   seabirds and, 186–87, 187n, 278
   sharks and, 178
   taro and, 176–77
   tattooing of, 176

Pompeii
　fictional account of slavery in, 203–12, 217–23, 228–32, 239–42
　fictional account of volcanic eruption of, 203–12, 228, 232, 239–42
　volcanic eruption destroying, 202
poop-knife, 15, 323, 329–30, 330n
population
　density of, 117, 201, 243, 347
　growth of, 134
　Spanish conquest and, 418
potatoes
　chaco or pasa recipe, 50–51
　chuño recipes, 49–50, 51, 52
　domestication in South America and, 49
　fictional account in South America, 47–48
　geophagy and, 50–51
　toxins in, 49–51
　varieties of, 49
pottery
　as diagnostic for different cultures, 336
　experimental archaeology and, 337–42, *338*
　features of cooking pots, 336–37
professions, development of, 134
Prussia, 367n
pyramid construction
　alien theories and, 153n
　experimental archaeology and, 150–58, 151n, *152*, 153n, *156*, 158n
　fictional account of, 120–22
　ramp theories and, 151–52, 153, 153n, 154–55, 154–55n, 158n
pyramidiots, 153n
pyramids
　contents of burial chambers for pharaohs, 158, 158n
　fictional account of robbing pharaoh's tomb, 119, 121, 123, 124–27, 137, 138, 146–48, 159–61
　security measures against tomb robbers, 158–59
　of Tenochtitlán, 381
Pythagoras, 374

queñuals (trees), 66
Quispe, Albino, 61

Ramses the Great, 141, 142
rattlesnakes, 272, *272*
recreation of past, and experimental archaeology, 3–7, 419, 420
Reinhardt, Django, 226
*Return to Oz* (film), 224n
Riday, Daniel, 251–52, 253n
Riegner, Dawn, 408–09, 412
Ritchie, Tessy, 408–09, 412
roasted wild nutmeg, as tattoo ink, 245
Rogers, Cliff, 408–10, 412
Roman Empire
　architecture of, 201
　bathhouses of, 212, 216
　beauty treatments of, 223
　Christians and, 210, 210n
　clothing fashions of, 223
　concrete of, 214–16
　Egypt compared to, 201, 212
　engineering of, 212
　fall of, 243, 277
　fictional account of food, 229–31
　fictional account of fullery in, 203–04, 209, 217, 219–22, 228, 231–32, 241
　fictional account of hairdressing, 219, 222–23, 228
　fictional account of slavery in Pompeii, 203–12, 217–23, 228–32, 239–42

# INDEX

fictional account of volcanic eruption of Pompeii, 203–12, 228, 232, 239–42
food of, 232–39, 237–38n, 239n
footwear of, 223
fullery of, 216, *220*
hairdressing of, 223–28, *225*, *226*
iron collar for punishing and tracking slaves, *204*
public toilets of, *210*, 213
road construction of, 212–14
Roman goddess Fortuna, *205*
slavery in, 201–02
social stratification in, 201
temple dedicated to Fortune, *205*
temples of, 216
togas of, 223, 233, 239
women's political influence and, 216
wool of, 216
Roy, Nathalie, 212–15, 213n, 216, 216n
Russia, land bridge to Alaska, 43

saber-tooth tigers, in North America, 75
*sala cattabia* (savory sourdough), 237–38n
salmon leather, 105n
Samoa, 167
Sánchez, Arturo, 393–96, 394n, 397, 398
sandcastle construction, 154
San people, in southern Africa, 9
Sasao, Mac, 178, 193–94
Scandinavia, 105n
seabirds
  as direct food sources, 187n
  navigation and, 186–87, 194, 278
  pollination and, 187n
  Polynesia and, 186–87, 187n, 278
sedentary living. *See also* Boncuklu, Turkey; Çatalhöyük, Turkey
  agriculture and, 117, 134

cities and, 81–82
fictional account of burial practices, 85–88, 90–92
transition to, 81
senilicide, 335n
Seuss, Dr., 53
Shakira, 394, 396
sharks
  fictional account of, 179–81, 197
  Polynesia and, 178
shell beads, *17*
Shishmaref, Alaska, 324–28
shock weapons
  of California native tribes, 264, 267–68
  experimental archaeology and, 264–67, *265*, 265n
  injuries due to, 303
shunting, 183n, 194
Siberia, 43, 148
silphium, 238–39
slavery
  fictional account of, 203–12, 217–23, 228–32, 239–42
  iron collar for punishing and tracking slaves in Rome, *204*
  in Roman Empire, 201–02
smallpox, 414
Smithsonian Institution, 30
snares
  fictional account of, 37–40
  illustration of, *37*
Solomon, 374
South America. *See also* Bolivia; Peru
  Andes highlands of, 44, *52*
  cultures of, 243
  fictional account of gathering in, 47–48, 48n
  fictional account of hunting in, 46, 47, 48, 48n, 56, 58–61, 69–70, 78–79

449

South America. (*Cont.*)
  fictional account of life in, 45–48, 54–61
  human colonization of, 43–44
  megafauna of, 75
  population of, 347
  potatoes domesticated in, 49
  slavery in, 201
South Seas tribes, plated trepanation holes and, 304n, 305
Spain
  Aztecs and, 381, 383, 386n, 389, 404–05, 407–08, 413–14
  cannons used by, 408–14
  diseases spread by, 414
  fictional accounts of Native allies, 385–89, 399–401, 405–07, 414–18
  horses of, 413–14, 414n
  Mesoamerican conquest of, 389, 407, 414, 418
  steel and, 407–08
spears
  atlatls compared to, 66, 67
  and experimental archaeology, 76
Spiritual Journey, Orange County, California, 252–57, 257n
steel blades, obsidian blades compared to, 19–20n
Stephens, Janet, 224–28, 224n, 228n
Stone Age people, pigments of, 112n
Stone Age technology, 20
stone tools. *See also* knapping stone tools
  collection of, *17*
  fictional accounts of, 45
  hammerstones and, 16–18
surfer's ear, 55n
survivalists, 4, 5
Sweden, 278

Taiwan, ancestral people of, 165–66
Tanahy, Dalani, 174–76
tanning animal hides, experimental archaeology and, 102–09, 105n, 176, 257, 306
tattoo needles, 250, 251
tattoos
  Arctic subdermal-sewing method for, 251, 319n
  bone needles for, 250, 250n, 251, 253, 253n
  Aaron Deter-Wolf on, 250–51, 250n, 252
  of Egyptian mummies, 250
  experimental archaeology and, 252–57
  fictional account of California native tribes and, 245, 258, 274
  fictional account of Polynesian culture and, 170, 171, 174, 198
  fictional account of Thule people and, 319–20
  flicking technique, 255, 256
  hand-poked tattoos, 251, 252–53, 255
  kapa art compared to, 176
  meaning of, 257
  Daniel Riday's reference set of, 251–52
  stabbing technique, 255, 256
  study of, 250
  tattoo needles, 250, 250n, 251, 253, 253n, 255
  tools for applying ink, 250, 250n, 251, 253, 255–56
*te lapa* (luminescence), 187
Tenochtitlán
  canals of, 381
  marketplace of, *382*
  Mexica people and, 381
  pyramids of, 381

Triple Alliance and, 382–83
view of, *382*
terns (seabird), 186, *186*
Texcoco, and Triple Alliance, 382–83
thermoplastic, 65
Thule people
   culture of, 315
   death by exposure and, 335n
   fictional account of birth, 317–24, 318n
   fictional account of hunting practices, 319, 320–21, 323–24, 332–33
   fictional account of rescue, 331–35, 342–46
   hunting practices of, 315
   semi-permanent settlements of, 315
   sharing practices of, 316
   summer chores of, 340
   technological sophistication of, 315–16
ticks, ash as preventive measure for, 12, 14–15
time travel, 7
Tlacopan, and Triple Alliance, 382–83
Tollund Man of Denmark (bog body), *296*
Tonga, 167
Treadway, Tiffany, 296–98
trebuchets, *362*, 363, 366–72, 367n, 408
tree sloths, 75
trepanations
   experimental archaeology and, 108, *304*, 305–10
   fictional account of, 310–14
   history of, 303, 304–05, 304–05n
   modern surgery compared to, 305
   procedure for, 303, 304, 305, 306
   self-trepanation, 309–10n
*tsama* watermelons, 11
turtles
   fictional account of hunting of, 34
   fictional account of use as bait, 39–40
Tutankamen, Pharaoh, tomb of, 158
Tu Youyou, 289n
*2001: A Space Odyssey* (film), 16, 18

ulu (curved knife), *322*
United States Military Academy, West Point, 408–11
University of Witwatersrand, 14

venoms, poisons distinguished, 273n
vicuñas
   appearance of, 52–53
   clay eaten by, 50
   depiction of female hunter from the Andes with, *52*
   dung as fuel source, 46, 53
   experimental archaeology and, 52–53
   fictional account of bones of, 47
   fictional account of dung of, 46
   fictional account of hides of, 45
   fictional account of hunting of, 46, 48, 48n, 56, 59–61, 70, 72
   hunting of, 64
Vikings
   blitzkrieg tactics of, 277
   female warriors of, 278
   fictional account of camp of, 289–90
   fictional account of siege of priory, 279–85, 289–94, 299–303, 310–14
   navigation and, 278
   pagan traditions of, 278
   poetry of, 278
   raids of, 277–78, 315
   range of, 278, 315
votive mummies, 145, 146

Wade, Ronn, 139–43
Wadley, Lyn, 14–15

wallata bird (Andean goose), 72
Walters Art Museum, 224–25
water root (*xwa*), 28, 29
Watson, Jim, 62–63, 63n
Wayne, John, 30
wheels, 150, 150n
Wilamaya Patjxa archaeological site, Peru, 61–64, 63n

wild potatoes, and fictional account of South America, 47
Wolfer, Laura, 306–08

yareta shrubs (cushion plants), 53–54, 65, 70, 78

Zheng He, 358–59

# ABOUT THE AUTHOR

**Sam Kean** is the *New York Times* best-selling author of *The Disappearing Spoon, The Icepick Surgeon, The Bastard Brigade, Caesar's Last Breath* (the *Guardian*'s Science Book of the Year), *The Tale of the Dueling Neurosurgeons,* and *The Violinist's Thumb.* He is a two-time finalist for the PEN / E. O. Wilson Literary Science Writing Award. His work has appeared in *The Best American Science and Nature Writing, The New Yorker, The Atlantic,* and the *New York Times Magazine,* among other publications, and he has been featured on NPR's *Radiolab, All Things Considered, Science Friday,* and *Fresh Air.* His podcast, *The Disappearing Spoon,* debuted at Number 1 on the iTunes science charts. Kean lives in Washington, DC.